NUMBERS AND FUNCTIONS
STEPS INTO ANALYSIS

Also by this Author:

Groups: A Path to Geometry

A Pathway into Number Theory, 2nd Edition

R. P. BURN

NUMBERS AND FUNCTIONS

STEPS INTO ANALYSIS

Third edition

CAMBRIDGE
UNIVERSITY PRESS

CAMBRIDGE
UNIVERSITY PRESS

University Printing House, Cambridge CB2 8BS, United Kingdom

One Liberty Plaza, 20th Floor, New York, NY 10006, USA

477 Williamstown Road, Port Melbourne, VIC 3207, Australia

314-321, 3rd Floor, Plot 3, Splendor Forum, Jasola District Centre, New Delhi - 110025, India

79 Anson Road, #06-04/06, Singapore 079906

Cambridge University Press is part of the University of Cambridge.

It furthers the University's mission by disseminating knowledge in the pursuit of education, learning and research at the highest international levels of excellence.

www.cambridge.org
Information on this title: www.cambridge.org/9781107444539

First published 1992
Second edition 2000
Reprinted with corrections 2006
Third edition 2015

A catalogue record for this publication is available from the British Library

Library of Congress Cataloging in Publication data
Burn, R. P.
Numbers and functions : steps into analysis / R.P. Burn. – Third edition.
pages cm
Includes bibliographical references and index.
ISBN 978-1-107-44453-9
1. Mathematical analysis. I. Title.
QA300.B88 2015
515–dc23
2014046495

ISBN 978-1-107-44453-9 Paperback

Contents

APPENDICES

*Preface to the first edition**

This text is written for those who have studied calculus in the sixth form at school, and are now ready to review that mathematics rigorously and to seek precision in its formulation. The question sequence given here tackles the key concepts and ideas one by one, and invites a self-imposed precision in each area. At the successful conclusion of the course, a student will have a view of the calculus which is in accord with modern standards of rigour, and a sound springboard from which to study metric spaces and point set topology, or multi-dimensional calculus.

Generations of students have found the study of the foundations of the calculus an uncomfortable business. The reasons for this discomfort are manifold.

(1) The student coming from the sixth form to university is already familiar with Newtonian calculus and has developed confidence in the subject by using it, and experiencing its power. Its validity has been established for him/her by reasonable argument and confirmed by its effectiveness. It is not a source of student uncertainty and this means that an axiomatic and rigorous presentation seems to make heavy weather of something which is believed to be sound, and criticisms of Newtonian calculus seem to be an irritating piece of intellectual nit-picking.

(2) At an age when a student's critical capacity is at its height, an axiomatic presentation can have a take-it-or-leave-it quality which *feels* humiliating: axioms for the real numbers have none of the 'let's-play-a-game' character by which some simpler systems appeal to the widespread interest in puzzles. The bald statement of axioms for the real numbers covers up a significant process of decision-making in their choice, and the Axiom of Completeness, which lies at the heart of most of the main results in analysis, seems superfluous at first sight, in whatever form it is expressed.

* The text of this preface was revised slightly for the second edition.

(3) Even when the axiom system has been accepted, proofs by contradiction can be a stumbling block, either because the results are unbelievable, as in the case of irrationality or uncountability, or because they make heavy weather of such seemingly obvious results as the theorem that a convergent sequence cannot have two limits.
(4) Definitions, particularly those of limits and continuity, appear strangely contrived and counter-intuitive.

There are also discomforts of lesser moment which none the less make the subject indigestible:

(5) the abstract definition of a function (when most students have only used the word function to mean a formula),
(6) the persistent use of inequalities in argument to tame infinity and infinitesimals,
(7) and proofs by induction (which play an incidental rôle in most school courses).
(8) The student who has overcome these hurdles will find that some of the best textbooks will present him/her with exercises at the end of each chapter which are so substantial that it could be a term's work to complete even those associated with three hours of lectures.
(9) The student who seeks help in the bibliography of his/her current text may find that the recommended literature is mostly for 'further reading'.

Most of these difficulties are well-attested in the literature on mathematical education. (See for example articles published in *Educational Studies in Mathematics* throughout the 1980s or the review article by David Tall (1992).)

With these difficulties in mind we may wonder how any students have survived such a course! The questions which they ask analysis lecturers may reveal their methods. Although I believed, as an undergraduate, that I was doing my best, I remember habitually asking questions about the details of the lecturer's exposition and never asking about the main ideas and results. I realise now that I was exercising only secretarial skills in the lecture room, and was not involved in the overall argument. A more participatory style of learning would have helped. Many of those who have just completed a degree in mathematics will affirm that 0.9 recurring is not the limit of a sequence, but an ordinary number less than 1! This is not evidence of any lack of intelligence: through the nineteenth century the best mathematicians stumbled because of the difficulty of imagining dense but incomplete sets of points and the seeming unreality of continuous but non-differentiable functions. There was real discomfort too in banning the language of infinitesimals from the discussion of limits. In their difficulties today, students have much in common with the best mathematicians of the nineteenth century.

It has been said that the most serious deficiency in undergraduate mathematics is the lack of an existence theorem for undergraduates! Put in other words, by an eminent educationalist, 'If I had to reduce all of educational psychology to just one principle, I would say this: the most important single factor influencing learning is what the learner already knows. Ascertain this, and teach him accordingly.' (Ausubel, 1968). There is a degree of recognition of this principle in virtually every elementary text on analysis, when, in the exercises at the end of a chapter, the strictly logical order of presentation is put aside and future results anticipated in order that the student should better understand the points at issue. The irrationality of π may be presumed before the number itself has been defined. The trigonometric, exponential and logarithmic functions almost invariably appear in exercises before they have been formally or analytically defined. In this book, my first concern, given the subject matter, has been to let students *use what they already know* to generate new concepts, and to explore situations which invite new definitions. In working through each chapter of the book the student will come to formulate, in a manner which respects modern standards of rigour, part of what is now the classical presentation of analysis. The formal achievements of each chapter are listed in summaries, but on the way there is no reluctance to use notions which will be familiar to students from their work in the sixth form. This, after all, is the way the subject developed historically. Sixth-form calculus operates with the standards of rigour which were current in the middle of the eighteenth century and it was from such a standpoint that the modern rigorous analysis of Bolzano, Cauchy, Riemann, Weierstress, Dedekind and Cantor grew.

So the first principle upon which this text has been constructed is that of involving the student in the generation of new concepts by using ideas and techniques which are already familiar. The second principle is that generalisation is one of the least difficult of the new notions which a student meets in university mathematics. The judiciously chosen special case which may be calculated or computed, provides the basis for a student's own formulation of a general theorem and this sequence of development (from special case to general theorem) keeps the student's understanding active when the formulation of a general theorem on its own would be opaque.

The third principle, on which the second is partly based, is that every student will have a pocket calculator with 'scientific' keys, and access to graph drawing facilities on a computer. A programmable calculator with graphic display will possess all the required facilities.

There is a fourth principle, which could perhaps be better called an ongoing tension for the teacher, of weighing the powerful definition and consequent easy theorem on the one hand, against the weak definition (which seems more meaningful) followed by the difficult theorem on the other. Which is the better teaching strategy? There is no absolute rule here. However

I have chosen 'every infinite decimal is convergent' as the axiom of completeness, and then established the convergence of monotonic bounded sequences adapting an argument given by P. du Bois-Reymond in 1882, and used by W. F. Osgood, 1907. Certainly to assume that monotonic bounded sequences are convergent and to deduce the convergence of the sequence for an infinite decimal takes less paper than the converse, but I believe that, more often than has usually been allowed, the combination of weaker definition and harder theorem keep the student's feet on the ground and his/her comprehension active.

Every mathematician knows theorems in which propositions are proved to be equivalent but in which one implication is established more easily than the other. A case in point is the neighbourhood definition of continuity compared with the convergent sequence definition. After considerable experience with both definitions it is clear to me that the convergent sequence definition provides a more effective teaching strategy, though it is arguable that this is only gained by covert use of the Axiom of Choice. (A detailed comparison of different limit definitions from the point of view of the learner is given in ch. 14 of Hauchart and Rouche.) As I said earlier, the issue is not one of principle, but simply an acknowledgement that the neat piece of logic which shortens a proof *may* make that proof and the result *less* comprehensible to a beginner. More research on optimal teaching strategies is needed.

It sometimes seems that those with pedagogical concerns are soft on mathematics. I hope that this book will contradict this impression. If anything, there is more insistence here than is usual that a student be aware of which parts of the axiomatic basis of the subject are needed at which juncture in the treatment.

The first two chapters are intended to enable the student who needs them to improve his/her technique and his/her confidence in two aspects of mathematics which need to become second nature for anyone studying university mathematics. The two areas are those of mathematical induction and of inequalities. Ironically, perhaps, in view of what I have written above, the majority of questions in chapter 2 develop rudimentary properties of the number system from stated axioms. These questions happen to be the most effective learning sequences I know for generating student skills in these areas. My debt to Landau's *Foundations of Analysis* and to Thurston's *The Number-system* will be evident. There is another skill which these exercises will foster, namely that of distinguishing between what is familiar and what has been proved. This is perhaps the key distinction to be drawn in the transition from school to university mathematics, where it is expected that everything which is to be assumed without proof is to be overtly stated. Commonly, a first course in analysis contains the postulational basis for most of a degree course in mathematics (see appendix 1). This justifies lecturers being particularly fussy about the reasoning used in proofs in analysis. In the second chapter we also establish various classical inequalities to use in later work on convergence.

In the third chapter we step into infinite processes and define the convergence of sequences. It is the definition of limit which is conventionally thought to be the greatest hurdle in starting analysis, and we define limits first in the context of null sequences, the preferred context in the treatments of Knopp (1928) and Burkill (1960). In order to establish basic theorems on convergence we assume Archimedean order. When the least upper bound postulate is used as a completeness axiom, it is common to deduce Archimedean order from this postulate. Unlike most properties established from a completeness axiom, Archimedean order holds for the rational numbers, and indeed for any subfield of the real numbers. So this proof can mislead. The distinctive function of Archimedean order in banishing infinite numbers and infinitesimals, whether the field is complete or not, is often missed. The Archimedean axiom expressly forbids ∞ being a member of the number field. Now that non-standard analysis is a live option, clarity is needed at this point. In any case, the notion of completeness is such a hurdle to students that there is good reason for proving as much as possible without it. We carry this idea through the book by studying sequences *without* completeness in chapter 3 and *with* completeness in chapters 4 and 5; continuous functions and limits *without* completeness in chapter 6 and *with* completeness in chapter 7; differentiation *without* completeness in chapter 8 and *with* completeness in chapter 9.

The fourth chapter is about the completeness of the real numbers. We identify irrational numbers and contrast the countability of the rationals with the uncountability of the set of infinite decimals. We adopt as a completeness axiom the property that every infinite decimal is convergent. We deduce that bounded monotonic sequences are convergent, and thereafter standard results follow one by one. Of the possible axioms for completeness, this is the only one which relates directly to the previous experience of the students. With completeness under our belt, we are ready to tackle the convergence of series in chapter 5.

The remaining chapters of the book are about real functions and start with a section which shows why the consideration of limiting processes requires the modern definition of a function and why formulae do not provide a sufficiently rich diet of possibilities. By adopting Cantor's sequential definition of continuity, a broad spectrum of results on continuous functions follows as a straightforward consequence of theorems about limits of sequences. The second half of chapter 6 is devoted to reconciling the sequential definitions of continuity and limit with Weierstrass' neighbourhood definitions, and deals with both one- and two-sided limits. These have been placed as far on in the course as possible. There is a covert appeal to the

Axiom of Choice in the harder proofs. Chapter 6 builds on chapter 3, but does not depend on completeness in any way. In chapter 7 we establish the difficult theorems about continuity on intervals, all of which depend on completeness. Taking advantage of the sequential definition of continuity, we use completeness in the proofs by claiming that a bounded sequence contains a convergent subsequence. There is again a covert appeal to the Axiom of Choice.

Chapters 8 and 9, on differentiation, are conventional in content, except in stressing the distinction between those properties which do not depend on completeness, in chapter 8 (the definition of derivative and the product, quotient and chain rules), and those which do, in chapter 9 (Rolle's Theorem to Taylor's Theorem). The differentiation of inverse functions appears out of place, in chapter 8. Chapter 10, on integration, starts with the computation of areas in ways which were, or could have been, used before Newton, and proceeds from these examples to the effective use of step functions in the theory of the Riemann integral. Completeness is used in the definition of upper and lower integrals.

The convention of defining logarithmic, exponential and circular functions either by neatly chosen integrals or by power series is almost universal. This seems to me to be an excellent procedure in a *second* course. But the origins of exponentials and logarithms lie in the use of indices and that is the starting point of our development in chapter 11. Likewise our development of circular functions starts by investigating the length of arc of a circle. The treatment is necessarily more lengthy than is usual, but offers some powerful applications of the theorems of chapters 1 to 10.

A chapter on uniform convergence completes the book. Some courses, with good reason, postpone such material to the second year. This chapter rounds off the problem sequence in two senses: firstly, by the discussion of term-by-term integration and differentiation, it completes a university-style treatment of sixth-form calculus; and secondly by discussing the convergence of functions it is possible to see the kind of questions which provoked the rigorous analysis of the late nineteenth century.

The interdependence of chapters is illustrated below.

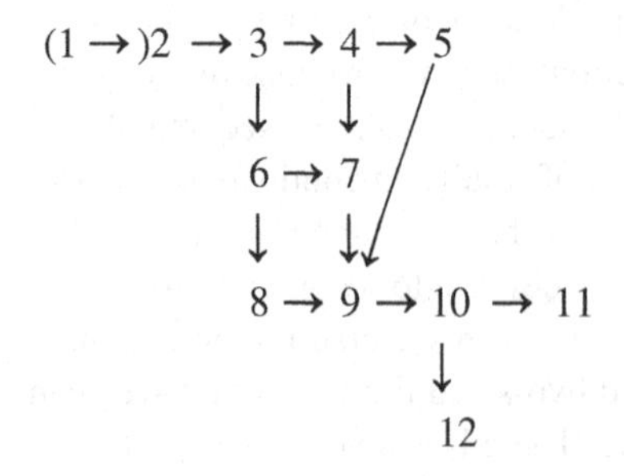

One review of my *Pathway into Number Theory* (*Times Higher Educational Supplement*, 3.12.82) suggested that a pathway to analysis would be of more value than a pathway to number theory. While not disagreeing with the reviewer, I could not, and still today cannot, see the two tasks as comparable. The subtlety of the concepts and definitions of undergraduate analysis is of a different order. But the reviewer's challenge has remained with me, and the success I have seen students achieve with my pathways to number theory and geometrical groups has spurred me on. None the less, I offer these steps (notice the cautious claim by comparison with the earlier books) aware that they contain more reversals of what I regard as an optimal teaching sequence than the earlier pathways.

It may be helpful to clarify the differences between the present text and other books on analysis which have the word 'Problem' in their titles. I refer firstly to the books of the Schaum series. Although their titles read *Theory and problems of*... the books consist for the most part of solutions. Secondly, a book with the title *Introductory Problem Course in Analysis and Topology* written by E. E. Moise consists of a list of theorems cited in logical sequence. Thirdly, the book *Problems and Propositions in Analysis* by G. Klambauer provides an enriching supplement to any analysis course, and, in my opinion, no lecturer in the subject should be without a copy. And finally there is the doyen of all problem books, that by Pólya and Szegö (1976), which expects greater maturity than the present text but, again, is a book no analysis lecturer should be without.

Sometimes authors of mathematics books claim that their publications are 'self-contained'. This is a coded claim which may be helpful to an experienced lecturer, but can be misleading to an undergraduate. It is never true that a book of university mathematics can be understood without experience of other mathematics. And when the concepts to be studied are counter-intuitive, or the proofs tricky, it is not just that one presentation is better than another, but that all presentations are problematic, and that whichever presentation a student meets *second* is more likely to be understood than the one met *first*. For these reasons I persistently encourage the consultation of other treatments. While it is highly desirable to recommend a priority course book (lest the student be entirely at the mercy of the lecturer) the lecturer needs to use the ideas of others to stimulate and improve his/her own teaching; and, because students are different from one another, no one lecturer or book is likely to supply quite what the student needs.

There is a distinction of nomenclature of books on this subject, with North American books tending to include the word 'calculus' in their titles and British books the word 'analysis'. In the nineteenth century the outstanding series of books by A. L. Cauchy clarified the distinction. His first volume was about convergence and continuity and entitled 'Course of Analysis' and his later volumes were on the differential and integral calculus.

The general study of convergence thus preceded its application in the context of differentiation (where limits are not reached) and in integration (where the limiting process occurs by the refinement of subdivisions of an interval). British university students have met a tool-kit approach to calculus at school and the change in name probably assists the change in attitude required in transferring from school to university. But the change of name is unhelpful to the extent that it is part of the purpose of every first course of analysis in British universities to clarify the concepts of school calculus and to put familiar results involving differentiation and integration on a more rigorous foundation.

I must express gratitude and indebtedness to many colleagues: to my own teacher Dr J. C. Burkill for his pursuit of simplicity and clarity of exposition; to Hilary Shuard my colleague at Homerton College whose capacity to put her finger on a difficulty and keep it there would put a terrier to shame (an unpublished text on analysis which she wrote holds an honoured place in my filing cabinet; time and again, when I have found all the standard texts unhelpful, Hilary has identified the dark place, and shown how to sweep away the cobwebs); to Alan Beardon for discussions about the subject over many years; to Dr D. J. H. Garling for redeeming my mistakes, and for invaluable advice on substantial points; to Dr T. W. Körner for clarifying the relationship between differentiability and invertibility of functions for me by constructing a differentiable bijection $\mathbb{Q} \to \mathbb{Q}$ whose inverse is not continuous; and to Dr F. Smithies for help with many historical points. At a fairly late stage in the book's production I received a mass of detailed and pertinent advice from Dr Tony Gardiner of Birmingham University which has led to many improvements. This book would not yet be finished but for the sustained encouragement and support of Cambridge University Press. The first book which made me believe that a humane approach to analysis might be possible was *The Calculus, a Genetic Approach* by O. Toeplitz. The historical key to the subject which he picked up can open more doors than we have yet seen. I still look forward to the publication of an undergraduate analysis book, structured by the historical development of the subject during the nineteenth century. [Added in 1995.] David Bressoud's new book is greatly to be welcomed, but we still await a text inspired by the history surrounding the insights of Weierstrass, and especially the development of the notion of completeness.

I would like to be told of any mistakes which students or lecturers find in the book.

School of Education, R. P. Burn
Exeter University, EX1 2LU
January 1991

Preface to the second edition

Under the inspiration of David Fowler, driven by the leadership of David Epstein, eased by the teaching ideas of Alyson Stibbard and challenged by the research of Lara Alcock, a remarkable transformation of the way in which students begin to study analysis took place at Warwick University. The division of teaching time between lectures and problem-solving by students in class changed to give student problem-solving pride of place. Many of the problems the students tackled in this approach to analysis were taken from the first five chapters of the first edition of this book. The experience has suggested a number of improvements to the original text. This new edition is intended to embody these improvements. The most obvious is the playing down of the Peano postulates for the natural numbers and algebraic axioms for the real numbers, which affects chapter 1 and the beginning of chapter 2. The second is in the display of 'summaries', which now appear when a major idea has been rounded off, not just at the end of the chapter. The third is an increase in the number of diagrams, and the fourth is the introduction of simple (and I hope evocative) names for small theorems, so that they may be cited more readily than when they only have a numerical reference. There are also numerous smaller points not just about individual questions. In particular, there is a suggestion of two column proofs in establishing the rudimentary properties of inequalities in chapter 2, and least upper bounds now figure more substantially in chapter 4. The stimulus of discussions with David Epstein both about overall strategy and about the details of question after question has been a rare and enriching experience.

I must also thank Paddy Paddam, previously of Cambridge, who has worked through every question in the text, and made invaluable comments to me from a student's eye view. I have also taken the opportunity to correct and improve the historical references and notes.

Kristiansand
March 2000

[illegible] second edition

Preface to the third edition

In addition to taking the opportunity to correct and improve details in the text and historical notes, this edition incorporates two new ideas. In chapter 3, the introduction to limits, especially in relation to null sequences, builds on the method of Archimedes for finding areas and volumes, and the early questions in chapter 10 investigate the determination of areas bounded by continuous monotonic curves without an appeal to completeness. This extends the structure of the book from limits, continuity and differentiation with and without completeness to consider integration also, with and without completeness. As in so many places in previous editions, the clarifying insights have emerged from seeking to understand the historical development. All the changes have been made in such a way as to maintain, so far as possible, the sequence and numbering of questions in the second edition. Only questions 3.24–26, 28 and 10.1, 3–9 have been substantially changed.

I gratefully acknowledge discussion of historical matters with Jeremy Gray, and also prolonged dialogue with Dirk Nelson which has improved the clarity and accuracy of the text.

Exeter
May 2014

Glossary

qn 27 refers to question 27 of the same chapter.
qn 6.27 refers to question 27 of chapter 6.
6.27^+ refers to what immediately follows qn 6.27.
6.27^- refers to what immediately precedes qn 6.27.

$x \in A$	x is an element of the set A; for example, $x \in \{x, \ldots\}$
$x \notin A$	x is not an element of the set A
$\{x \mid x \in A\}$ or $\{x : x \in A\}$	the set A
$A \subseteq B$	A is a subset of B every member of A is a member of B $x \in A \Rightarrow x \in B$
$A \cup B$	the union of A and B $\{x \mid x \in A \textit{ or } x \in B\}$
$A \cap B$	the intersection of A and B $\{x \mid x \in A \textit{ and } x \in B\}$
$A \backslash B$	$\{x \mid x \in A \textit{ and } x \notin B\}$
$A \times B$	the cartesian product $\{(a, b) \mid a \in A, b \in B\}$
$f : A \to B$	the function f, a subset of $A \times B$, $\{(a, f(a)) \mid a \in A, f(a) \in B\}$ A is the domain and B is the co-domain of the function f
$f(A)$	$\{f(x) \mid x \in A\} \subseteq B$
$x \mapsto f(x)$	the function f, x is an element of the domain of f
$\mathbb{N}$	the set of natural numbers or counting numbers $\{1, 2, 3, \ldots\}$

$\mathbb{Z}$	the set of integers $\{0, \pm 1, \pm 2, \ldots\}$
$\mathbb{Z}^+$	the set of positive integers $\{1, 2, 3, \ldots\}$
$\mathbb{Q}$	the set of rational numbers $\{p/q \mid p \in \mathbb{Z}, q \in \mathbb{Z}\backslash\{0\}\}$
$\mathbb{Q}^+$	the set of positive rationals $\{x \mid x \in \mathbb{Q}, 0 < x\}$
$\mathbb{R}$	the set of real numbers, or infinite decimals
$\mathbb{R}^+$	the set of positive real numbers $\{x \mid x \in \mathbb{R}, 0 < x\}$
$\lfloor x \rfloor$	the integral part of x, also called floor x $\lfloor x \rfloor \leqslant x < \lfloor x \rfloor + 1$
$[a, b]$	closed interval $\{x \mid a \leqslant x \leqslant b, x, a, b \in \mathbb{R}\}$
(a, b)	open interval $\{x \mid a < x < b, x, a, b \in \mathbb{R}\}$, the symbol may also denote the ordered pair, (a, b), or the two coordinates of a point in $\mathbb{R}^2$
$[a, b)$	half-open interval $\{x \mid a \leqslant x < b, x, a, b \in \mathbb{R}\}$
$[a, \infty)$	closed half-ray $\{x \mid a \leqslant x, a, x \in \mathbb{R}\}$
(a, ∞)	open half-ray $\{x \mid a < x, a, x \in \mathbb{R}\}$
(a_n)	the sequence $\{a_n \mid n \in \mathbb{N}, a_n \in \mathbb{R}\}$ that is, a function: $\mathbb{N} \to \mathbb{R}$
$(a_n) \to a$ as $n \to \infty$	For any $\varepsilon > 0$, $\lvert a_n - a \rvert < \varepsilon$ for all $n > N$
$\binom{n}{r}$	$\dfrac{n!}{(n-r)!r!}$ when $n \in \mathbb{Z}^+$
$\binom{a}{r}$	$\dfrac{a(a-1)(a-2)\ldots(a-r+1)}{r!}$ the binomial coefficient when $a \in \mathbb{R}$
$\sum_{n=1}^{n=N} a_n$	$a_1 + a_2 + a_3 + \ldots + a_N$
$\ln x$	the natural logarithm of x $\log_e x$

PART I

Numbers

1

Mathematical induction

Preliminary reading: Rosenbaum, Polya ch. 7.
Concurrent reading: Sominskii.
Further reading: Thurston chs 1 and 2, Ledermann and Weir.

The set of all whole numbers $\{1, 2, 3, \ldots\}$ is often denoted by $\mathbb{N}$. We will usually call $\mathbb{N}$ the set of *natural numbers*, and, sometimes, the set of *positive integers*. Note that $\mathbb{N}$ excludes 0.

1 $1^2 + 2^2 + 3^2 + \ldots + n^2 = \dfrac{n(n+1)(2n+1)}{6}$.

Is this proposition true or false?
Test it when $n = 1$, $n = 2$ and $n = 3$.
How many values of n should you test if you want to be sure it is true for all natural numbers n?

If we write $f(n) = \dfrac{n(n+1)(2n+1)}{6}$, show that

$$f(n) + (n+1)^2 = f(n+1).$$

Now suppose that the proposition with which we started holds for some particular value of n: add $(n+1)^2$ to both sides of the equation and deduce that the proposition holds for the next value of n.
Since we have already established that the proposition holds for $n = 1, 2$, and 3, the argument we have just formulated shows that it must hold for $n = 4$, and then by the same argument for $n = 5$, and so on.

2 When n is a natural number, $6^n - 5n + 4$ is divisible by 5.
Check this proposition for $n = 1, 2$ and 3.
By examining the difference between this number and $6^{n+1} - 5(n+1) + 4$, show that if the proposition holds for one value of n, it holds for the succeeding value of n. When you have done this you have

established the two components of the proof of the proposition by mathematical induction.

3 Prove the following propositions by induction (some have easy alternative proofs which do not use induction):

(i) (*Triangular numbers*) $1+2+3+\ldots+n=\frac{1}{2}n(n+1)$,
(ii) $a+(a+d)+(a+2d)+\ldots+(a+(n-1)d)=\frac{1}{2}n(2a+(n-1)d)$,
(iii) $1^3+2^3+3^3+\ldots+n^3=[\frac{1}{2}n(n+1)]^2$,
(iv) $1\cdot 2+2\cdot 3+3\cdot 4+\ldots+n(n+1)=\frac{1}{3}n(n+1)(n+2)$,
(v) $1+2+4+\ldots+2^{n-1}=2^n-1$,
(vi) $1+x+x^2+\ldots+x^{n-1}=\dfrac{x^n-1}{x-1}$, provided $x\neq 1$.
(vii) (optional) $1+2x+3x^2+\ldots+nx^{n-1}=\dfrac{nx^n}{x-1}-\dfrac{x^n-1}{(x-1)^2}$, provided $x\neq 1$.

4 Pascal's triangle, shown here, is defined inductively, each entry being the sum of the two (or one) entries in the preceding row nearest to the new entry. The $(r+1)$th entry in the nth row is denoted by $\binom{n}{r}$

$$
\begin{array}{ccccccccccc}
 & & & & 1 & & 1 & & & & \\
 & & & 1 & & 2 & & 1 & & & \\
 & & 1 & & 3 & & 3 & & 1 & & \\
 & 1 & & 4 & & 6 & & 4 & & 1 & \\
1 & & 5 & & 10 & & 10 & & 5 & & 1 \\
\end{array}
$$
$$1 \quad \ldots \quad \ldots \quad \ldots \quad \ldots \quad \ldots \quad 1$$

and called 'n choose r' because it happens to count the number of ways of choosing r objects from a set of n objects. Notice that we always have $0\leq r\leq n$. From the portion of Pascal's triangle which has been shown, for example,

$$\binom{3}{0}=1,\ \binom{3}{1}=3,\ \binom{3}{2}=3 \text{ and } \binom{3}{3}=1.$$

From the definition of Pascal's triangle we have

$$\binom{n}{r-1}+\binom{n}{r}=\binom{n+1}{r}, \text{ when } 0<r<n.$$

Prove by induction that $\binom{n}{r}=\dfrac{n!}{r!(n-r)!}$, taking $0!=1$.

Verify that this formula gives $\binom{n}{r}=\binom{n}{n-r}$.

What aspect of Pascal's triangle does this reflect?

5 *The Binomial Theorem for positive integral index*
Prove by induction that

$$(1+x)^n = \binom{n}{0} + \binom{n}{1}x + \binom{n}{2}x^2 + \ldots + \binom{n}{r}x^r + \ldots + \binom{n}{n}x^n.$$

6 Examine each of the propositions:

(i) $2^{n-1} \leq n^2$,
(ii) $n^2 + n + 41$ is a prime number.

Find values of n for which these propositions hold and also a value of n for which each is false.
If you were to attempt a proof by induction of either of these propositions, where would the proof break down?

7 Examine the numbers $6^n - 5n + 1$ and $6^{n+1} - 5(n+1) + 1$.
Find their difference.
Deduce that if the first of these numbers were divisible by 5 then the second would also be divisible by 5. Deduce also that if the second were divisible by 5 then the first would also be divisible by 5.
Is the first number divisible by 5 when $n = 1$?
Are there any values of n for which these numbers are divisible by 5?
If you were to attempt a proof by induction that the first number was divisible by 5, where would the proof break down?

8 Define $y^{(0)} = y$, $y^{(1)} = \dfrac{dy}{dx}$, and $y^{(n+1)} = \dfrac{dy^{(n)}}{dx}$.

(i) Let $y = \ln(1+x)$, so $\dfrac{dy}{dx} = \dfrac{1}{1+x}$.
Prove that for $n \geq 1$, $(1+x)y^{(n+1)} + ny^{(n)} = 0$.
Deduce that when $x = 0$, $y^{(n+1)} = (-1)^n n!$

(ii) Let $y = \arctan x$, so $\dfrac{dy}{dx} = \dfrac{1}{1+x^2}$.
Prove that for $n \geq 1$,
$(1+x^2)y^{(n+2)} + 2(n+1)xy^{(n+1)} + n(n+1)y^{(n)} = 0$.
Deduce that when $x = 0$, $y^{(2n)} = 0$ and $y^{(2n+1)} = (-1)^n(2n)!$

There is ambiguity in the literature on the question of whether $\mathbb{N}$ contains 0 or not. If mathematics starts with sets, and particularly the empty set, the equipment for counting the elements of finite sets must include 0. The psychological origins of counting, however, start with 1. It is this convention we have called *natural*. Both conventions are well-established.

THE PRINCIPLE OF MATHEMATICAL INDUCTION

A proposition, $P(n)$, relating to a natural number n, is valid for all natural numbers n, if

(a) $P(1)$ is true; the proposition is valid for $n = 1$ and
(b) $P(n) \Rightarrow P(n+1)$; the proposition for n implies the proposition for $n+1$.

Historical Note

The study of triangular numbers is said to go back to the Pythagoreans (6th century BC). The sum of the first n squares was known to Archimedes (c. 250 BC) and the sum of the first n cubes to the Arabs (c. AD 1010). The justification of these forms used induction implicitly.

The earliest use of proof by mathematical induction in the literature is by Maurolycus in his study of polygonal numbers published in Venice in 1575. Pascal knew the method of Maurolycus and used it for work on the binomial coefficients (c. 1657). In 1713, Jacques Bernoulli used an inductive proof to make rigorous the claim of Wallis (1656) that

$$\sum_{r=0}^{r=n} \frac{r^k}{n^{k+1}} \to \frac{1}{k+1} \text{ as } n \to \infty.$$

By the latter part of the eighteenth century, induction was being used by several authors. The name 'mathematical induction' as distinct from scientific induction is due to De Morgan (1838). The use of inductive definitions long pre-dates this method of proof.

The well-ordering principle, that every non-empty set of natural numbers has a least member, is equivalent to mathematical induction (see Ledermann and Weir, 1996). Well-ordering was used by Euclid (VII.31) to show that every integer has a prime factor and by Fermat (1637) in his proof of the non-existence of integral solutions to $x^4 + y^4 = z^2$. However, the recognition of the relationship between well-ordering and induction is recent, and from an historical point of view, awareness of the two principles developed independently.

During the nineteenth century there were several attempts to give a formal description of the natural numbers, notably by H. Grassmann (1861) and H. von Helmholtz (1887) both of whom assumed the Principle of Induction. R. Dedekind (1888) defined $\mathbb{N}$ as an arbitrary infinite chain and was able to prove the Principle of Induction. But in 1889, G. Peano, working with Dedekind's material, adopted the Principle of Mathematical Induction as one of his five postulates for defining the natural numbers, and his account remains standard to this day though Peano's Postulates admit non-standard models, which we exclude with Archimedean order, in chapter 3.

Answers and comments

1 Let $P(n)$ mean '$1^2 + 2^2 + 3^2 + \ldots + n^2 = n(n+1)(2n+1)/6$'.
You checked that $P(1)$, $P(2)$ and $P(3)$ were all true.
The algebra you did showed that $P(n) \Rightarrow P(n+1)$.
So (by induction) $P(n)$ is true for all natural numbers n.

2 Let $P(n)$ mean '$6^n - 5n + 4$ is divisible by 5'.
Because $5|5$, $P(1)$ is true. Because $5|30$, $P(2)$ is true. Because $5|205$, $P(3)$ is true.
The difference between the expression for $n+1$ and the expression for n is $5 \cdot 6^n - 5$, which is divisible by 5, so $P(n) \Rightarrow P(n+1)$. And (by induction) $P(n)$ is true for all natural numbers n.

3 Take each of the six propositions in turn as $P(n)$. Verify the truth of $P(1)$ in each case. The algebra behind the proof of $P(n) \Rightarrow P(n+1)$ is as follows:

(i) $\frac{1}{2}n(n+1) + (n+1) = \frac{1}{2}(n+1)(n+2)$,
(ii) $\frac{1}{2}n[2a + (n-1)d] + (a + nd) = \frac{1}{2}(n+1)(2a + nd)$,
(iii) $[\frac{1}{2}n(n+1)]^2 + (n+1)^3 = [\frac{1}{2}(n+1)(n+2)]^2$,
(iv) $\frac{1}{3}n(n+1)(n+2) + (n+1)(n+2) = \frac{1}{3}(n+1)(n+2)(n+3)$,
(v) $(2^n - 1) + 2^n = 2^{n+1} - 1$,
(vi) $\dfrac{x^n - 1}{x - 1} + x^n = \dfrac{x^{n+1} - 1}{x - 1}$,
(vii) $$\frac{nx^n}{x-1} - \frac{x^n - 1}{(x-1)^2} + (n+1)x^n = \frac{nx^n + (n+1)x^n(x-1)}{x-1} - \frac{x^n - 1}{(x-1)^2}$$
$$= \frac{(n+1)x^{n+1}}{x-1} - \frac{x^{n+1} - 1}{(x-1)^2}.$$

4 If $P(n)$ is the proposition we are asked to prove for $0 \le r \le n$,
$$\binom{n}{0} = \binom{n}{n} = 1$$
is easily checked, and incorporates the truth of $P(1)$. If we assume $P(n)$ and apply it to the Pascal triangle property, we get $P(n+1)$.
Symmetry about a vertical axis.

5 For the inductive step, one must show that the coefficient of x^r in $(1+x)^{n+1} = (1+x)^n(1+x)$ is $\binom{n}{r-1} + \binom{n}{r}$.

6 (i) True for $n \le 6$, false for $7 \le n$.
(ii) True for $n \le 39$, false for $n = 40$, or a multiple of 41, or $40 + k^2$.

$P(n)$ does not imply $P(n+1)$.

7 $6^n - 5n + 1$ divisible by $5 \Leftrightarrow 6^{n+1} - 5(n+1) + 1$ divisible by 5.
But both statements are false because the first statement is false when $n = 1$. $P(1)$ is false.

8 (i) $(1+x)dy/dx = 1$, or $(1+x)y^{(1)} = 1$, so $(1+x)y^{(2)} + y^{(1)} = 0$, and the proposition holds for $n = 1$.
Suppose $(1+x)y^{(n+1)} + ny^{(n)} = 0$, differentiating we have
$(1+x)y^{(n+2)} + (n+1)y^{(n+1)} = 0$, and the proposition for n implies the proposition for $n+1$.
When $x = 0$, $y^{(1)} = 1$ and $y^{(n+1)} = -ny^{(n)}$. So
$y^{(n+1)} = (-1)^n n! \Rightarrow y^{(n+2)} = (-1)^{n+1}(n+1)!$
and the result holds by induction.

(ii) Differentiate $(1+x^2)y^{(1)} = 1$ twice to get result for $n = 1$.
Differentiate result for n to obtain result for $n+1$. Result follows by induction.
When $x = 0$, $y^{(1)} = 1$, $y^{(2)} = 0$ and $y^{(n+2)} + n(n+1)y^{(n)} = 0$. So result holds for $n = 1$.
Also $y^{(2n)} = 0 \Rightarrow y^{(2n+2)} = 0$, and $y^{(2n+1)} = (-1)^n(2n)!$
$\Rightarrow y^{(2n+3)} = -(2n+1)(2n+2)(-1)^n(2n)! = (-1)^{n+1}(2n+2)!$
So result holds by induction.

2

Inequalities

Preliminary reading: Beckenbach and Bellman.
Concurrent reading: Korovkin, Kazarinoff.
Further reading: Thurston ch. F, Ivanov ch. 4.

Positive numbers and their properties

The addition, subtraction, multiplication and division of numbers will work in this course the way they did for you in school. Their essential algebraic properties are listed in appendix 1 under the heading 'Algebraic properties of a field of numbers'. Look at this list briefly. Examine which of these algebraic properties hold for the set of integers

$$\mathbb{Z} = \{0, \pm 1, \pm 2, \pm 3, \ldots\},$$

and then check that the set of rational numbers

$$\mathbb{Q} = \left\{ \frac{p}{q} \mid p, q \in \mathbb{Z}, q \neq 0 \right\}$$

satisfies *all* these algebraic properties. There is no need to remember the list for this course. The letter $\mathbb{Z}$ is the first letter of the German word *Zahl*, meaning *number*. The letter $\mathbb{Q}$ is the first letter of *quotient*; each rational number being a quotient of integers. The letter $\mathbb{R}$, the first letter of the word *real*, is used to denote the set of real numbers, the numbers needed for measuring distances along a line. We will presume that all the algebraic properties of numbers mentioned in appendix 1 hold for the set of real numbers. The difference between $\mathbb{Q}$ and $\mathbb{R}$ will be examined in chapter 4.

It is questions of convergence and the finding of limits which characterise a course in analysis. It is necessary to use inequalities in order to define these infinite processes. The purpose of this chapter is to sharpen your awareness of inequalities so that you know how to argue with them and what you need to be careful about when you are using them in an argument.

1 To prepare for a formal treatment of inequalities, determine for what numbers x you want to claim the inequality $2x < 3x$. Check whether you expect it to hold when $x = 1$, when $x = 0$ and when $x = -1$.

FORMAL PROPERTIES OF POSITIVE NUMBERS

1. If a is a number, then *either* $a = 0$, *or* a is positive *or* $-a$ is positive, and only one of these is true. When $-a$ is positive, a is said to be negative. This property is called the *trichotomy law* because of the *three* possibilities.
2. The sum of two positive numbers is positive. This is also described by saying that the positive numbers are *closed under addition.*
3. The product of two positive numbers is positive. This is also described by saying that the positive numbers are *closed under multiplication.*

We introduce the inequality '$<$' with the following

DEFINITION OF 'LESS THAN'
We say $a < b$ if and only if $b - a$ is positive.

The subset of positive numbers in $\mathbb{Z}$ is denoted by $\mathbb{Z}^+$, the subset of positive numbers in $\mathbb{Q}$ is denoted by $\mathbb{Q}^+$, and the subset of positive numbers in $\mathbb{R}$ is denoted by $\mathbb{R}^+$.

We can go on from here to define $b > a$ if and only if $a < b$, and to modify these definitions for $\leq$ and $\geq$. We will keep to '$<$' until some elementary properties have been established. In the proofs, be careful only to use the properties of '$<$' which have been given, or which you have successfully deduced from them. The answers to these questions have been set out so as to highlight the reasons for each step.

2 Use the *definition of less than* to give an inequality equivalent to the proposition 'b is positive'.

3 If $0 < -a$, use the definition to prove that $a < 0$.

4 If $a < 0$, use the definition to prove that $0 < -a$.

5 Use the law of trichotomy to prove that for any number a, either $a = 0$ or $0 < a$ or $a < 0$, and only one of these is true.

6 If $-b < -a$, use the definition to prove that $a < b$.

7 *Extended trichotomy law*
Apply the trichotomy law to the number $b - a$ to prove that for any two numbers a and b, *either* $a = b$, *or* $a < b$, *or* $b < a$, and exactly one of these is true.

8 If $a < b$, use the definition to prove $a + c < b + c$.

9 *Transitive law*
If $a < b$ and $b < c$, use closure under addition to prove that $a < c$.

When both $a < b$ and $b < c$, we usually write $a < b < c$.
It is the combination of *extended trichotomy* with the *transitive law* that makes it so helpful to mark numbers, in order, along a straight line.

10 If $a < b$ and $c < d$, prove that $a + c < b + d$.

11 (i) If $a < b$ and $0 < c$, prove that $a \cdot c < b \cdot c$.
(ii) Deduce that if $a < 0$ and $0 < c$, then $a \cdot c < 0$.
(iii) If $0 < a < b$ and $0 < c < d$, prove that $ac < bd$.

12 If $a < b$ and $a \cdot c < b \cdot c$, prove that $0 < c$.

Question 12 is closely related to qn 11(i). But the constructive methods we have used so far do not provide a proof. If you suppose that the conclusion is *wrong*, you can contradict what has been given. The contradiction is the means of showing that in fact the conclusion must be right. Such a proof is called a *proof by contradiction*.

13 Sketch a graph of the line $y = cx$ for positive c to illustrate the consequent relationship between $a < b$ and $ca < cb$, as in qns 11 and 12. Sketch a graph of the line $y = cx$ for negative c to illustrate the consequent relationship between $a < b$ and $cb < ca$.

14 Give an example to show that it is possible to have $b^2 < a^2$ when $a < b$. If $0 < a < b$, prove that $a^2 < b^2$.

15 If $a \neq 0$, prove that $0 < a^2$.

16 Use qn 15 to prove that $0 < 1$. Deduce that there is no number a such that $a^2 = -1$ in a number system with inequalities.

17 If $0 < a, 0 < b$ and $a^2 < b^2$, prove that $a < b$. Make a proof by contradiction. Use trichotomy and qn 14.

A consequence of this is that if $0 < a < b$, then $0 < \surd a < \surd b$. Note that $\surd(a^2)$ is always positive, or zero.

18 Sketch a graph of $y = x^2$ and illustrate on it how $a < b$ may be compatible with any one of $a^2 < b^2$, $a^2 = b^2$ and $b^2 < a^2$.

19 (i) With an argument like that of qn 14, show that if $0 < a < b$, then $a^3 < b^3$.
(ii) With an argument like that of qn 17, show that if $0 < a,\ 0 < b$ and $a^3 < b^3$, then $a < b$.

20 (i) If $0 < a < b$, prove by induction that $a^n < b^n$ for all positive integers, n.
(ii) If $0 < a$ and $0 < b$, and $a^n < b^n$ for some positive integer n, prove that $a < b$. A consequence of this is that if $0 < a < b$, then $0 < \sqrt[n]{a} < \sqrt[n]{b}$.

21 Sketch graphs of $y = x^n$ for various positive integers n, and decide for which positive integers n, $a < b \Leftrightarrow a^n < b^n$.

22 If $0 < a$, prove that $0 < 1/a$. Use qns 12 and 16.

23 If $0 < a < b$, prove that $0 < 1/b < 1/a$.

24 If $a < 0$, prove that $1/a < 0$.

25 If $a < b < 0$, prove that $1/b < 1/a < 0$.

26 Sketch a graph of $y = 1/x$ and illustrate on it how, provided neither a nor b is 0, $a < b$ may be compatible either with $1/b < 1/a$ or with $1/a < 1/b$.

27 Sketch a graph of $y = 1/(1 + x)$. Decide for what values of x, $1/(1 + x) > 1$, and then prove your claim formally.

$a \leq b$ is defined to mean *either* $a < b$ *or* $a = b$. Thus both $2 \leq 3$ and $2 \leq 2$ are true. Likewise $a \geq b$ is defined to mean *either* $a > b$ *or* $a = b$.

28 True or false?

(i) $a \geq b \Rightarrow a \geq b - 1$.
(ii) $a \geq b \Rightarrow a \geq b + 1$.
(iii) $a < b \Rightarrow a \leq b$.
(iv) $a \leq b \Rightarrow a < b$.

29 *Jakob Bernoulli's inequality* (1689)
Use qn 11 to prove by induction that if $-1 < x$, then $1 + nx \leq (1 + x)^n$ for all positive integers n.

For positive values of x, this can also be deduced from the binomial theorem. Sketch graphs of $y = 1 + nx$ and $y = (1 + x)^n$ for $n = 2$ and $n = 3$ to illustrate Bernoulli's inequality. Put the graphs for $n = 2$ and $n = 3$ on separate diagrams. Use a computer graphing package or a graphics calculator if possible.

Summary: Properties of order

Trichotomy For any two numbers, a and b, either $a < b$ or $b < a$ or $a = b$, and only one of these holds.

Transitivity If $a < b$ and $b < c$, then $a < c$.

Addition	If $a < b$ and $c < d$, then $a + c < b + d$.
Multiplication	If $a < b$ and $0 < c$, then $ac < bc$.
	If $a < b$ and $c < 0$, then $bc < ac$.
Squares	$0 \le a^2$, for all a.
Powers	If $0 < a < b$, then $a^n < b^n$ for all positive integers n.
	Suppose $0 < a$ and $0 < b$. If $a^n < b^n$ for some positive integer n, then $a < b$.
Reciprocals	If $0 < a < b$, then $0 < 1/b < 1/a$.
	If $a < b < 0$, then $1/b < 1/a < 0$.
Bernoulli's inequality	If $-1 < x$, then $(1 + x)^n \ge 1 + nx$, for all positive integers n.

Arithmetic mean and geometric mean

30 If 2, A and 8 are three consecutive terms of an arithmetic progression, what is the number A? A is called the *arithmetic mean* of 2 and 8.
If 2, G, 8 are three consecutive terms of a geometric progression of positive terms, what is the number G? G is called the *geometric mean* of 2 and 8.
Which is the greater, A or G?

31 Write down the arithmetic mean and the geometric mean of the numbers 2 and 3. Check that $(2\frac{1}{2})^2 = 6\frac{1}{4}$, and deduce that the arithmetic mean of 2 and 3 is greater than their geometric mean. Obtain the same result by using the inequality $0 < (\sqrt{3} - \sqrt{2})^2$.

32 (*Euclid*) If a and b are positive, find the arithmetic mean and the geometric mean of the numbers a^2 and b^2. Which of the means is greater? Under what circumstances are the two means equal?

33 If a and b are positive, prove that
$\sqrt{(ab)} \le \frac{1}{2}(a + b)$.

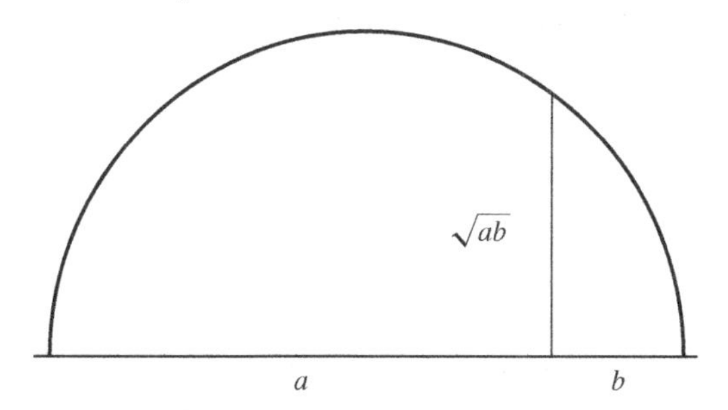

Figure 2.1

Under what circumstances can there be equality?

34 For any positive number a, prove that $2 \le a + (1/a)$.

35 In a remote village, the (only) greengrocer uses a scale-balance and a 1 kg weight. Unfortunately, as a result of an accident, the balance breaks and the greengrocer, in repairing it, does not get the point of balance exactly in the middle, but slightly offset.

A customer enters and asks for 2 kg of apples. The greengrocer places his 1 kg weight in the left-hand pan and fills the right-hand pan with apples until the scales balance. He then empties the apples into a brown paper bag. Now he puts the 1 kg weight to the right-hand pan and fills up the left-hand pan with apples until the scales balance. He then adds these apples to those in the brown paper bag and gives the lot to the customer. (Law of moments: when the scale pans balance, the weight in one pan times the length of its scale arm is equal to the weight in the other pan times the length of its scale arm.) Decide between

(1) The customer gets 2 kg of apples.
(2) The customer gets *more* than 2 kg of apples.
(3) The customer gets *less* than 2 kg of apples.

36 For a given perimeter, what shape of rectangle gives the greatest area?

37 (*Heron,* AD 250) If x and a are both positive numbers, what are the arithmetic and geometric means of the numbers x and a^2/x?
If $x_1 > 0$ and x_n is defined by $x_{n+1} = \frac{1}{2}(x_n + a^2/x_n)$, show that $a \le x_{n+1} \le x_n$, for $n \ge 2$.

38 If a_1 and b_1 are given positive numbers with $a_1 < b_1$, and two sequences of positive numbers are defined by
$a_{n+1} = \sqrt{(a_n b_n)}$ and $b_{n+1} = \frac{1}{2}(a_n + b_n)$,
prove that

(i) $0 < a_n < a_{n+1} < b_{n+1} < b_n$,
(ii) $0 < b_{n+1} - a_{n+1} < b_{n+1} - a_n = \frac{1}{2}(b_n - a_n)$,
(iii) $0 < b_{n+1} - a_{n+1} < (\frac{1}{2})^n(b_1 - a_1)$.

So the two means get closer together as n increases.

(iv) Prove that $b_{n+1} - a_{n+1} = \frac{1}{2}(\sqrt{b_n} - \sqrt{a_n})^2$ and that when $1 \le a < b$, $\sqrt{b} - \sqrt{a} < \frac{1}{2}(b - a)$.
(v) If $a_1 = 1$ and $b_1 = 2$, show that $b_5 - a_5 < 1/2^{45}$, so the two sequences close in very fast.

39 (Optional) If a regular n-gon is inscribed in a circle of unit radius, each side has length $2 \sin \pi/n$, so its perimeter is $2n \sin \pi/n$. Denote this perimeter by I_n. If a regular n-gon is circumscribed about a circle of unit radius, each side has length $2 \tan \pi/n$, so the perimeter is $2n \tan \pi/n$. Denote this perimeter by C_n.

Prove that $I_{2n} = \surd(I_n C_{2n})$ and that $C_{2n} = \dfrac{2I_n C_n}{I_n + C_n}$.

Let $a_n = \dfrac{1}{C_{2^n}}$ and $b_n = \dfrac{1}{I_{2^n}}$.

Deduce that $a_{n+1} = \frac{1}{2}(a_n + b_n)$ and $b_{n+1} = \surd(b_n a_{n+1})$.

Thus $0 \le a_n < a_{n+1} < b_{n+1} < b_n$.

Prove that $b_{n+1} - a_{n+1} < \frac{1}{4}(b_n - a_n)$, so the two sequences close in on each other. Check that $a_2 = \frac{1}{8}$ and $b_2 = 1/(4\surd 2)$. What number do you think a_n and b_n approach as n increases? (If you run a computer program which gives successive values of a_n and b_n you may be surprised to see that the terms of the two sequences approach each other much more slowly than the terms of the sequences in qn 38.)

Completing the square

40 (i) For what value of c is $x^2 + 6x + c$ a perfect square?

(ii) For what value of d is $4x^2 + 6x + d$ a perfect square?

(iii) What expression for e will make $a^2x^2 + abx + e$ a perfect square?

(iv) Assuming that $a \neq 0$, find a formula for f such that the equation $ax^2 + bx + c = 0$ is equivalent to the equation $(ax + f)^2 = f^2 - ac$. Solve this equation for x, when $f^2 \ge ac$.

41 Use the identity $a^2x^2 + abx + ac = (ax + \frac{1}{2}b)^2 + ac - \frac{1}{4}b^2$ to show that when $0 < a$, $c - b^2/4a \le ax^2 + bx + c$, for all values of x.
Is equality possible? Illustrate on a graph.

42 If $0 < a$, prove that $0 \le ax^2 + bx + c$, for all values of x, if and only if $b^2 \le 4ac$.
If $b^2 < 4ac$, is it possible to have $ax^2 + bx + c = 0$?
What is the analogous claim if $a < 0$?

The sequence $(1 + 1/n)^n$

43 Simplify the product $(b - a)(b^n + b^{n-1}a + b^{n-2}a^2 + \ldots + a^n)$.
Check that your answer will imply qn 1.3(vi).
Let $0 < a < b$. Show that

$$(n+1)a^n < \frac{b^{n+1} - a^{n+1}}{b - a} < (n+1)b^n.$$

44 Use a calculator to find

$$\left(1+\frac{1}{2}\right)^2, \left(1+\frac{1}{3}\right)^3, \left(1+\frac{1}{4}\right)^4, \left(1+\frac{1}{5}\right)^5$$

to two places of decimals. Denote these numbers by a_2, a_3, a_4 and a_5 respectively. And in general let

$$a_n = \left(1 + \frac{1}{n}\right)^n.$$

45 Put $b = 1 + \dfrac{1}{n}$ and $a = 1 + \dfrac{1}{n+1}$
in the right-hand inequality of qn 43 to prove that $a_n < a_{n+1}$ (with the notation of qn 44). Thus $2 \leq a_n$, for all n.

46 Use a calculator to find

$$\left(1 + \frac{1}{2}\right)^3, \left(1 + \frac{1}{3}\right)^4, \left(1 + \frac{1}{4}\right)^5, \left(1 + \frac{1}{5}\right)^6$$

to two places of decimals. Denote these numbers by b_2, b_3, b_4 and b_5 respectively. And in general let

$$b_n = \left(1 + \frac{1}{n}\right)^{n+1}.$$

47 If $(1 + 1/n)(1 - 1/x) = 1$, prove that $x = n + 1$.
Deduce that

$$b_n = \left(1 + \frac{1}{n}\right)^{n+1} = \left(1 - \frac{1}{n+1}\right)^{-(n+1)},$$

with the notation of qn 46.

48 Put $b = 1 - \dfrac{1}{n+1}$ and $a = 1 - \dfrac{1}{n}$ in the left-hand inequality of qn 43 to prove that $b_n < b_{n-1}$ (with the notation of qn 46).

49 Justify the inequalities $a_n < a_{n+1} < b_{n+1} < b_n$, with the notation of qns 44 and 46, and use qn 46 to show that $a_n < 3$ for all positive integers n. (This result will be used in qn 4.36.)

nth roots

50 If $1 < a$, prove that

(i) $1 < a < a^2 < \ldots < a^n$,

(ii) $\dfrac{1 + a + a^2 + \ldots + a^{n-1}}{n} < \dfrac{1 + a + a^2 + \ldots + a^{n-1} + a^n}{n+1}$,

(iii) $\dfrac{a^n - 1}{n} < \dfrac{a^{n+1} - 1}{n+1}$ (from qn 1.3(vi)),

(iv) if $a = b^{1/n(n+1)}$, $(n+1)(\sqrt[n+1]{b} - 1) < n(\sqrt[n]{b} - 1)$.

51 Deduce from qn 49 that if $n \geq 3$, then $(1 + 1/n)^n < n$, and so $\sqrt[n+1]{(n+1)} < \sqrt[n]{n}$.

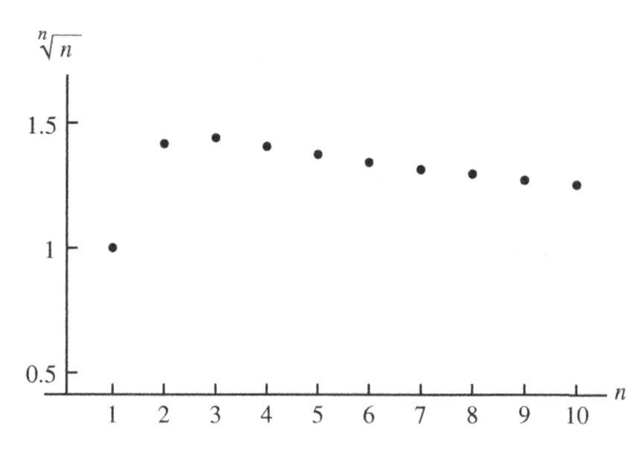

Figure 2.2

Summary: Results on inequalities

Arithmetic and geometric means

If a and b are positive, then $\sqrt{(ab)} \leq \frac{1}{2}(a + b)$.
For positive a, $a + 1/a \geq 2$.

Completing the square

Provided $a \neq 0$,
$ax^2 + bx + c = (ax + b/2)^2/a + c - b^2/4a$.
Provided a is positive, $ax^2 + bx + c \geq 0$ for all values of $x \Leftrightarrow b^2 \leq 4ac$.

$(1 + 1/n)^n$ — $(1 + 1/n)^n$ increases with n and $2 \leq (1 + 1/n)^n < 3$.

nth roots — When $a > 1$, $(n + 1)(\sqrt[n+1]{a} - 1) < n(\sqrt[n]{a} - 1)$.
$\sqrt[n]{n}$ decreases with n, when $n > 3$.

Absolute value

DEFINITION

We define $|a| = \begin{cases} a & \textbf{when } 0 \leq a, \textbf{ and} \\ -a & \textbf{when } a < 0. \end{cases}$

52 Sketch the graph of the absolute value function $y = |x|$.

The results that follow on absolute value may be scanned rather than proved. The results are essential for the rest of the book, but the proofs are not notably instructive. We will repeatedly appeal to qns 61 and 63.

53 Prove that $\max(-a, a) = |a|$.

54 Prove that $a \le |a|$ and $-|a| \le a$, for all a. Thus $-|a| \le a \le |a|$, for all a.

55 Prove that $0 \le |a|$ for all a. Deal with a positive and a negative separately.

56 Prove that $|a|^2 = a^2$ for all a.

57 Prove that $|a| = |-a|$ for all a.
Beware of presuming that $-a$ has to be negative. -2 must be negative, but $-a$ need not be. Another way of writing this result is $|\pm a| = |a|$.

58 Prove that $|ab| = |a| \cdot |b|$ for all a and b.

59 Prove that $|a/b| = |a|/|b|$, provided $b \ne 0$.

60 On the graph of $y = |x|$ (qn 52) draw the line $y = b$, for some positive b. For what values of x do you get $|x| \le b$?
Use the equivalence of $a < b$ with $-b < -a$, and of $-b < a$ with $-a < b$ (qn 6) to show that $-b \le a \le b$ implies $-b \le -a \le b$, and deduce that $-b \le a \le b$ implies $|a| \le b$.
Conversely, show that if $|a| \le b$, or $\max(-a,\ a) \le b$, then $-b \le a \le b$.

61 If a and b are two numbers of the same sign verify that $|a + b| = |a| + |b|$.
If a and b are two numbers of different sign verify that $|a + b| < |a| + |b|$.
From qn 54 we have $-|a| \le a \le |a|$ and $-|b| \le b \le |b|$ for all numbers a and b. Now use qn 10 to obtain $-|a| - |b| \le a + b \le |a| + |b|$, and then qn 60 to prove that $|a + b| \le |a| + |b|$ for any two numbers a and b. If one substitutes $a = x - y$ and $b = y - z$ in this inequality, we get $|x - z| \le |x - y| + |y - z|$. Now $|x - y|$ is the distance from x to y on the number line, so this inequality says that for any three points on the number line, the sum of the distances between two pairs cannot be less than the distance between the third pair, and it is from this that the name *triangle inequality* stems.

62 Deduce from qn 61 that $|a - b| \le |a| + |b|$.
Deduce also from qn 61 that $|a + b + c| \le |a| + |b| + |c|$.

63 By putting $b = c - a$ in the triangle inequality, prove that $|c| - |a| \le |c - a|$, and deduce that $|a| - |c| \le |a - c| = |c - a|$.
Prove further that $||c| - |a|| \le |c - a|$. This result can be pictured by marking two points on the real line drawn on tracing paper. If the paper is

folded at 0 so that the line falls onto itself, the left-hand side of the inequality is the folded distance between a and c, while the right-hand side gives the unfolded distance between a and c.

64 *The triangle inequality in two dimensions*
The strange name given to the inequality $|x+y| \leq |x| + |y|$ is more understandable in terms of the inequality for complex numbers $|z+w| \leq |z| + |w|$. This result is the claim that for a triangle in the Argand diagram, the sum of the lengths of two sides is greater than the third. We prove this by considering the triangle in $\mathbb{R} \times \mathbb{R}$ with vertices $O = (0, 0)$, $A = (a, b)$ and $P = (p, q)$.

(i) Show that for any real numbers a, b, p and q, $(ap + bq)^2 \leq (a^2 + b^2)(p^2 + q^2)$.
(ii) By taking the square root of the inequality in (i) deduce that $\sqrt{(a-p)^2 + (b-q)^2} \leq \sqrt{a^2 + b^2} + \sqrt{p^2 + q^2}$. Thus $|AP| \leq |OA| + |OP|$.

65 Can you find more than one number a which satisfies

(a) $-\varepsilon < a < \varepsilon$, for all positive ε;
(b) $|a| < \varepsilon$, for all positive ε?
[Argue by contradiction. If $a \neq 0$, try $\varepsilon = \frac{1}{2}|a|$.]

66 (*Newton*, 1687) If a and b are two numbers such that $|a - b| < \varepsilon$, for all positive ε, what can you say about a and b?

Summary: Results on absolute value

Definition	When $0 \leq a$, $\lvert a\rvert = a$. When $a < 0$, $\lvert a\rvert = -a$.
Multiplication qn 58	$\lvert ab\rvert = \lvert a\rvert \cdot \lvert b\rvert$
Triangle inequality qn 61	$\lvert a+b\rvert \leq \lvert a\rvert + \lvert b\rvert$
Theorem qn 63	$\lvert\lvert a\rvert - \lvert b\rvert\rvert \leq \lvert a-b\rvert$.

This chapter on inequalities has come at the front of this book because, in what follows, a pair of inequalities will be used repeatedly like a carpenter's vice to pin down exactly one number.

Historical Note

Early steps towards the arithmetical axiomatisation of the number system were taken by G. Peacock in 1830 and 1845, with his principle of permanence of forms. Precise arguments assuming the total ordering of the positive real numbers pervade Euclid's *Elements*. The notation $<$ and $>$ was introduced by Harriot and first appeared in print in 1631. The symbols $\leq$ and $\geq$ were first used in France about one hundred years later. In the seventeenth century Bernoulli's inequality was only considered for $x > 0$. An early version of it appeared in the work of Gregory of St Vincent (1647) where he showed that an increasing geometric progression always eventually exceeded an arithmetic progression and used this to prove that a decreasing geometric progression must be a null sequence. See qn 3.39. The arithmetic mean – geometric mean inequality appears in Euclid's *Elements* (II.5, V.25 and lemma between X.59 and X.60). The general form of this inequality, for n positive numbers, was established by C. Maclaurin in 1729. The sequence $(1 + 1/n)^n$ was first investigated by L. Euler (1736) who named its limit with his own initial. The notation for absolute value $|x|$ was due to Weierstrass who used it in his lectures in Berlin from 1859. It did not appear in print until 1877. Bolzano (1817) wrote $x < \pm 1$, where we would write $|x| < 1$. Harnack (1881) wrote abs m and Jordan (1882) wrote mod m for $|m|$. Question 65 encapsulates an idea used by Archimedes to obtain areas and volumes by contradiction.

Answers and comments

1 Holds for x positive, not for $x = 0$ or for x negative.

2

	proposition	justification
	b is positive	given
$\Leftrightarrow$	$b - 0$ is positive	algebra
$\Leftrightarrow$	$0 < b$	definition

3

	proposition	justification
	$0 < -a$	given
$\Rightarrow$	$-a - 0$ is positive	definition
$\Rightarrow$	$0 - a$ is positive	algebra
$\Rightarrow$	$a < 0$	definition

4

	proposition	justification
	$a < 0$	given
$\Rightarrow$	$0 - a$ is positive	definition
$\Rightarrow$	$-a - 0$ is positive	algebra
$\Rightarrow$	$0 < -a$	definition

Notice the difference between qns 3 and 4. Together they show that $0 < -a \Leftrightarrow a < 0$. Question 4 is the *converse* of qn 3. Question 3 is the *converse* of qn 4.

5

	proposition	justification
	a is positive	
$\Leftrightarrow$	$0 < a$	qn 2
	$-a$ is positive	
$\Leftrightarrow$	$a < 0$	qns 2, 3 and 4
	$a = 0$ or a is positive or $-a$ is positive	trichotomy
$\Leftrightarrow$	$a = 0$ or $0 < a$ or $a < 0$	from above

6

	proposition	justification
	$-b < -a$	given
$\Rightarrow$	$-a - (-b)$ is positive	definition
$\Rightarrow$	$b - a$ is positive	algebra
$\Rightarrow$	$a < b$	definition

7

	proposition	justification
	$b - a = 0$	
$\Leftrightarrow$	$a = b$	algebra
	$b - a$ is positive	
$\Leftrightarrow$	$a < b$	definition
	$-(b - a)$ is positive	
$\Leftrightarrow$	$a - b$ is positive	algebra
$\Leftrightarrow$	$b < a$	definition
	$b - a = 0$ or $b - a$ is positive or $-(b - a)$ is positive	trichotomy
$\Leftrightarrow$	$b - a = 0$ or $a < b$ or $b < a$	from above

8

	proposition	justification
	$a < b$	given
$\Rightarrow$	$b - a$ is positive	definition
$\Rightarrow$	$(b + c) - (a + c)$ is positive	algebra
$\Rightarrow$	$a + c < b + c$	definition

9

	proposition	justification
	$a < b$	given
$\Rightarrow$	$b - a$ is positive	definition
	$b < c$	given
$\Rightarrow$	$c - b$ is positive	definition
$\Rightarrow$	$(b - a) + (c - b)$ is positive	closure under addition
$\Rightarrow$	$c - a$ is positive	algebra
$\Rightarrow$	$a < c$	definition

10

	proposition	justification
	$a < b$	given
$\Rightarrow$	$b - a$ is positive	definition
	$c < d$	given
$\Rightarrow$	$d - c$ is positive	definition
$\Rightarrow$	$(b - a) + (d - c)$ is positive	closure under addition
$\Rightarrow$	$(b + d) - (a + c)$ is positive	algebra
$\Rightarrow$	$a + c < b + d$	definition

11 (i)

	proposition	justification
	$a < b$	given
$\Rightarrow$	$b - a$ is positive	definition
	$0 < c$	given
$\Rightarrow$	c is positive	definition
$\Rightarrow$	$(b - a) \cdot c$ is positive	closure under multiplication
$\Rightarrow$	$b \cdot c - a \cdot c$ is positive	algebra
$\Rightarrow$	$a \cdot c < b \cdot c$	definition

(ii) Put $b = 0$. (iii) $ac < bc < bd$.

12

	proposition	justification
	$a < b$	given
	$b - a$ positive	definition
	$a \cdot c < b \cdot c$	given
	$c = 0$	hypothesis
$\Rightarrow$	$a \cdot c = b \cdot c$	algebra

This contradicts what is given so hypothesis is false.

	proposition	justification
	$c < 0$	hypothesis
$\Rightarrow$	$0 - c$ is positive	definition
$\Rightarrow$	$(b - a) \cdot (-c)$ is positive	closure under multiplication
$\Rightarrow$	$a \cdot c - b \cdot c$ is positive	algebra
$\Rightarrow$	$b \cdot c < a \cdot c$	definition

This contradicts what is given so hypothesis is false.

	proposition	justification
$\Rightarrow$	$0 < c$	trichotomy and qn 11

Sherlock Holmes said that when what is impossible has been eliminated, what remains, however improbable, must be the truth. The only difference between the 'Sherlock Holmes method' and a mathematical proof by contradiction, is that in mathematics the final conclusion is what we were really expecting all along.

14 $-2 < 1$, but $1^2 < (-2)^2 = 4$.

	proposition	justification
	$a < b$ and $0 < a$	given
$\Rightarrow$	$a^2 < ab$	qn 11
	$a < b$ and $0 < b$	given, transitivity
$\Rightarrow$	$ab < b^2$	qn 11
$\Rightarrow$	$a^2 < ab$ and $ab < b^2$	above
$\Rightarrow$	$a^2 < b^2$	transitivity

15

	proposition	justification
	$a \neq 0$	given
$\Rightarrow$	$0 < a$ or $0 < -a$	trichotomy, qn 5
$\Rightarrow$	$0 < a^2$ or $0 < (-a)^2$	positive closure for multiplication
	$a^2 = (-a)^2$	algebra
$\Rightarrow$	$0 < a^2$	

16 $0 < 1^2 = 1$. So -1 is negative, and no square is negative from qn 15.

17 Given $0 < a$ and $0 < b$; $a = b \Rightarrow a^2 = b^2$; $b < a \Rightarrow b^2 < a^2$ from qn 14. Both these conclusions conflict with what is given so $a < b$, by trichotomy.

19 (i) $0 < a < b \Rightarrow a^3 < a^2b < ab^2 < b^3$.
(ii) Given $0 < a$ and $0 < b$; $a = b \Rightarrow a^3 = b^3$; $b < a \Rightarrow b^3 < a^3$ from part (i). Both these conclusions conflict with what is given, so $a < b$, by trichotomy.

20 (i) The result has already been established for $n = 1$ (trivial), $n = 2$ (qn 14) and $n = 3$ (qn 19(i)). $a^n < b^n \Rightarrow a^{n+1} < ab^n$.
$a < b \Rightarrow ab^n < b^{n+1}$. So $a^{n+1} < b^{n+1}$ by transitivity. The result follows by induction.
(ii) Given $0 < a$ and $0 < b$; $a = b \Rightarrow a^n = b^n$; $b < a \Rightarrow b^n < a^n$ by part (i). Both these conclusions conflict with what is given, so $a < b$, by trichotomy.

21 The equivalence holds for odd integers n.

22 $0 < a$ and $0 \cdot (1/a) = 0 < 1 = a \cdot (1/a)$, from qn 16, implies $0 < 1/a$ from qn 12.

23 $0 < 1/a$ and $0 < 1/b$ from qn 22. So $a \cdot (1/a) \cdot (1/b) < b \cdot (1/a) \cdot (1/b)$, from qn 11. Now $0 < 1/b < 1/a$, claiming transitivity.

24 $a < 0$ and $a \cdot (-1/a) = -1 < 0 = 0 \cdot (-1/a)$ (from qn 16) implies $0 < -1/a$ (qn 12). So $1/a < 0$ from qn 3.

25 $1/a < 0$ and $1/b < 0$ from qn 24. So $0 < (-1/a)(-1/b) = 1/ab$ from qn 4 and positive closure. Thus $a < b$ implies $a \cdot (1/ab) < b \cdot (1/ab)$ or $1/b < 1/a$ and by transitivity $1/b < 1/a < 0$.

27 Apply trichotomy to $1 + x$. $1 + x = 0$ is inadmissible.
$1 + x < 0 \Leftrightarrow 1/(1 + x) < 0$.
$1 + x > 0$ and $1/(1 + x) > 1 \Rightarrow 1 > 1 + x \Leftrightarrow 0 > x$, so $-1 < x < 0$.
Now $-1 < x < 0 \Leftrightarrow 0 < 1 + x < 1 \Rightarrow 1 < 1/(1 + x)$.

28 (i) True.
(ii) False, try $a = 3$ and $b = 2\frac{1}{2}$.

(iii) True.
(iv) False, try $a = b = 2$.

29 Result trivial for $n = 1$.

$$-1 < x \Rightarrow 0 < 1 + x, \text{ so}$$
$$1 + nx \le (1+x)^n \Rightarrow (1+nx)(1+x) \le (1+x)^{n+1}$$
$$\Rightarrow 1 + (n+1)x + nx^2 \le (1+x)^{n+1}$$
$$\Rightarrow 1 + (n+1)x \le (1+x)^{n+1}.$$

30 $A = 5, G = 4$.

31 $A = 2\frac{1}{2}, G = \sqrt{6}$. $(2\frac{1}{2})^2 > 6$. Also $2\sqrt{6} < 5$.

32 $0 \le (a-b)^2 \Rightarrow 2ab \le a^2 + b^2$. Equality only when $a = b$.

33 $0 \le (\sqrt{a} - \sqrt{b})^2 \Rightarrow 2\sqrt{a}\sqrt{b} \le a + b \Rightarrow \sqrt{a}\sqrt{b} \le \frac{1}{2}(a+b)$. Equality only when $\sqrt{a} = \sqrt{b}$.

34 Put $b = 1/a$ in qn 33.

35 If the arms of the scale are l (on the left) and r (on the right), and the two amounts of apples weighed out are v (on the left) and w (on the right), $lv = r \cdot 1$ and $rw = l \cdot 1$.
Now $v + w = r/l + l/r > 2$, from qn 34 unless $l = r$.

36 If the sides of the rectangle are of length a and b, the constant perimeter $= 2(a+b)$, while the area is ab. $\sqrt{ab} \le \frac{1}{2}(a+b)$ which is constant, so $\sqrt{ab}$ and thus the area is maximum when $a = b$.

37 $G = a$, $A = \frac{1}{2}(x + a^2/x)$. GM $\le$ AM $\Rightarrow a \le x_{n+1} \le x_n$, except possibly for $n = 1$. This is the basis of Heron's method for finding square roots.

38 (i) needs GM < AM and induction.
(iii) needs (ii) and induction.
(v) $b_{n+1} - a_{n+1} < \frac{1}{8}(b_n - a_n)^2$.
Such pairs of sequences were investigated by Gauss.

39 Both a_n and b_n approach $1/2\pi$ as n increases.

40 (i) $c = 9$,
(ii) $d = (3/2)^2$,
(iii) $e = (b/2)^2$,
(iv) $f = b/2$,
$x = (-f \pm \sqrt{(f^2 - ac)})/a = (-b \pm \sqrt{(b^2 - 4ac)})/2a$.

41 From the identity, $a^2x^2 + abx + ac \ge ac - \frac{1}{4}b^2$. Since $a > 0$, $ax^2 + bx + c \ge (ac - \frac{1}{4}b^2)/a$. Equality is possible only when $ax + \frac{1}{2}b = 0$.

42 $ax^2 + bx + c = (ax + \frac{1}{2}b)^2/a + (4ac - b^2)/4a$. For $0 < a$, the right-hand side is positive or 0 for all x precisely when $4ac - b^2$ is positive or zero. If $b^2 < 4ac$, the right-hand side is always positive, and so cannot $= 0$.
For $a < 0$, $ax^2 + bx + c \leq 0$ provided $b^2 \leq 4ac$.

44 $a_2 = 2.25$, $a_3 = 2.37$, $a_4 = 2.44$, $a_5 = 2.49$, increasing.

46 $b_2 = 3.37$, $b_3 = 3.16$, $b_4 = 3.05$, $b_5 = 2.99$, decreasing.

49 The first and last inequalities were established in qns 45 and 48. The middle one only needs $1 < 1 + 1/n$. As a consequence any a_i is less than any b_j.
$b_5 = (1 + 1/5)^6 = (6/5)^6 = 46\,656/15\,625 < 46\,875/15\,625 = 3$.

50 (i) Since $0 < 1$, this involves repeated use of qn 11(i).

51 From qn 49 $(1 + 1/n)^n < 3$, so for $n \geq 3$, $(1 + 1/n)^n < n$.
$((n + 1)/n)^n < n \Leftrightarrow (n + 1)^n < n^{n+1}$.

52

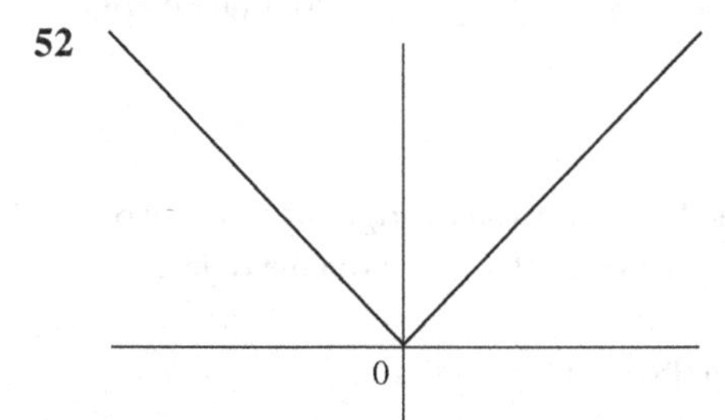

Figure 2.3

53 When $a = 0$, $a = -a = |a|$. When a is positive, $-a < 0 < a = |a|$. When a is negative, $a < 0 < -a = |a|$.

54 $a \leq |a|$ follows from qn 53. So also $-a \leq |a| \Rightarrow -|a| \leq a$, from qn 6.

55 The result follows from the definition when $a \geq 0$. When $a < 0$, use qn 4.

56 $|a|^2 = (+a)(+a) = a^2 = (-a)(-a)$.

57 $|a| = \max(-a,\ a) = \max(a,\ -a) = |-a|$.

58 $(\pm a)(\pm b) = \pm(ab)$ sums up the four non-zero cases.

59 Provided $b \neq 0$, $|a| = |(a/b) \cdot b| = |(a/b)| \cdot |b|$ by qn 58.

60 When x lies between $\pm b$.
If $-b \leq a \leq b$ and $-b \leq -a \leq b$, then $\max(-a,\ a) \leq b$ and $|a| \leq b$.
If $|a| \leq b$, then $\max(-a,\ a) \leq b$, so $a \leq b$ and $-a \leq b$. This last inequality implies $-b \leq a$, so together we have $-b \leq a \leq b$.
The work in this question proves that $|a| \leq b \Leftrightarrow -b \leq a \leq b$.

61 $-|a| \leq a \leq |a|$ and $-|b| \leq b \leq |b|$ together imply
$-|a| - |b| \leq a + b \leq |a| + |b|$, so from qn 60, $|a + b| \leq |a| + |b|$.

62 For the first inequality put $-b$ for b in the triangle inequality.
$|a+b+c| \leq |a|+|b+c| \leq |a|+|b|+|c|$.

63 The result $||c|-|a|| \leq |c-a|$ follows from qn 60.

64 (i) $0 \leq (aq-bp)^2$.
(ii) Keep the negative sign on square rooting the left-hand side; multiply by 2 and add $a^2+b^2+p^2+q^2$, to both sides.

3

Sequences

A first bite at infinity

Preliminary reading: Hemmings and Tahta, O'Brien.
Concurrent reading: Hart.
Further reading: Knopp.

Introduction

The notion of a sequence is a familiar and intuitive one. The following examples of sequences of numbers illustrate this notion.

$1, 2, 3, 4, 5, \ldots$

$2, 4, 6, 8, 10, \ldots$

$\frac{1}{2}, \frac{1}{4}, \frac{1}{8}, \frac{1}{16}, \frac{1}{32}, \ldots$

$3, 3.1, 3.14, 3.141, 3.1415, \ldots$

$-1, 1, -1, 1, -1, \ldots$

$\frac{1}{2}, \frac{2}{3}, \frac{3}{4}, \frac{4}{5}, \frac{5}{6}, \ldots$

The general way of writing down a sequence is

$a_1, a_2, a_3, \ldots, a_n, \ldots$

so that for each natural number n, there is an element of the sequence, a_n. This means that a sequence must be an infinite (not finite) list of terms, though repetition is allowed as in the fifth example above. Such a sequence is denoted by (a_n). a_n is the single number which appears as the nth term of the sequence. (a_n) refers to an infinite list of numbers, in order; (a_n) is the sequence as a whole.

1 For each of the sequences listed above, except the fourth, suggest a formula for the nth term, a_n. (A sequence may or may not have a formula to define it.)

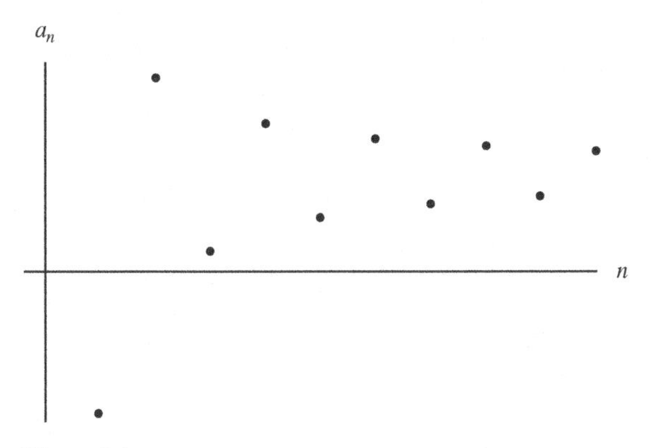

Figure 3.1

2 It is possible to draw a graph of the sequence (a_n) by plotting the points (n, a_n) on a conventional cartesian graph of $\mathbb{R} \times \mathbb{R}$.
Draw the graphs of the first five terms of the six sequences given above.

Although graph drawing provides a useful picture of the first few terms of the sequence, it does not illustrate what happens in the long run. Most of the important questions about sequences, which go on for ever, are not decided by the first few terms but depend on the behaviour of the nth term a_n as n gets larger and larger, and these terms are not illustrated on the graphs you have drawn.

3 (a) According to the definition at the start of this chapter, why do the integers from 1 to 10, in order, *not* form a sequence? Or the integers from 1 to a million, in order?
(b) Construct a sequence whose terms take only one value. Such a sequence is called a *constant sequence*.
(c) Construct a sequence whose terms take exactly three values.
(d) Construct a sequence whose terms, a_n, take exactly three values but which becomes constant for large n.

The central property of sequences we will study is that of convergence. This turns out to be surprisingly difficult to define. In order to become familiar with sequences, we first examine a number of more straightforward properties which we will find useful.

Monotonic sequences

4 Test each of the sequences (i)–(vii), defined below, to determine whether any one or more of the following four properties, (a)–(d), holds for all values of n.

(a) $a_n < a_{n+1}$.
(b) $a_n \leq a_{n+1}$.
(c) $a_n > a_{n+1}$.
(d) $a_n \geq a_{n+1}$.

(i) $a_n = n$,
(ii) $a_n = 1/n$,
(iii) $a_n = (-1)^n$,
(iv) $a_{2n-1} = n, \ a_{2n} = n$,
(v) $a_n = -1/n$,
(vi) $a_n = 1$,
(vii) $a_n = 2^{-n}$.

When property (a) in qn 4 holds, the sequence is said to be *strictly monotonic increasing* or just *strictly increasing*.

When property (b) holds, the sequence is said to be *monotonic increasing* or just *increasing*.

When property (c) holds, the sequence is said to be *strictly monotonic decreasing* or *strictly decreasing*.

When property (d) holds, the sequence is said to be *monotonic decreasing* or *decreasing*.

When any one of (a), (b), (c) or (d) holds, the sequence is said to be *monotonic*.

The seemingly bizarre status of sequence 4(vi) is a consequence of wanting useful definitions. The fact that constant sequences satisfy conditions (b) and (d) simultaneously is not a sufficient reason to revise our definitions, because experience has shown that these definitions, as they stand, lead to simple statements of theorems and proofs.

Bounded sequences

5 Test each of the sequences (a_n) (i)–(iv) defined here, to determine whether either or both of the following properties applies. Sketch a graph of the first few terms.

(a) There is a number U such that $a_n \leq U$ for all n.
(b) There is a number L such that $L \leq a_n$ for all n.

(i) $a_n = (-2)^n$,
(ii) $a_n = (-1)^n/n$,
(iii) $a_n = \sin n$,
(iv) $a_n = \sqrt{n}$.

When property (a) in qn 5 holds, the sequence (a_n) is said to be *bounded above*, and the number U is called an *upper bound* of the sequence.

When property (b) holds, the sequence is said to be *bounded below*, and the number L is called a *lower bound* of the sequence.

When both properties (a) and (b) hold, so that there are numbers L and U such that $L \le a_n \le U$, for all values of n, the sequence (a_n) is said to be *bounded*.

A sequence which is neither bounded above nor bounded below is said to be *unbounded*. (Some authors say that a sequence is unbounded when it is either not bounded above, or not bounded below, but differing conventions in this matter will not affect the formulation of any theorems in this book.) Notice that upper bounds and lower bounds, if they exist, are not unique: for if U is an upper bound, then so is $U + 1$, and if L is a lower bound, then so is $L - 1$.

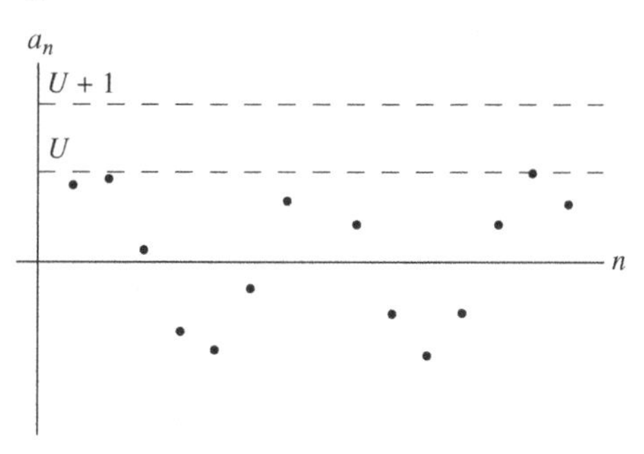

Figure 3.2

Beware of the language here. The sequence $1, 2, 3, \ldots, n, \ldots$ is neither bounded nor unbounded! But the definitions, as we have given them, lead to simple statements of theorems and proofs.

6 A sequence (a_n) is known to be monotonic increasing.

(i) Might it have an upper bound?
(ii) Might it have a lower bound?
(iii) Must it have an upper bound?
(iv) Must it have a lower bound?

Give an example to show each possibility or impossibility. How must your answers be revised if the same questions are asked about an arbitrary monotonic *decreasing* sequence?

7 If a sequence is not bounded above, must it contain

(i) a positive term,
(ii) an infinity of positive terms?

Subsequences

The sequences

$$2, 4, 6, 8, \ldots, 2n, \ldots,$$
$$1, 3, 5, 7, \ldots, 2n - 1, \ldots,$$
$$1, 4, 9, 16, \ldots, n^2, \ldots,$$
$$2, 4, 8, 16, \ldots, 2^n, \ldots,$$
$$1, 3, 6, 10, \ldots, \tfrac{1}{2}n(n + 1), \ldots,$$
$$5, 6, 7, 8, \ldots, n + 4, \ldots,$$

or generally, the strictly increasing sequence

$$n_1, n_2, n_3, n_4, \ \ldots, n_i, \ \ldots,$$

all provide examples (the last is a thoroughly general example) of infinite subsets of the natural numbers, $\mathbb{N}$.

With any such infinite subset, a subsequence of a sequence may be constructed.

Thus from the sequence $a_1, a_2, a_3, \ldots, a_n, \ldots$ various subsequences, such as

$$a_2, a_4, a_6, \ \ldots, a_{2n}, \ \ldots$$
$$a_1, a_4, a_9, \ \ldots, a_{n^2}, \ \ldots$$
$$a_5, a_6, a_7, \ \ldots, a_{n+4}, \ \ldots,$$

may be constructed (using the first, the third and the sixth of the subsets of $\mathbb{N}$ given above), or in general

$$a_{n_1}, a_{n_2}, a_{n_3}, \ldots, a_{n_i}, \ldots.$$

You should think of n_1 as the first suffix (or value of n) used in the subsequence, n_2 as the second suffix used in the subsequence, and so on.

One erases some (or none!) of the terms of a sequence to obtain a subsequence.

A *subsequence* of a sequence (a_n) is an infinite subset of the a_n with the terms of the subsequence occurring in the same order as in the original sequence.

In the definition of a sequence (a_n), n runs through all of the natural numbers, $\mathbb{N}$, in order.

With the general notation for a *subsequence* (a_{n_i}), n_i runs through some (or all) of the natural numbers in order (and i runs through them all).

8 Why are none of the following sets, as they stand, subsequences of (a_n)?

(i) $a_{10}, a_{20}, a_{30}, a_{40}, a_{50}, a_{60}, a_{70}, a_{80}, a_{90}, a_{100}$;
(ii) $a_5, a_4, a_{10}, a_9, \ldots, a_{5n}, a_{5n-1}, \ldots$;
(iii) $(a_{n/2})$;
(iv) (a_{2n-3}).

9 For the sequence $(a_n) = 10, -11, -12, 13, 14, -15, -16, 17, 18, -19, -20, 21, 22, \ldots$ identify n_1, n_2, n_3, n_4 and n_5 for

(i) the subsequence of positive terms;
(ii) the subsequence of negative terms;
(iii) the subsequence of even terms;
(iv) the subsequence of odd terms.

10 (a_{n_i}) is a subsequence of (a_n).

(i) (a) Is it possible to have $n_6 < n_5$? (b) Is it possible to have $n_6 = n_5$? (c) What is the least possible value of n_5 for any subsequence? (d) Name a term of (a_{n_i}) which you know comes later in the sequence (a_n) than a_{100}.
(ii) If $i < j$, does it follow that $n_i < n_j$ and that $a_{n_i} < a_{n_j}$?

11 A sequence (a_n) is known to be monotonic increasing, but not strictly monotonic increasing.

(i) Might there be a strictly monotonic increasing subsequence of (a_n)?
(ii) Must there be a strictly monotonic increasing subsequence of (a_n)?

12 If a sequence is bounded, must each of its subsequences be bounded?

13 (i) If the subsequence $a_2, a_3, a_4, \ldots, a_{n+1}, \ldots$ is bounded, does it follow that the sequence (a_n) is bounded?
(ii) If the subsequence $a_3, a_4, a_5, \ldots, a_{n+2}, \ldots$ is bounded, does it follow that the sequence (a_n) is bounded?
(iii) If the subsequence $a_{1+k}, a_{2+k}, a_{3+k}, \ldots, a_{n+k}, \ldots$ is bounded, does it follow that the sequence (a_n) is bounded?

We describe the result of qn 13 by saying that if a sequence is *eventually bounded*, then it is *bounded*. Each of the sequences of qn 13 is called a *shift* of the sequence (a_n). Obviously every shift of a bounded sequence is bounded (qn 12), and from qn 13, if a shift is bounded, then the whole sequence is bounded.

14 (a) If the subsequence (a_{2n}) is bounded, does it follow that (a_n) is bounded?

(b) A sequence (a_n) is known to be unbounded.

(i) Might it contain a bounded subsequence?

(ii) Must it contain a bounded subsequence?

Look back to the definition of *bounded* and *unbounded.*

Rather unexpectedly, it is possible to show that *any* sequence has a monotonic subsequence. To prove this we will use the notion of a *floor term* of a sequence. We will call a term of a sequence a 'floor term' when none of its successors is strictly less. The term a_f is a 'floor term' if $a_n \geq a_f$ when $n \geq f$. A 'floor term' is a lower bound for the rest of the sequence. The phrase 'floor term' is used in qns 15 and 16, and not again in the course.

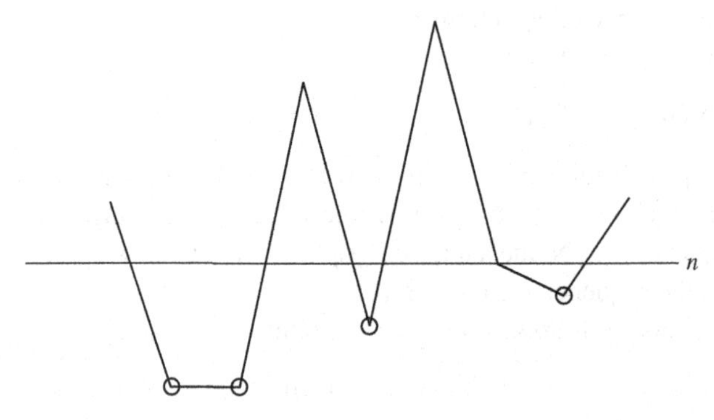

Figure 3.3

15 Underline the floor terms in each of the following sequences.

(i) $-1, \frac{1}{2}, -\frac{1}{3}, \frac{1}{4}, -\frac{1}{5}, \frac{1}{6}, \ldots, (-1)^n/n, \ldots$

(ii) $1, 0, 1, 0, 1, \frac{3}{4}, 1, \frac{5}{8}, 1, \frac{7}{12}, \ldots, 1, \frac{2n+1}{4n}, \ldots$

(iii) $-1, 2, -3, 4, -5, 6, \ldots, (-1)^n n, \ldots$

Sketch a graph of the first six terms of each of these sequences and circle the points corresponding to floor terms.

16 (*Erdős and Szekeres*, 1935)

(i) If an arbitrary sequence contains an infinite number of floor terms, show that they form a monotonic increasing subsequence.

(ii) If an arbitrary sequence contains just a finite number of floor terms, and the last one is a_F, construct a strictly decreasing subsequence with a_{F+1} as its first term.

(iii) If an arbitrary sequence contains no floor terms at all, construct a strictly decreasing subsequence with a_1 as its first term.

The theorem proved in qn 16 guarantees the existence of a monotonic subsequence even for a sequence such as $(\sin n)$.

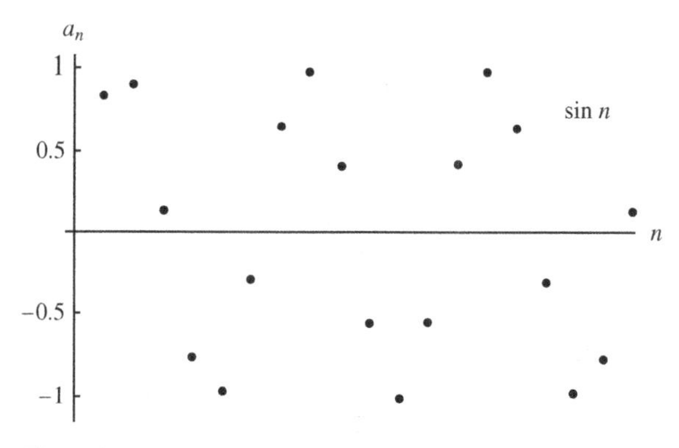

Figure 3.4

Sequences tending to infinity

17 Describe the following four sequences saying whether or not they are monotonic and whether or not they are bounded above or below.

(i) $1, 1, 2, 1, 3, 1, \ldots, n, 1, \ldots$
(ii) $-1, 2, -3, 4, -5, \ldots, (-1)^n n, \ldots$
(iii) $2, 1, 4, 3, \ldots, 2n, 2n-1, \ldots$
(iv) $11, 12, 11, 12, \ldots, 11, 12, \ldots$

In each case decide whether there is a stage beyond which all terms in the sequence are

(a) greater than 1, in which case we say that the sequence is eventually greater than 1,
(b) greater than 10, so that the sequence is eventually greater than 10,
(c) greater than 100, so that the sequence is eventually greater than 100.

Only sequence (iii) is said to tend to $+\infty$ (plus infinity).

We say that a sequence *tends to* $+\infty$ when any number, C, however chosen, is eventually a lower bound for the sequence. So a sequence must pass infinitely many tests if it is to be said to tend to $+\infty$ because there are infinitely many choices for C, and the terms of the sequence must eventually exceed each of them.

For each value of C, we look for a value N of n sufficiently great to make $a_n \geq C$, for all $n > N$.

18 Select values of C to demonstrate that qn 17(i), (ii) and (iv) do not tend to $+\infty$.

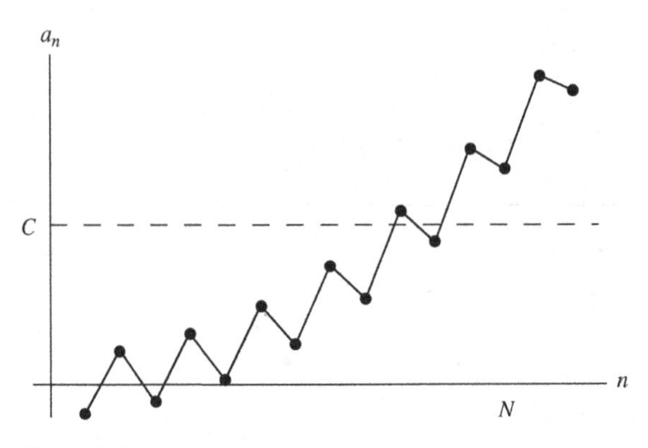

Figure 3.5

In relation to a sequence (a_n), the term 'eventually', the phrase 'provided n is large enough', and the phrase 'for sufficiently large n' each mean **'for all n greater than some fixed number N'**.

We can now gather the discussion into a formal definition and say that

> **the sequence (a_n) tends to $+\infty$,**
> **if and only if, given any number C,**
> **there is a number N such that $n > N \Rightarrow a_n \geq C$.**

Yet another way to express this definition is to say that, for each number C, $a_n \geq C$ for all but a finite number of terms of the sequence.

'(a_n) tends to infinity' is often written $(a_n) \to +\infty$ as $n \to \infty$.

Archimedean order and the integer function

An important property of the real numbers is

given any number A, there is an integer n, which is greater than A

This is called the *Archimedean property*. It was first pointed out by the Greeks, Archimedes among them, who probably wanted to say that if you repeatedly combine together straight line segments of length 1, you will never cover an area A, no matter how many times (n times) you do so. In contrast, when working with real numbers only, a number A can *always* be exceeded by an integer.

An equivalent way of expressing the property of Archimedean order is to say that for any two positive numbers a and b, there is a term in the sequence

$a, 2a, 3a, 4a, \ldots$ which exceeds b; that is, for some positive integer n, $na > b$. It is worth checking for yourself that each of these two descriptions of the property implies the other.

By making our assumptions explicit and precise, we can see the logical relations between different properties. For this reason you should cite the Archimedean property, when you use it, at least in this chapter. Later on, the Archimedean property can be assumed.

To see that the sequence of natural numbers tends to infinity we argue as follows. If C is any number, by the Archimedean property, we can choose a positive integer n such that $C < n$. Then $n < n+1$, so $C < n+1$. By induction all subsequent natural numbers are also greater than C. So C is an eventual lower bound for the sequence of natural numbers.

It also follows that there is an integer *less than* any given number A: if we apply the Archimedean property to $-A$ there is an integer n such that $-A < n$, and so $-n < A$. So any real number A divides the set of integers into two sets, those less than or equal to A and those greater than A. If for a given number A, m and n are integers such that $m < A < n$, we need only scrutinise the $n - m$ integers, $m, m+1, \ldots, n-1$ to determine the greatest integer less than or equal to A. We denote this integer by $\lfloor A \rfloor$.

$\lfloor \pi \rfloor = 3$, $\lfloor 2 \rfloor = 2$ and $\lfloor -2.1 \rfloor = -3$. $\lfloor A \rfloor \le A < \lfloor A \rfloor + 1$.

-1 0 1 2 · C $\lfloor C \rfloor$ $\lfloor C+1 \rfloor$

19 (i) Rewrite $\lfloor A \rfloor \le A < \lfloor A \rfloor + 1$ in the equivalent form $\cdots < \lfloor A \rfloor \le \ldots$.

(ii) (optional)

(a) For each real number x, could there be an integer $N(x)$ such that $N(x) < x < N(x) + 1$?

(b) For each real number x, could there be an integer $M(x)$ such that $M(x) < x < M(x+1)$?

(c) For each real number x, you know how to find an integer $K(x)$ such that $K(x) \le x \le K(x+1)$. Can you find more than one way of defining $K(x)$ to satisfy this condition?

20 Prove that $(\sqrt{n}) \to +\infty$ as $n \to \infty$.

21 (a) Let y be a fixed positive number. Show that the sequence $(ny) \to +\infty$.

(b) Let $x > 1$, so that for some positive y, $x = 1 + y$. Use Bernoulli's inequality (qn 2.29), to show that $x^n \ge 1 + ny$, and deduce that $(x^n) \to +\infty$

22 If a sequence tends to $+\infty$, must it be monotonic increasing?

23 Construct a definition for '$(a_n) \to -\infty$ as $n \to \infty$'.
Check that your definition guarantees that the sequence defined by $a_n = -n$ tends to $-\infty$.

Summary: The language of sequences

The symbol for a sequence, (a_n), denotes the infinite list of numbers $a_1, a_2, a_3, \ldots, a_n, \ldots$ in this order. $n \in \mathbb{N}$.

Definition qn 4 — A sequence (a_n) is said to be *increasing* when $a_n \leq a_{n+1}$ for all n, and is said to be *decreasing* when $a_{n+1} \leq a_n$ for all n. Both increasing and decreasing sequences are called *monotonic.*

Definition qn 5 — A sequence (a_n) is said to be *bounded above* when there is a number U such that $a_n \leq U$ for all n.
A sequence (a_n) is said to be *bounded below* when there is a number L such that $L \leq a_n$ for all n.
A sequence which is bounded above and bounded below is said to be *bounded.*

Definition qn 8 — A sequence (a_{n_i}) is called a *subsequence* of (a_n), if $n_i < n_j$ whenever $i < j$.

Theorem qn 16 — Every sequence has a monotonic subsequence.

Definition qn 18 — A sequence (a_n) is said to *tend to infinity* if and only if for any number C, there is an N, such that $n > N \Rightarrow a_n \geq C$. This is written $(a_n) \to +\infty$ as $n \to \infty$.

Property of Archimedean order
Given a number, there is an integer which is greater. Equivalently, the positive integers are not bounded above.

Theorem — For every real number A, there is a unique integer $\lfloor A \rfloor$ such that $\lfloor A \rfloor \leq A < \lfloor A \rfloor + 1$.

Null sequences

Even to the untutored eye, the harmonic sequence

$$1, \frac{1}{2}, \frac{1}{3}, \frac{1}{4}, \ldots, \frac{1}{n}, \ldots$$

will be seen to 'tend to 0'. We work towards a precise definition of this phrase and its synonym 'converges to 0'.

24 (i) Given a positive number ε, use the property of Archimedean order to claim an integer greater than $1/\varepsilon$, and deduce that there is a positive integer n such that $1/n < \varepsilon$.

(ii) Use qn 2.65(a) to show that if $-1/n < a < 1/n$ for all positive integers n, then $a = 0$.

The two inequalities here and in qn 2.65(a) act like a carpenter's vice.

(iii) If b and c are given positive numbers and $-b/n < a < c/n$ for all positive integers n, prove that $a = 0$. Apply the property of Archimedean order to c/ε and b/ε respectively and for b/ε apply qn 2.6.

(iv) If b and c are given positive numbers and $-b/n < a - k < c/n$ for all positive integers n, prove that $a = k$. Apply part (iii).

To see the power of qn 24, we adapt a method of proof due to Archimedes in qn 25.

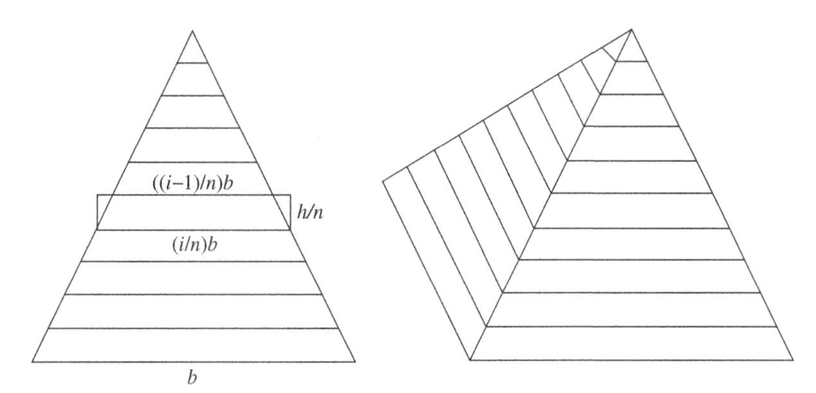

Figure 3.6

25 (*After Archimedes' quadrature of the spiral*) We will find the volume of a pyramid with a $b \times b$ square base and height h.
We start by slicing the pyramid parallel to the base with n slices of thickness h/n. A typical slice has one face as a square $(i/n)b \times (i/n)b$ and the parallel face a square $((i-1)/n)b \times ((i-1)/n)b$, so the volume of the slice is more than $h/n \times ((i-1)/n)b \times ((i-1)/n)b$ but less than $h/n \times (i/n)b \times (i/n)b$.

(i) Add the slices to show that

$$\frac{hb^2}{n^3}\left(1^2 + 2^2 + \ldots + (n-1)^2\right) < \textit{pyramid} < \frac{hb^2}{n^3}(1^2 + 2^2 + \ldots + n^2).$$

Now use qn 1.1 to obtain

$$\frac{hb^2}{6}\left(1-\frac{1}{n}\right)\left(2-\frac{1}{n}\right) < pyramid < \frac{hb^2}{6}\left(1+\frac{1}{n}\right)\left(2+\frac{1}{n}\right).$$

(ii) The expressions on either side approach $hb^2/3$ for large n. Subtract $hb^2/3$ from each of the three terms to obtain

$$\frac{hb^2}{6}\left(-\frac{3}{n}+\frac{1}{n^2}\right) < pyramid - \frac{hb^2}{3} < \frac{hb^2}{6}\left(\frac{3}{n}+\frac{1}{n^2}\right).$$

Deduce that $-\dfrac{hb^2}{2n} < pyramid - \dfrac{hb^2}{3} < \dfrac{2hb^2}{3n}$, since $\dfrac{1}{n^2} < \dfrac{1}{n}$.

This holds for every positive integer n. Finally apply qn 24(iv) to show the volume of the pyramid is $hb^2/3$.

If you look at qns 10.1 and 10.2 you will see how this method may be used (as it was by Fermat, 1636) to obtain the area under a parabola and other curves. The results of qn 24 depend upon the fact that there are terms of the harmonic sequence which are less than any given positive number. The work in qns 24 and 25 does not use limits. The argument depends instead on the contradiction in qn 2.65.

We have seen that $-1/n < a < 1/n$ for all positive integers n, implies that $a = 0$. This argument works because a is constant. We now ask what happens if a is not constant, and we have a sequence (a_n) such that $-1/n < a_n < 1/n$ for all positive integers n.

26 Give some examples of sequences (a_n) such that $-1/n < a_n < 1/n$ for all positive integers n. Give monotonic and non-monotonic examples. Give examples including positive, negative and zero terms. Might such a sequence be constant? What can you say about values of a_n far along the sequence?

Mathematicians say that every sequence satisfying the conditions of qn 26 'tends to 0' or 'converges to 0'. The *commonsense* meaning of 'tend to' and 'converge' suggest getting nearer *from one side*, and *never quite reaching* like the harmonic sequence which never reaches 0. The examples of convergence studied in the seventeenth and eighteenth centuries had both these properties. But when we consider convergence today *nearness to the limit* is the only issue, and whether from above or below and whether the limit is reached occasionally or frequently is of secondary concern.

27 Examine the sequence

$$-1, \frac{1}{2}, -\frac{1}{3}, -\frac{1}{4}, -\frac{1}{5}, \ldots, (-1)^n/n, \ldots$$

(i) Beyond what stage in the sequence are the remaining terms between -0.1 and 0.1?
(ii) Beyond what stage in the sequence are the remaining terms between -0.01 and 0.01?
(iii) Beyond what stage in the sequence are the remaining terms between -0.001 and 0.001?
(iv) Beyond what stage in the sequence are the remaining terms between $-\varepsilon$ and ε, where ε is a given positive number? The choice of ε, the Greek *e*, here, is to stand for 'error', where the terms of the sequence are thought of as successive attempts to hit the target 0.

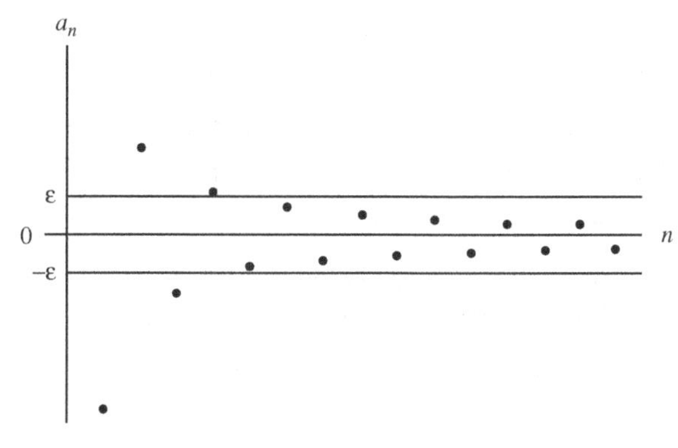

Figure 3.7

A sequence (a_n) is said to be a *null sequence*, or to tend to 0, or converge to 0, when,

> for any positive number ε, however small, there is a stage in the sequence beyond which the remaining terms of the sequence are between $-\varepsilon$ and ε;

or, equivalently, for any given positive ε,

> ε is eventually an upper bound for the sequence and $-\varepsilon$ is eventually a lower bound for the sequence:

or, equivalently,

> all but a finite number of terms of the sequence lie between $\pm\varepsilon$, or satisfy $|a_n| < \varepsilon$.

These conditions must hold for every possible choice of ε. So, to be called a null sequence, the sequence must satisfy an infinity of conditions.

DEFINITION OF A NULL SEQUENCE

Formally we write $(a_n) \to 0$ as $n \to \infty$ if and only if, given $\varepsilon > 0$, there is an N such that $n > N \Rightarrow |a_n| < \varepsilon$.

This definition could also have been given by saying that the subsequence of all terms after a_N is bounded by $\pm\varepsilon$. If a sequence is null, there are many acceptable Ns for each ε.

28 For each of the sequences below, prove that it is not null by naming an ε for which no related N exists.

(i) A sequence in which each term is strictly less than its predecessor:

$$2, 1, 0, -1, -2, -3, \ldots, -n, \ldots$$

(ii) A sequence in which each term is strictly less than its predecessor while remaining positive:

$$2, \frac{3}{2}, \frac{4}{3}, \frac{5}{4}, \ldots, \frac{n+1}{n}, \ldots$$

(iii) A sequence in which, for sufficiently large n, each term is less than some small positive number:

$$2, 1, 0, -0.1, -0.1, -0.1, \ldots, -0.1, \ldots$$

(iv) A sequence in which, for sufficiently large n, the absolute value of each term is less than some small positive number:

$$2, 1, 0, -0.1, 0.01, -0.1, 0.01, \ldots, -0.1, 0.01, \ldots$$

(v) A sequence with arbitrarily small terms:

$$1, \frac{1}{2}, 1, \frac{1}{4}, 1, \frac{1}{8}, \ldots$$

(vi) The constant sequence

$$1, 1, 1, \ldots$$

See Figure 3.8 which illustrates the sequence in qn 27.

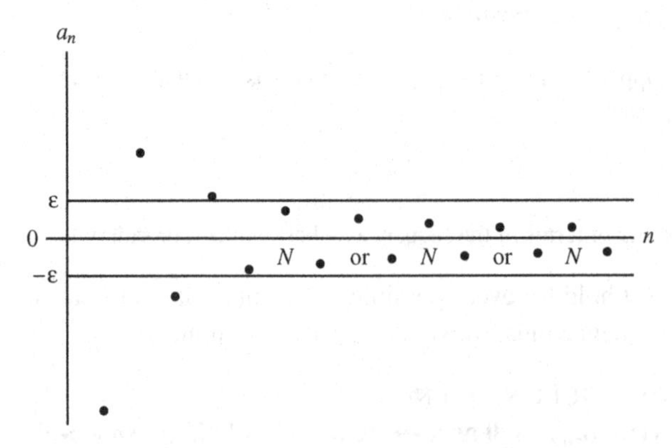

Figure 3.8

29 Prove that the sequence $(1/n)$ is a null sequence. Use qn 24(i).

30 Prove that the sequence $(1/\sqrt{n})$ is a null sequence.

31 If $(a_n) \to +\infty$, prove that $(1/a_n)$ is a null sequence.
If (a_n) is a null sequence and $a_n > 0$ for all n, prove that $(1/a_n) \to +\infty$.
Give an example to show that it is possible to have a null sequence (a_n) without having $(1/a_n) \to +\infty$.

32 *The scalar rule for null sequences*
Let (a_n) be a null sequence and c a constant number.
Prove that $(c \cdot a_n)$ is a null sequence.
Consider the cases $c \neq 0$ and $c = 0$, in turn.
Deduce that $(10/\sqrt{n})$ is a null sequence.

33 *The absolute value rule for null sequences*

(a) Let (a_n) be a null sequence. Prove that $(|a_n|)$ is a null sequence.
(b) Conversely, let $(|a_n|)$ be a null sequence. Prove that (a_n) is a null sequence.

34 *A squeeze rule or sandwich theorem for null sequences* I
Let (a_n) be a null sequence, and $0 \leq b_n \leq a_n$, for all n. Prove that (b_n) is a null sequence.

35 Prove that each of the following sequences is null:

(i) $(1/(n+1))$,
(ii) $(10/(n+1))$,
(iii) $(20/(7n+3))$,
(iv) $(1/(n^2+1))$,
(v) $(\sqrt{(n+1)} - \sqrt{n})$,
(vi) $((\sin n)/n)$,
(vii) $(\sin n\pi)$,
(viii) $(\sin(n!\pi \cdot c))$ where c is a rational number,
(ix) (a_n) where $a_{2n-1} = 1/(2n-1)$ and $a_{2n} = 1/n$.

Standard properties of the sine function should be used: they will be proved in chapter 11.

36 (a) Must every subsequence of a null sequence be null? (The answer here gives *the subsequence rule for null sequences.*)
(b) Check the converse. If a sequence has a subsequence which is null, must the original sequence have been null?

(c) *The shift rule for null sequences*
For some fixed positive integer k, the sequence (a_{n+k}) is null. Prove that (a_n) is null. The sequence (a_{n+k}) is a special kind of subsequence of (a_n) for which the converse of the result in (a) *is* valid. The result may also be expressed by saying that a sequence which is eventually null is a null sequence.

(d) *A squeeze rule for null sequences, with a shift* I
Let (a_n) be a null sequence, and $0 \leq b_n \leq a_n$, for all $n > k$. Prove that (b_n) is a null sequence.

37 Prove that each of the following sequences is null by building on qn 36:

(i) $(1/(2n-1))$,
(ii) $(1/(2n^2-1))$,
(iii) (a_n), where $a_n = n$ for $n \leq 10$, and $a_n = 1/n$ for $n > 10$,
(iv) $(1/(3n-100))$.

38 If a constant sequence (c) is a null sequence, what can be said about c?

39 (*Gregory of St Vincent*, 1647) A particularly important family of null sequences is formed by geometric progressions. Suppose $0 < x < 1$.

(a) Prove that $1 < 1/x$, using qn 2.23, and let $1/x = 1 + y$, so $y > 0$.
(b) Show that $(1/x)^n \geq 1 + ny$, using Bernoulli's inequality, qn 2.29.
(c) Prove that $0 < x^n \leq 1/(yn+1)$.
(d) Use $1/(yn+1) < (1/y)(1/n)$, the scalar rule and the squeeze rule to prove that (x^n) is a null sequence.
(e) Prove that for any constant c, $(c \cdot x^n)$ is a null sequence.

40 (a) Use a calculator or a computer to evaluate terms of the sequence $(n^2/2^n)$. Calculate the values for $n = 1, 2, 5, 10, 20, 50$. Would you conjecture that the sequence is null?
(b) Let $a_n = n^2/2^n$. Verify that $\dfrac{a_{n+1}}{a_n} = \dfrac{1}{2}\left(1 + \dfrac{1}{n}\right)^2$.
(c) Find an integer k, such that if $n \geq k$, then $\dfrac{a_{n+1}}{a_n} < \dfrac{3}{4}$.
(d) By considering that $a_{k+1} < \frac{3}{4}a_k$ and that $a_{k+2} < \frac{3}{4}a_{k+1}$, etc., prove that $a_{k+n} < (\frac{3}{4})^n a_k$.
(e) Why is $((\frac{3}{4})^n a_k)$ a null sequence? You have chosen k in part (c) and it remains fixed.
(f) Use the squeeze rule (qn 36(d)) to show that $(n^2/2^n)$ is a null sequence.

(g) What numbers might have been used in place of $\frac{3}{4}$ in part (c) (with perhaps a different k) which would still have led to a proof that the sequence was null?

41 Use the method of qn 40 to establish that the following sequences are null:

(i) $(n^4/2^n)$,
(ii) $(2^n/n!)$,
(iii) $(n!/n^n)$ using qn 2.45,
(iv) $(n^3(0.9)^n)$.

42 Use the absolute value rule (qn 33) to extend qn 39 and establish that $(c \cdot x^n)$ is a null sequence when $-1 < x < 1$.

43 *The product rule for null sequences*
Let both (a_n) and (b_n) be null sequences, and suppose $\varepsilon > 0$ is given.

(i) Must there be an N' such that $|a_n| < \varepsilon$ when $n > N'$?
(ii) Must there be an N'' such that $|b_n| < 1$ when $n > N''$?
(iii) Can you find an N such that when $n > N$, then both $n > N'$ and $n > N''$?
(iv) If $n > N$, must $|a_n b_n| < \varepsilon$?

You have proved that the termwise product $(a_n b_n)$ of two null sequences is null.

(v) If the sequence (c_n) is also null, what about $(a_n b_n c_n)$? And the termwise product of k null sequences?

44 *The sum rule for null sequences*
Let both (a_n) and (b_n) be null sequences, and suppose $\varepsilon > 0$ is given.

(i) Must there be an N' such that $|a_n| < \varepsilon/2$ when $n > N'$?
(ii) Must there be an N'' such that $|b_n| < \varepsilon/2$ when $n > N''$?
(iii) Is there an N such that when $n > N$, then both $n > N'$ and $n > N''$?
(iv) If $n > N$, must $|a_n + b_n| < \varepsilon$? Use the triangle inequality (qn 2.61).

You have proved that the termwise sum $(a_n + b_n)$ of two null sequences is null.

(v) If the sequence (c_n) is also null, what about $(a_n + b_n + c_n)$? And the termwise sum of k null sequences?

45 Let both (a_n) and (b_n) be null sequences. Show that $(c \cdot a_n + d \cdot b_n)$ is a null sequence, where c and d are constants. Use the scalar rule and the sum rule.

The difference rule
By choosing suitable values for c and d, show that $(a_n - b_n)$ is a null sequence.

46 (i) *A squeeze rule or sandwich theorem for null sequences* II
Let both (a_n) and (c_n) be null sequences, and $a_n \leq b_n \leq c_n$, for all n.
(a) Say why $(c_n - a_n)$ is a null sequence.
(b) Say why $(b_n - a_n)$ is a null sequence.
(c) Finally say why (b_n) is a null sequence.
(ii) *The squeeze rule with a shift* II
Let both (a_n) and (c_n) be null sequences, and $a_n \leq b_n \leq c_n$, for $n > k$. Say why (b_n) must be a null sequence.

47 Let $A_n = a_1 + a_2 + \ldots + a_n$.
Give an example of a sequence (a_n) which is not null, but for which the sequence (A_n/n) is null.

Summary: Null sequences

Definition A sequence (a_n) is said to tend to zero or to be a null sequence, if and only if, given any $\varepsilon > 0$, there exists an N such that $n > N \Rightarrow |a_n| < \varepsilon$. This is expressed symbolically by writing $(a_n) \to 0$ as $n \to \infty$.

Theorem qn 29 $(1/n)$ is a null sequence.

Theorem qns 39, 42 (x^n) is a null sequence, when $-1 < x < 1$.

The scalar rule for null sequences
qn 32 If (a_n) is a null sequence, then $(c \cdot a_n)$ is a null sequence.

The absolute value rule for null sequences
qn 33 (a_n) is a null sequence $\Leftrightarrow$ $(|a_n|)$ is a null sequence.

The subsequence rule for null sequences
qn 36(a) Every subsequence of a null sequence is null.

The shift rule for null sequences
qn 36(c) If for some fixed positive integer k, (a_{n+k}) is a null sequence, then (a_n) is a null sequence.

If (a_n) and (b_n) are null sequences, then

The sum rule for null sequences
qn 44 (i) $(a_n + b_n)$ is a null sequence;

The difference rule for null sequences
qn 45 (ii) $(a_n - b_n)$ is a null sequence;
qn 45 (iii) $(c \cdot a_n + d \cdot b_n)$ is a null sequence;

The product rule for null sequences
qn 43 (iv) $(a_n b_n)$ is a null sequence;
The squeeze rule for null sequences
qn 46 (v) when $a_n \le c_n \le b_n$, eventually,
(c_n) is a null sequence.

Convergent sequences and their limits

48 Examine the terms of the sequence given by
$a_n = \frac{n}{n+1}$.
To what limit do you think this sequence tends?
What can you say about the sequence $(a_n - 1)$?

Up to this point we have avoided the term 'limit' because in non-mathematical usage (*e.g.* a speed limit) it is synonymous with 'bound'. However, for mathematicians, bounds need not be approached, whereas limits are numbers to which the terms of sequences and the values of functions get close.

FIRST DEFINITION OF CONVERGENCE:
EXTENSION OF NULL SEQUENCE DEFINITION
The sequence (a_n) tends to the limit a, or converges to a, if and only if $(a_n - a)$ is a null sequence.

This is expressed symbolically by writing
either $(a_n) \to a$ as $n \to \infty$ *or* $\lim_{n\to\infty} a_n = a$.
We read each of the arrows here as *tends to*.
A sequence of numbers with a number as limit is said to be *convergent*.

49 Prove

(i) $(n/(2n-1)) \to \frac{1}{2}$;
(ii) $((2n^2+1)/(n^2+1)) \to 2$.

50 The sequence $((-1)^n)$, for example, does not have two limits, $+1$ and -1. The definition of a limit rules out ambiguity. This is quite easy to prove. Suppose (a_n) is a sequence and both $(a_n) \to a$ and $(a_n) \to b$. By definition $(a_n - a)$ is a null sequence and $(a_n - b)$ is a null sequence. Now use the difference rule (qn 45) to show that the constant sequence $(b - a)$ is a null sequence and deduce that $a = b$ (qn 38). In fact, the sequence $((-1)^n)$ is not convergent.

51 Look back to qn 19 and the discussion preceding it, for the definition of $\lfloor x \rfloor$. Write down the first four terms of the sequence $(\lfloor 10^n a \rfloor / 10^n)$

(i) when $a = 6\frac{1}{4}$, (ii) when $a = \frac{1}{3}$, (iii) when $a = \sqrt{2}$.

Prove that, for *any* a,

$$0 \le a - \frac{\lfloor 10^n a \rfloor}{10^n} < \frac{1}{10^n}.$$

Deduce that for any number a there is a sequence of rational numbers which tends to it.

Question 51 shows both the power and the limitation of decimal expressions. Each of the terms of the sequence $(\lfloor 10^n a \rfloor / 10^n)$ is a *terminating decimal* and if the sequence is eventually constant, then the number a is a *terminating decimal*. Terminating decimals provide an excellent way of approximating to the points of a number line, but there are many points on a number line which may only be pinpointed as the *limit* of a sequence of terminating decimals and not by a single terminating decimal. An infinite decimal is the limit of a sequence of terminating decimals, presuming that sequence converges. The infinite decimal 1.4142... denotes the limit of the sequence 1, 1.4, 1.41, 1.414, ... and this is what is meant by writing $\sqrt{2} = 1.4142\ldots$. The sequence determines a unique number – the limit of the sequence.

Many of the theorems we have established for null sequences can be extended to give theorems about convergent sequences whatever their limits.

52 *The shift rule*
Use the shift rule for null sequences (qn 36(c)) to prove that if $(a_{n+k}) \to a$ for some fixed positive integer k, then $(a_n) \to a$. This result may also be described by saying that a sequence which eventually converges to a, converges to a.

53 Let (a_n) be a sequence.

(i) We know that $a_n = 0$ for $n < 10^6$; what do you think the limit is?
(ii) We know that $a_n = 0$ for $n < 10^6$, and $a_n = 1$ for $10^6 \le n < 10^9$; what do you think the limit is?
(iii) The sequence is given by $a_n = 0$ for $n < 10^6$, $a_n = 1$ for $10^6 \le n < 10^9$, and $a_n = 2$ for $10^9 \le n$. What is the limit?

54 Let $(a_n) \to a$ and $(b_n) \to b$.

(i) *The scalar rule*
Use the scalar rule for null sequences (qn 32) to show that $(c \cdot a_n) \to c \cdot a$.
(ii) *The subsequence rule*
Use the subsequence rule for null sequences (qn 36(a)) to show that every subsequence of (a_n) converges to a.

(iii) *The sum rule*
Use the sum rule for null sequences (qn 44) to show that $(a_n + b_n) \to a + b$.

(iv) Use the scalar rule and sum rule to show that $(c \cdot a_n + d \cdot b_n) \to c \cdot a + d \cdot b$.

(v) *The difference rule*
Apply (iv) to show that $(a_n - b_n) \to a - b$.

(vi) *The product rule*
Use the scalar rule, the sum rule and the product rule for null sequences (qn 43) and the equation
$a_n b_n - ab = (a_n - a)(b_n - b) + a(b_n - b) + b(a_n - a)$
to show that $(a_n b_n) \to ab$.

(vii) *The absolute value rule*
Use the absolute value rule for null sequences (qn 33), the reverse triangle inequality (qn 2.63) and the squeeze rule for null sequences (qn 34) to show that $(|a_n|) \to |a|$.

(viii) *The squeeze rule or sandwich theorem*
When $a_n \le c_n \le b_n$, for $n > k$, and $a = b$, use the squeeze rule for null sequences (qn 46(ii)) to show that $(c_n) \to a$.

55 (i) Extend the sum rule (qn 54(iii)) to a sum of three convergent sequences and then to a sum of k convergent sequences.

(ii) Extend the product rule (qn 54(vi)) to a product of three convergent sequences, and then to a product of k convergent sequences.

(iii) If $(a_n) \to a$, what is the limit of the sequence $(1 + 2a_n + 3a_n^2 + 4a_n^3 + 5a_n^4)$?

It was mentioned in chapter 2 that many square roots, cube roots, etc., of integers are not rational numbers. So if we only knew about rational numbers, the square root of 2 'would not exist'. Roots of numbers are problematic. For example, $\sqrt{(-1)}$ 'does not exist' if we are thinking only of real numbers, as we are in this book. In chapter 4 we will meet this issue head-on and show that nth roots of positive numbers always exist and are unique, something which you probably always took for granted. In the mean time you should proceed as if this is not a problem.

56 Use a calculator to explore the sequences $(\sqrt[n]{2})$, $(\sqrt[n]{10})$ and $(\sqrt[n]{1000})$. Repeated use of the square root button gives a subsequence in each case.

57 (a) Let $1 < b$. Use qn 2.20 to show that $1 < \sqrt[n]{b}$.

(b) Let $1 + a_n = \sqrt[n]{b}$. Use Bernoulli's inequality (qn 2.29) to show that $1 + na_n \le b$.

(c) Prove that (a_n) is a null sequence.

(d) Deduce that $(\sqrt[n]{b}) \to 1$ as $n \to \infty$.

58 (a) Do you think the sequence $(\sqrt[n]{(2^n + 3^n)})$ is convergent?
(b) Use a calculator to explore the sequence $(\sqrt[n]{(2^n + 3^n)})$.
(c) Find the limit of the sequence $(\sqrt[n]{(a^n + b^n)})$ when $0 \le a \le b$.

59 (a) For $n \ge 2$, check that $\sqrt[n]{n} > 1$. See qn 2.51.
(b) Let $\sqrt[n]{n} = 1 + a_n$. Use the binomial theorem to show
$$n > \frac{n(n-1)}{2}(a_n)^2.$$
(c) Show that
$$a_n < \sqrt{\frac{2}{n-1}},$$ and deduce that (a_n) is a null sequence.
(d) Deduce that $(\sqrt[n]{n}) \to 1$.

STANDARD DEFINITION OF CONVERGENCE
The sequence (a_n) tends to the limit a, or converges to a, and we write $(a_n) \to a$ as $n \to \infty$ if and only if, given any $\varepsilon > 0$, there exists an N such that $n > N \Rightarrow |a_n - a| < \varepsilon$.

60 Show that the two definitions of convergence are equivalent.

Boundedness of convergent sequences

61 For what values of L and U is the inequality $L < x < U$ equivalent to the inequality $|x - a| < \varepsilon$? Mark a, ε, L and U on a number line.
For these values of L and U the interval $\{x | L < x < U\}$ is called an ε-*neighbourhood* of a.

62 Let $(a_n) \to a$. Prove that the sequence (a_n) is eventually bounded above by $a + 1$ and eventually bounded below by $a - 1$. (*Hint*. Take $\varepsilon = 1$.)
Illustrate this proof with a graph.
Deduce from qn 13 that every convergent sequence is bounded. Give an example to show that a bounded sequence need not be convergent.

63 Let $(a_n) \to a$, and $a > 0$. Identify a positive number which is eventually a lower bound for the sequence (a_n).

Quotients of convergent sequences

64 What is the relationship between the limits of the sequences $(2n/(n+1))$ and $((n+1)/2n)$?

65 *The reciprocal rule for non-null sequences*
Let (a_n) be a sequence of non-zero terms with $(a_n) \to a$ and $a > 0$. We wish to show that $(1/a_n) \to 1/a$, and so we must examine

$$\left|\frac{1}{a_n} - \frac{1}{a}\right| = \frac{|a - a_n|}{|a_n a|} = \left|\frac{1}{a_n}\right| \cdot \frac{|a - a_n|}{a}.$$

(a) How do you know that $\left(\frac{|a - a_n|}{a}\right)$ is a null sequence?

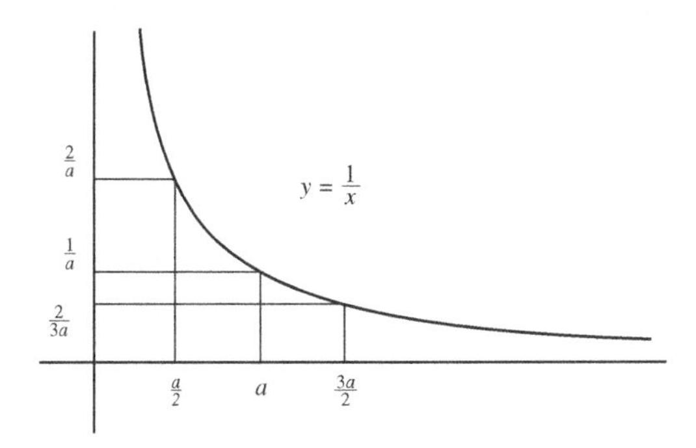

(b) As in qn 63, show that, eventually,

$$\frac{a}{2} < a_n < \frac{3a}{2},$$

and hence

$$\frac{2}{3a} < \frac{1}{a_n} < \frac{2}{a}.$$

(c) Deduce that, eventually,

$$\left|\frac{1}{a_n} - \frac{1}{a}\right| < \frac{2}{a} \cdot \frac{|a - a_n|}{a}.$$

(d) Now use (a), the scalar rule for null sequences (qn 32) and the squeeze rule with a shift (qn 46(ii)) to prove that $(1/a_n) \to 1/a$.

66 Let (a_n) be a sequence of non-zero terms with $(a_n) \to a$ and $a < 0$. Prove that $(1/a_n) \to 1/a$, by applying the result of qn 65 to the sequence $(-a_n)$.

67 *The quotient rule*
Let $(a_n) \to a$, $(b_n) \to b$, and suppose that all of the b_n and b are non-zero. Prove that $(a_n/b_n) \to a/b$.

68 Find the limits of the sequences with nth terms given here, stating which rules you are using.

(i) $\dfrac{1 + 1/n}{2 + 1/n}$,

(ii) $\dfrac{3n + 2}{4n - 3}$,

(iii) $\dfrac{n^2-1}{3n^2+n}$,

(iv) $\dfrac{(\frac{1}{2})^n-1}{(\frac{1}{2})^n+1}$,

(v) $\dfrac{2^n-1}{2^n+1}$,

(vi) $\dfrac{n^3+n^2+n+1}{3n^4+5}$,

(vii) $\dfrac{(\sqrt{n}+3)(\sqrt{n}-1)}{3\sqrt{n}-5n}$,

(viii) $\dfrac{a^n-1}{a^n+1}$, $0 < a$,

(ix) $\dfrac{1+2+\ldots+n}{n^2}$,

(x) $\sqrt[n]{b}$, where $0 < b < 1$.

69 Let $a_n = \sqrt{(n+8)} - \sqrt{n}$, $b_n = \sqrt{(n+\sqrt{n})} - \sqrt{n}$ and $c_n = \sqrt{(n+n/8)} - \sqrt{n}$.
Show that $a_n > b_n > c_n$ when $n < 64$.
Find the limits of the sequences (a_n), (b_n) and (c_n) if they exist.
(*Hint*. For positive numbers, x and y,
$x - y = (\sqrt{x} - \sqrt{y})(\sqrt{x} + \sqrt{y})$.)

d'Alembert's ratio test

70 Give examples of sequences (a_n) of positive terms for which

$$\left(\frac{a_{n+1}}{a_n}\right) \to \frac{1}{2}.$$

You will find one example in qn 40. Are all the sequences which you have found convergent? To what limit?
By choosing $\varepsilon = \frac{1}{4}$ for the sequence of ratios, show that the method of qn 40 can be applied here to prove that any such sequence is null.

71 Let (a_n) be a sequence of positive terms such that

$$\left(\frac{a_{n+1}}{a_n}\right) \to l < 1.$$

Show that, for sufficiently large n, the sequence (a_n) is squeezed between 0 and an appropriately chosen null geometric progression. Deduce that (a_n) is a null sequence.

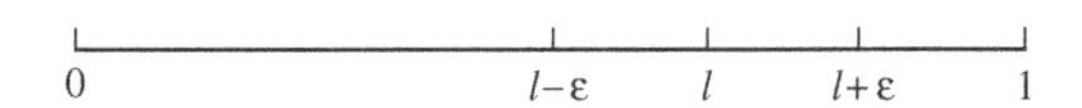

72 Let (a_n) be a sequence of positive terms such that

$$\left(\frac{a_{n+1}}{a_n}\right) \to l > 1.$$

Show that, for sufficiently large n, the terms of the sequence (a_n) are greater than the corresponding terms of an appropriately chosen geometric progression, which from qn 21 tends to $+\infty$.
Deduce that $(a_n) \to +\infty$.

73 Give an example of a null sequence of positive terms (a_n) for which

$$\left(\frac{a_{n+1}}{a_n}\right) \to 1.$$

Also give an example of an increasing non-convergent sequence of positive terms (a_n) for which

$$\left(\frac{a_{n+1}}{a_n}\right) \to 1.$$

This means that the condition that the ratio of successive terms tend to 1 does not determine the convergence of the original sequence one way or the other.

74 Let k be a given integer. Determine for what values of x the sequences with nth terms

(i) $n^k x^n$ and
(ii) $\dfrac{x^n}{n!}$ are null.

Can the value of k affect the answer?

Convergent sequences in closed intervals

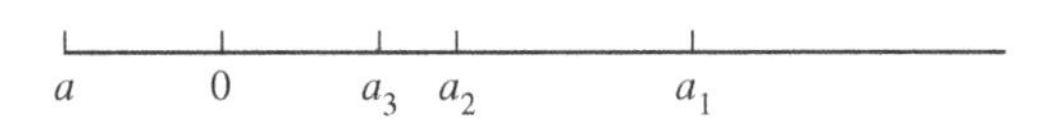

Figure 3.9

75 We investigate the question 'Can a sequence of positive terms have a negative limit?'
'Obviously not,' you think. Here is a more reasoned response.

Let $(a_n) \to a, 0 \leq a_n$ and $a \leq 0$. Use the inequality $0 \leq a_n \leq a_n - a$ to prove that $a = 0$.
Deduce that, if $(a_n) \to a$ and $0 \leq a_n$, then necessarily $0 \leq a$.

76 *The inequality rule*
Let $(a_n) \to a$, $(b_n) \to b$ and $a_n \leq b_n$ for all n. By considering the sequence $(b_n - a_n)$, and using the difference rule, prove that $a \leq b$.

77 By considering the sequences defined by $a_n = -1/n$ and $b_n = 1/n$, show that even if $a_n < b_n$ for all n, $(a_n) \to a$ and $(b_n) \to b$ do *not* imply $a < b$. This gives a *counter-example* to the proposition that if $a_n < b_n$ for all n, $(a_n) \to a$ and $(b_n) \to b$ imply $a < b$.

78 *The closed interval property*
Let $(a_n) \to a$ and $A \leq a_n \leq B$ for all n. Prove that $A \leq a \leq B$. The reason for calling the interval $\{x | A \leq x \leq B\}$ a *closed* interval is that no convergent sequence in the interval can have a limit which escapes from the interval.
Give an example to show that if $(a_n) \to a$ and $A < a_n < B$ for all n, it does not necessarily follow that $A < a < B$.
The interval $\{x | A < x < B\}$ is called an *open* interval, partly because a convergent sequence in the interval may have a limit outside the interval.

79 Let (a_n) be a sequence of positive terms such that $a_{n+1}^2 = a_n + 2$, and $(a_n) \to a$. Prove that $a = 2$. Use the sum rule (qn 54(iii)), the product rule (qn 54(vi)) and the closed interval property.
(*Optional*) Prove that (a_n) is monotonic.
(*Viète*, 1593) As an example, $(2 \cos \pi/2^n)$ is a special case of this sequence.

80 Let (a_n) be a monotonic increasing sequence with a convergent subsequence $(a_{n_i}) \to A$.

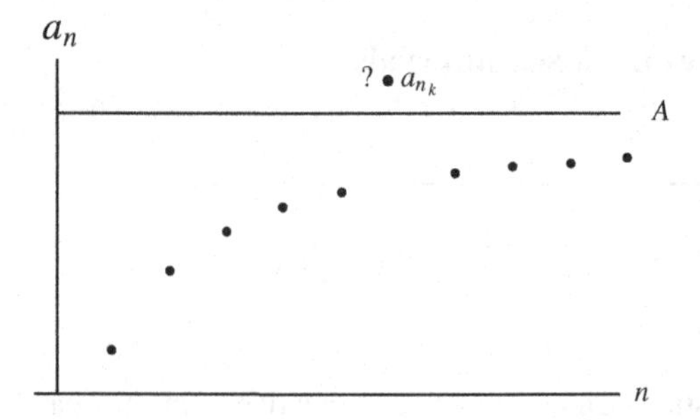

Figure 3.10

(a) Show that $a_{n_i} \leq A$ for all n_i, by contradiction, supposing that $A < a_{n_k} = A + \varepsilon$, for some positive integer k, and showing that the subsequence (beyond the n_kth term) could not then be *both* monotonic increasing *and* convergent to A.
(b) Use the fact that every term of the sequence (a_n) is followed eventually by a term of the convergent subsequence to prove that $a_n \leq A$ for all n.
(c) For any positive ε, use the convergence of the subsequence to A to find a term of the subsequence between $A - \varepsilon$ and A.
(d) Name an N such that $n > N \Rightarrow |a_n - A| < \varepsilon$, so that $(a_n) \to A$.

State the theorem you have proved through (a),(b),(c) and (d). State an analogous theorem for monotonic decreasing sequences. Combine the statements of these two theorems into a single theorem about the convergence of monotonic sequences.

Intuition and convergence

81 An isosceles right-angled triangle is constructed with a given line segment l as hypotenuse. It has perimeter p_1 and area A_1. *Two* isosceles right-angled triangles are constructed each with hypotenuse on half of the line segment l, but with the two hypotenuses occupying the whole of l. The two triangles, which only overlap at one vertex, have combined perimeter p_2 and combined area A_2.

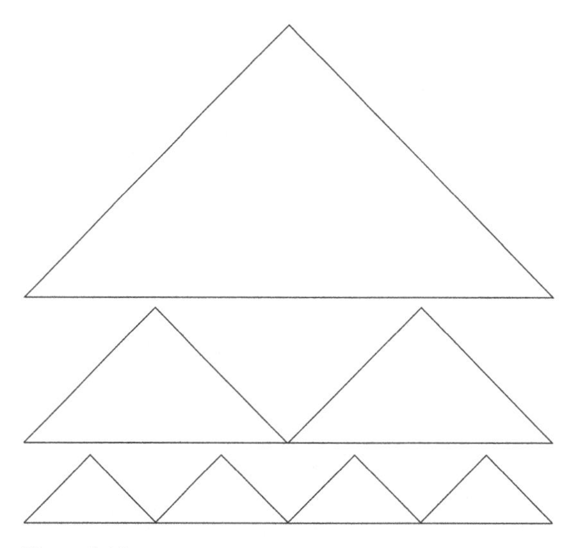

Figure 3.11

This replacement of a triangle by two smaller triangles is repeated indefinitely, so that at each stage of replacement there are 1, 2, 4, 8, ..., 2^n, ... congruent triangles with hypotenuses filling the line segment l, and at the nth stage, the combined perimeter of these triangles is p_n and their combined area is A_n.

(a) What is the relation between p_n and p_{n+1}?
(b) What is the relation between A_n and A_{n+1}?
(c) What can you say about the two sequences (p_n) and (A_n)?

82 (*Koch*, 1904) An equilateral triangle is given with perimeter p_1 and area A_1. This is stage 1.
Now an equivalent triangle is added to each side on the outside of the original figure, at its points of trisection. The result is a polygon with 12 sides, perimeter p_2 and area A_2. This is stage 2.
The same kind of extension is now made on each of the 12 sides of this polygon, resulting in a further polygon with 48 sides, perimeter p_3 and area A_3. This is stage 3.
This process is continued indefinitely.

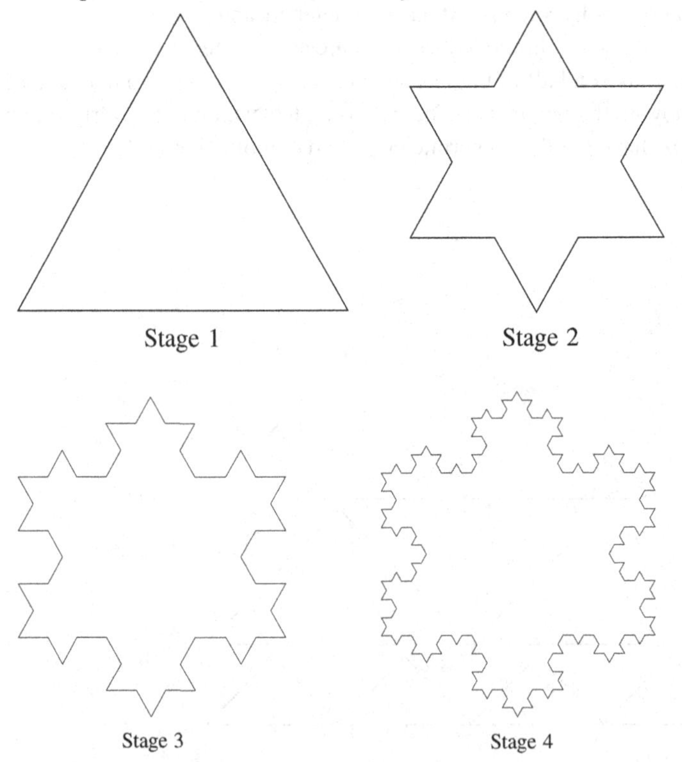

Figure 3.12

(a) What is the number of sides of the nth polygon?
(b) Show that $p_{n+1} = \frac{4}{3} p_n$.
(c) Show that, in making the $(n+1)$th polygon, each little triangle being adjoined has area $(\frac{1}{9})^n A_1$.
(d) Show that $A_{n+1} = A_n + (\frac{3}{9})(\frac{4}{9})^{n-1} A_1$.
(e) Prove (from chapter 1 qn 3(vi)) that

$$1 + \tfrac{4}{9} + (\tfrac{4}{9})^2 + \ldots + (\tfrac{4}{9})^{n-1} = \frac{(\frac{4}{9})^n - 1}{\frac{4}{9} - 1}.$$

(f) Prove that $(A_n) \to \frac{8}{5} A_1$ as $n \to \infty$.
(g) Prove that $(p_n) \to +\infty$ as $n \to \infty$.

83 The two big ideas of this chapter have been null sequences and convergence. For each of these notions write down (a) some (varied) examples, (b) some theorems and (c) some related ideas. Then look back at the summaries to see what you left out, and whether you wish to revise your opinion as to which ideas are the most important.

Summary: Convergent sequences

Definition A sequence (a_n) is said to tend to a limit a or to be convergent, if and only if, given $\varepsilon > 0$, there exists an N such that $n > N \Rightarrow |a_n - a| < \varepsilon$. This is expressed symbolically by writing $(a_n) \to a$ as $n \to \infty$.

The subsequence rule
qn 54(ii) If $(a_n) \to a$ then every subsequence of $(a_n) \to a$.

The shift rule
qn 52 If, for some fixed integer k, $(a_{n+k}) \to a$, then $(a_n) \to a$.

The scalar rule
qn 54(i) If $(a_n) \to a$, then $(c \cdot a_n) \to c \cdot a$.

The absolute value rule
qn 54(vii) If $(a_n) \to a$, then $(|a_n|) \to |a|$.

The reciprocal rule for non-null sequences
qns 64, 65 If $(a_n) \to a$, and $a_n, a \neq 0$, then $(1/a_n) \to 1/a$.

Theorem Every convergent sequence is bounded.
qn 62

The closed interval rule
qn78 If $(a_n) \to a$ and $A \leq a_n \leq B$, then $A \leq a \leq B$.

	If $(a_n) \to a$ and $(b_n) \to b$ then
The sum rule	
qn 54(iii)	(i) $(a_n + b_n) \to a + b$;
The difference rule	
qn 54(v)	(ii) $(a_n - b_n) \to a - b$;
qn 54(iv)	(iii) $(c \cdot a_n + d \cdot b_n) \to c \cdot a + d \cdot b$;
The product rule	
qn 54(vi)	(iv) $(a_n b_n) \to ab$;
The squeeze rule	
qn 54(viii)	(v) when $a_n \le c_n \le b_n$, eventually, and $a = b$, $(c_n) \to a$;
The quotient rule	
qn 67	(vi) when $b_n, b \neq 0,\ (a_n/b_n) \to a/b$;
The inequality rule	
qn 76	(vii) when $a_n \le b_n,\ a \le b$.
Theorem	Every real number is the limit of some sequence
qn 51	of terminating decimals.
Theorem	If $0 < a,\ (\sqrt[n]{a}) \to 1$.
qns 57, 68(x)	
Theorem	$(\sqrt[n]{n}) \to 1$.
qn 59	

Historical Note

The 'Archimedean property' appears in Euclid in the form of a definition of ratio (Book V, Def. 4), and thus pre-dates Archimedes (250 BC). It was important to Euclid because it precluded comparing a measurement of length with a measurement of area or a measurement of area with a measurement of volume. There is an equivalent proposition in Euclid Book X.1 that if a half or more of a quantity is removed, then half or more of what is left, and so on, then after a finite number of steps what remains is less than any predetermined quantity. This proposition was used to great effect by Euclid (Book XII) in relation to the area of a circle, the volume of a pyramid and cone, and the volume of a sphere, and the Archimedean property itself by Archimedes in his calculations of areas and volumes. In modern terms, Euclid X.1 established that $(1/2^n)$ tends to 0. There are no limit arguments in classical Greek mathematics. Both Euclid and Archimedes used inequalities and their contradiction to determine areas and volumes of non-rectilinear figures. To prove $A = K$, they contradicted $|A - K| > 0$ by repeated halvings. The importance of the Archimedean property in modern times dates from 1891, when G. Veronese recognised that it eliminated constant infinitesimals. In Hilbert's *Foundations of Geometry* (1899) it is referred to as the 'axiom of continuity'.

The ε in our definition of limit is foreshadowed in the 'predetermined quantity' of Euclid X.1, and also by Newton who wrote 'Quantities, and the ratio of quantities, which in any finite time converge continually to equality, and before the end of that time approach *nearer to each other than by any given difference*, become ultimately equal.' (*Principia*, Book 1, Lemma 1, 1687). See question 2.66. During the eighteenth century a great deal of work was done on particular sequences and series, and sophisticated methods for approximating and finding limits were developed. D'Alembert had studied the binomial expansion of $(1 + x)^m$ when m is rational (1768) and had established the boundedness of the series by comparison with two geometric series when $|x| < 1$. Newton's statement on limits was refined by d'Alembert (1765) who wrote 'One says that a quantity is the limit of another quantity, if the second approaches the first closer than any given quantity, however small.' There are indications of how to prove that the limit of a product is equal to the product of the limits in d'Alembert and de la Chapelle (1789) and of how to prove that the limit of a quotient is the quotient of the limits (without attention to the possibility of a zero denominator) in L'Huilier (1795). But formal proofs, as in this chapter, had to await the methods of Cauchy's *Analyse Algébrique* (1821) in his proof that $(a_{n+1} - a_n) \to k$ implies $(a_n/n) \to k$. The expression 'for sufficiently large n' and its interpretation 'there exists an N such that when $n > N \ldots$' we owe to Cauchy. The suffix notation for sequences stems from Lagrange (1759) who actually used superscripts and not subscripts. The suffix notation proper is due to Lacroix (1800), but it was in Cauchy's hands that it became the powerful tool we know today. Most of the limits which were studied in the eighteenth century were one-sided, as was mentioned after qn 26, and there was also a presumption (explicit in d'Alembert) that limits were not to be reached.

The notation for absolute value was not available to Cauchy, though he sometimes wrote $\sqrt{(x^2)}$ where we would write $|x|$. Where absolute value was pertinent, early-nineteenth-century mathematicians would declare that an inequality holds 'without regard to sign'. The symbolism of absolute value and the notion of neighbourhood were introduced by K. Weierstrass in his lectures in 1859.

The symbol ∞ was introduced by J. Wallis in 1655, but he manipulated it like a number. The 'lim' notation was due to L'Huilier in 1786, and mid-nineteenth-century mathematicians, including Weierstrass, used expressions such as

$$\lim_{n=\infty} n^2 = \infty.$$

This only became unacceptable with the greater precision about what is and what is not a real number in the latter part of the nineteenth century, and it was in 1905 that the Cambridge mathematician J. G. Leathem proposed the use of '$n \to \infty$'. G. H. Hardy (1908) vehemently exhorted his readers to forgo writing $n = \infty$, though even he allowed the statement $\lim a_n = \infty$.

The language of bounds and boundedness, introduced by Pasch (1882), was adopted at the beginning of the twentieth century. For most of the nineteenth century a bound was called a limit (as in common non-mathematical usage) and the boundedness of a set described by saying that its elements were finite.

The integral part of x was used for certain special cases by Legendre in 1808 and Dirichlet in 1849, and in the sense that we now use $\lfloor x \rfloor$ by F. Mertens in 1874. G. Peano wrote Ex for $\lfloor x \rfloor$ in 1899 (E being the first letter of *entier*, the word for *whole number* in French). Through most of the twentieth century, the notation $[x]$ was used.

Both recurring (1657) and non-recurring (1693) infinite decimals appear first in the work of J. Wallis.

Answers and comments

1 $n, 2n, 1/2^n, \pi$ to $n-1$ places of decimals, $(-1)^n, n/(n+1)$.

4

	(a)	(b)	(c)	(d)
(i)	✓	✓	×	×
(ii)	×	×	✓	✓
(iii)	×	×	×	×
(iv)	×	✓	×	×
(v)	✓	✓	×	×
(vi)	×	✓	×	✓
(vii)	×	×	✓	✓

5

	(a)	(b)
(i)	×	×
(ii)	✓	✓
(iii)	✓	✓
(iv)	×	✓

6 (i) Yes, $(-1/n)$; (ii) yes; (iii) no, (n); (iv) yes, always, a_1.
Monotonic decreasing: (i) yes; (ii) yes $(1/n)$; (iii) yes, always, a_1; (iv) no, $(-n)$.

7 (i) Yes; (ii) yes, because if there were only a finite number, the greatest would be an upper bound.

8 (i) A finite list, (ii) not in order, (iii) not in original sequence when n is odd, (iv) not in original sequence when $n = 1$.

9 (i) $1, 4, 5, 8, 9, \ldots$, (ii) $2, 3, 6, 7, 10, \ldots$, (iii) $1, 3, 5, 7, 9, \ldots$, (iv) $2, 4, 6, 8, 10, \ldots$.

10 (i) (a) No. The n_i form an increasing sequence. (b) No. The n_i form a strictly increasing sequence. (c) 5. (d) $a_{n_{101}}$, an integer greater than 101 may also be used.
(ii) (n_i) is strictly increasing, so $i < j \Rightarrow n_i < n_j$. But the sequence (a_n) is arbitrary, so the subsequence need not be increasing.

11 (i) Yes, see qn 4(iv); (ii) not for a constant sequence.

12 Yes, the bounds for the sequence are certainly bounds for any subsequence.

13 (i) If the subsequence is bounded above by U and bounded below by L, then the sequence has $\max(a_1, U)$ as an upper bound and $\min(a_1, L)$ as a lower bound. 'max' is formally defined at qn 6.44.
(ii) As (i) with upper bound $\max(a_1, a_2, U)$, lower bound $\min(a_1, a_2, L)$.
(iii) As (i) with upper bound $\max(a_1, a_2, \ldots, a_k, U)$ and lower bound $\min(a_1, a_2, \ldots, a_k, L)$.

14 (a) No. Take the sequence (a_n) defined by $a_{2n} = 1$ and $a_{2n-1} = n$, for example. (b) (i) Yes, $a_{2n-1} = 0$, $a_{2n} = (-2)^n$; (ii) no, $(-2)^n$.

15 (i) $-1, -1/3, -1/5, \ldots$; (ii) 0, 0; (iii) none.

16 (i) If the floor term a_{f_j} succeeds the floor term a_{f_i}, then by the definition of a_{f_i}, $a_{f_i} \leq a_{f_j}$, and the subsequence (a_{f_i}) is monotonic increasing.
(ii) Since no term after a_F is a floor term, every term thereafter is eventually succeeded by a lesser term, giving a strictly decreasing subsequence.
(iii) Since there are no floor terms, every term is eventually succeeded by a lesser term, again giving a strictly decreasing subsequence.

17 (i) Not monotonic; bounded below; all terms ≥ 1.
(ii) Not monotonic; unbounded.
(iii) Not monotonic; bounded below; all terms ≥ 1; all terms past the 10th ≥ 10; all terms past the 100th ≥ 100.
(iv) Not monotonic; bounded; all terms ≥ 10.

18 (i) $1 < C$; (ii) any C; (iv) $11 < C$.

19 (i) $A - 1 < \lfloor A \rfloor \leq A$. (ii) (a) $0 < x - N(x) < 1$ which is impossible when x is an integer. (b) $x < M(x+1) < x+1$, so not possible when x is an integer. (c) Of course $K(x) = \lfloor x \rfloor$ is possible, and is inevitable when x is not an integer, but when x is an integer we may have $x = K(x)$ or $K(x+1)$.

20 Take $N \geq C^2$.

21 (a) Take $N \geq C/y$.

22 Although in the common-sense use of the word, a sequence which tends to $+\infty$ is 'increasing', qn 17(iii) shows that such a sequence need not be monotonic increasing, and therefore not, in the mathematically precise sense, increasing.

23 $(a_n) \to -\infty$ when, given any number C, there exists an N such that $n > N \Rightarrow a_n \leq C$.

24 (ii) $-1/n < a < 1/n$ for all positive integers n implies $-\varepsilon < a < \varepsilon$ for all positive ε by (i).
(iii) For a given ε, by Archimedean order, there is an $n > c/\varepsilon$, so $c/n < \varepsilon$ and $a < c/n < \varepsilon$. Similarly there is an $n > b/\varepsilon$, so $b/n < \varepsilon$ and so $-\varepsilon < -b/n < a$. The result follows from qn 2.65(a).
(iv) From (iii), $a - k = 0$.

26 $(1/2n)$, $(1/2^n)$,$(-1/2n)$, monotonic.
$((-1)^n/2n)$,(a^n) where $a_n = 0$ when n has a factor 3, and $a_n = (-1)^n/2n$ otherwise, not monotonic.

27 (i) $n > 10$; (ii) $n > 100$; (iii) $n > 1000$; (iv) $n > 1/\varepsilon$.

28 For (i), any ε will do. For (ii) take $\varepsilon = 1$ or less. For (iii) take $\varepsilon = 0.1$ or less. For (iv) take $\varepsilon = 0.01$ or less. For (v) take $\varepsilon = 1$ or less.
For the constant sequence (vi) $1, 1, 1, \ldots$, take $\varepsilon = \frac{1}{2}$. For no term in this sequence is $|a_n| < \frac{1}{2}$.

29 Given $\varepsilon > 0$, let N be an integer greater than $1/\varepsilon$. Then
$n > N \Rightarrow n > 1/\varepsilon \Rightarrow 1/n < \varepsilon$.

30 Given $\varepsilon > 0$, let N be an integer greater than $1/\varepsilon^2$ (which exists by the property of Archimedean order). Then
$n > N \Rightarrow n > 1/\varepsilon^2 \Rightarrow 1/\sqrt{n} < \varepsilon$.

31 Given $\varepsilon > 0$, there exists an N such that
$n > N \Rightarrow a_n > 1/\varepsilon \Rightarrow 0 < 1/a_n < \varepsilon$.
For given $C > 0$, there exists N such that
$n > N \Rightarrow |a_n| < 1/C \Rightarrow 0 < a_n < 1/C \Rightarrow C < 1/a_n$,
so $(1/a_n) \to +\infty$.
In general $(a_n) \to 0$ does not imply $(1/a_n) \to +\infty$. Put $a_n = (-1)^n/n$, for example.

32 Case (i), $c \neq 0$. Given $\varepsilon > 0$, choose N so that $n > N \Rightarrow |a_n| < \varepsilon/|c|$. Then $|c \cdot a_n| < \varepsilon$.
Case (ii), $c = 0$. Given $\varepsilon > 0$, $|c \cdot a_n| = 0 < \varepsilon$, for all n.
Use qn 30 and put $c = 10$.

33 Since $\big||a_n|\big| = |a_n|$, the same N is sufficient for either sequence.

34 If $|a_n| < \varepsilon$ and $0 \leq b_n \leq a_n$, then $0 \leq b_n < \varepsilon$, and so $|b_n| < \varepsilon$. So
$n > N \Rightarrow |a_n| < \varepsilon \Rightarrow |b_n| < \varepsilon$.

35
(i) Use qn 29 and the squeeze rule (qn 34).
(ii) Use the scalar rule (qn 32) and part (i).
(iii) Show that $1/(7n + 3) < 1/7n$. Then use qn 29, the scalar rule (qn 32) and the squeeze rule (qn 34).
(iv) Show that $1/(n^2 + 1) < 1/n$. Then use qn 29 and the squeeze rule (qn 34).
(v) nth term $= \dfrac{1}{\sqrt{n+1} + \sqrt{n}} < \dfrac{1}{2\sqrt{n}}$. Use qn 30, the scalar rule (qn 32) and the squeeze rule (qn 34).
(vi) $\left|\dfrac{\sin n}{n}\right| \leq \dfrac{1}{n}$. Use qn 29, the squeeze rule (qn 34) and the absolute value rule (qn 33).
(vii) All terms are 0.
(viii) If $c = p/q$, where p and q are integers, q positive, then when $n \geq q$, the terms are 0.

(ix) $0 \le a_n \le 2/n$. Use qn 29, the scalar rule (qn 32) and the squeeze rule (qn 34).

36 (a) Yes. If $n > N \Rightarrow |a_n| < \varepsilon$, then $n_i > N \Rightarrow |a_{n_i}| < \varepsilon$.
(b) No. See qn 26(v).
(c) $[n > N \Rightarrow |a_{n+k}| < \varepsilon] \Leftrightarrow [n > N + k \Rightarrow |a_n| < \varepsilon]$.
(d) (a_n) null $\Rightarrow (a_{n+k})$ null, by (a),
$0 \le b_n \le a_n$ for $n > k \Rightarrow 0 \le b_{n+k} \le a_{n+k}$ for all n.
So (b_{n+k}) is a null sequence by the squeeze rule (qn 34), and (b_n) is a null sequence by (c).

37 (i) and (ii) are subsequences of $(1/n)$. Use qn 36(a), the subsequence rule.
(ii) for $n > 10$, terms $\le 1/n$, use qn 36(d) with $k = 10$.
(iii) is a subsequence of $(1/n)$ for $n \ge 34$. Use qn 36(a) to claim that the subsequence is null and the shift rule (qn 36(c)) with $k = 34$.

38 See qn 2.65(b).

39 (d) $(1/n)$ is a null sequence by qn 29, so $(1/yn)$ is a null sequence by the scalar rule (qn 32), so $(1/(yn + 1))$ is a null sequence by the squeeze rule (qn 34) and finally (x^n) is a null sequence by the squeeze rule (qn 34).
(e) Use the scalar rule (qn 32).

40 (c) $k \ge 5$.
(d) For example, $a_{k+2} < (\frac{3}{4})^2 a_k$.
(e) From qn 39(e).
(f) From (d) and the squeeze rule (qn 34) (a_{n+k}) is a null sequence, so from the shift rule (qn 36(c))(a_n) is a null sequence.
(g) In order to appeal to qn 39, the number used must be < 1. Because of part (b) the number must be greater than $\frac{1}{2}$. If we want to use $\frac{3}{5}$, then $k \ge 11$.

41 (i) $a_{n+1}/a_n = \frac{1}{2}(1 + 1/n)^4 < \frac{3}{4}$, when $n > 10$.
(ii) $a_{n+1}/a_n = 2/(n + 1) < \frac{3}{4}$, when $n > 2$.
(iii) $a_{n+1}/a_n = 1/(1 + 1/n)^n < \frac{1}{2}$, for all $n > 1$.
(iv) $a_{n+1}/a_n = (0.9)(1 + 1/n)^3 < 0.95$, when $n > 55$.

42 When $-1 < x < 1$, $(|x|^n)$ is a null sequence by qn 39(d). Now $|x^n| = |x|^n$, so $(|x^n|)$ is a null sequence, and using the absolute value rule (qn 33), (x^n) is also a null sequence.

43 (iii) Take $N = \max(N', N'')$.
(iv) $|a_n b_n| = |a_n| \cdot |b_n| < |a_n| < \varepsilon$, from qns 2.58 and 2.11.
(v) $a_n b_n c_n$ is the product of $a_n b_n$ and c_n. Product of k terms by induction.

44 (iii) Take $N = \max(N',\ N'')$.
(iv) $|a_n + b_n| \le |a_n| + |b_n| < \frac{1}{2}\varepsilon + \frac{1}{2}\varepsilon = \varepsilon$, from qns 2.61 and 2.10.
(v) $a_n + b_n + c_n$ is the sum of $a_n + b_n$ and c_n. Sum of k terms by induction.

45 $(c \cdot a_n)$ and $(d \cdot b_n)$ are both null sequences from the scalar rule, qn 32. Their sum is a null sequence from the sum rule, qn 44.
The difference rule follows by putting $c = 1$ and $d = -1$.

46 (i) (a) From the difference rule (qn 45).
(b) $0 \le b_n - a_n \le c_n - a_n$ and the squeeze rule, qn 34.
(c) $b_n = (b_n - a_n) + a_n$. Now use the sum rule, qn 44.
(ii) (a_{n+k}) and (c_{n+k}) are null sequences by the subsequence rule, qn 36(a).
So (b_{n+k}) is a null sequence from part (i). Now (b_n) is a null sequence by the shift rule, qn 36(c).

47 For example $a_n = (-1)^n$.

48 Terms tend to 1. $(a_n - 1)$ is null.

49 (i) $n/(2n-1) - \frac{1}{2} = 1/2(2n-1)$, so $(a_n - \frac{1}{2})$ is a subsequence of $(1/n)$.
(ii) $(2n^2+1)/(n^2+1) - 2 = -1/(n^2+1)$, so $(a_n - 2)$ is a scalar multiple of a subsequence of $(1/n)$.

50 The constant sequence $(b - a)$ is null. Now $b - a = 0$, for if not $\varepsilon = \frac{1}{2}|b - a|$ gives a contradiction.

51 (i) 6.2, 6.25, 6.25, 6.25.
(ii) 0.3, 0.33, 0.333, 0.3333.
(iii) 1.4, 1.41, 1.414, 1.4142.
$\lfloor 10^n a \rfloor \le 10^n a < \lfloor 10^n a \rfloor + 1$ gives the inequality.
$(1/10^n)$ is null by qn 39, so $(\lfloor 10^n a \rfloor / 10^n) \to a$ by the squeeze rule, qn 34.
Every term of $(\lfloor 10^n a \rfloor / 10^n)$ is rational.

52 $(a_{n+k}) \to a \Rightarrow (a_{n+k} - a)$ is null (by definition) $\Rightarrow (a_n - a)$ is null (by the shift rule, qn 36(c)) $\Rightarrow (a_n) \to a$ (by definition).

53 (i) and (ii) give no basis for a guess.
(iii) With $k = 10^9$, $(a_{n+k} - 2)$ is null, so $(a_n - 2)$ is null by the shift rule, and $(a_n) \to 2$.

54 (i) $(a_n) \to a \Rightarrow (a_n - a)$ null $\Rightarrow (c \cdot (a_n - a))$ null $\Rightarrow (c \cdot a_n - c \cdot a)$ null $\Rightarrow (c \cdot a_n) \to c \cdot a$.
(ii) $(a_n) \to a \Rightarrow (a_n - a)$ null $\Rightarrow$ every subsequence of $(a_n - a)$ is null, by the subsequence rule, qn 36(a) $\Rightarrow$ every subsequence of $(a_n) \to a$.
(v) Put $c = 1$ and $d = -1$ in the linear rule.

(vi) $(a(b_n - b))$ and $(b(a_n - a))$ are both null sequences by the scalar rule (qn 32), and $((a_n - a)(b_n - b))$ is a null sequence by the product rule (qn 43). Now use the sum rule (qn 44).

(vii) $(a_n - a)$ null $\Rightarrow (|a_n - a|)$ null, by the absolute value rule (qn 33). Now $0 \le \big||a_n| - |a|\big| \le |a_n - a|$ from qn 2.63, and so by the squeeze rule (qn 34), $(\big||a_n| - |a|\big|)$ is null, and by the absolute value rule (qn 33), $(|a_n| - |a|)$ is null, so $(|a_n|) \to |a|$.

(viii) $a_n \le c_n \le b_n \Rightarrow a_n - a \le c_n - a \le b_n - a$. Now the squeeze rule with a shift (qn 46(ii)) gives $(c_n) \to a$.

55 (i) $a_n + b_n + c_n$ is the sum of $a_n + b_n$ and c_n. The sum of k terms by induction.

(ii) $a_n b_n c_n$ is the product of $a_n b_n$ and c_n. The product of k terms by induction.

(iii) By repeated use the product rule, the scalar rule and the sum rule, the limit is $1 + 2a + 3a^2 + 4a^3 + 5a^4$.

56 Each tends to 1.

57 (c) $0 < a_n \le (b - 1)/n$. Now use the scalar rule (qn 32) and the squeeze rule (qn 34).

58 (c) $\sqrt[n]{(a^n + b^n)} = b \cdot \sqrt[n]{((a/b)^n + 1)}$, so
$b < \sqrt[n]{(a^n + b^n)} = b \cdot \sqrt[n]{(a/b + 1)} \le b \cdot \sqrt[n]{2}$.
Now use qn 57, the scalar rule (qn 32) and the squeeze rule (qn 54(viii)).

59 (b) For $n > 2$, the third term in the expansion of $(1 + a_n)^n$ is $\frac{1}{2}n(n - 1)(a_n)^2$.

(c) For $n > 2$, $(1/\sqrt{(n - 1)})$ is a subsequence of a null sequence by qn 30. So (a_n) is null by the scalar rule and the squeeze rule.

60 $(a_n - a)$ is a null sequence $\Leftrightarrow$ given $\varepsilon > 0$, there exists an N such that $n > N \Rightarrow |a_n - a| < \varepsilon$.

61 $L = a - \varepsilon,\ U = a + \varepsilon$.

62 From the definition there exists an N such that
$n > N \Rightarrow |a_n - a| < 1 \Rightarrow a - 1 < a_n < a + 1$.
So every convergent sequence is eventually bounded. $((-1)^n)$ is bounded but not convergent.

63 $\frac{1}{2}a$, by taking $\varepsilon = \frac{1}{2}a$.

64 2 and $\frac{1}{2}$ are reciprocal.

65 (a) $(a_n - a)$ is a null sequence, given. So $(|a_n - a|)$ is a null sequence by the absolute value rule (qn 33). Then $(|a_n - a|/a)$ is a null sequence by the scalar rule (qn 32), since $a \ne 0$.

(b) The first pair of inequalities comes from taking $\varepsilon = \frac{1}{2}a$ in the standard definition of $(a_n) \to a$, the second from reciprocating as in qn 2.23.
(c) Apply the second inequality in (b) to the equation at the head of the question.

66 If (a_n) is a sequence of non-zero terms with $(a_n) \to a$, and $a < 0$, then $(-a_n)$ is a sequence of non-zero terms with $(-a_n) \to -a$ by the scalar rule (qn 54(i)), and $0 < -a$. So by qn 65, $(-1/a_n) \to -1/a$, and by the scalar rule (qn 54(i)), $(1/a_n) \to 1/a$.

67 Use the reciprocal rule (qns 65, 66) and the product rule (qn 54(vi)).

68 In parts (i) to (ix) the form of the expression for the nth term is adjusted until the quotient rule can be applied.

(i) $\frac{1}{2}$.
(ii) nth term $= \dfrac{3 + 2/n}{4 - 3/n}$, limit $\frac{3}{4}$.
(iii) nth term $= \dfrac{1 - 1/n^2}{3 + 3/n}$, limit $\frac{1}{3}$.
(iv) Limit -1.
(v) Limit 1.
(vi) Limit 0, divide top and bottom by n^4.
(vii) Limit $-\frac{1}{5}$.
(viii) Limit -1 when $a < 1$; limit 0 when $a = 1$; limit 1 when $1 < a$.
(ix) Limit $\frac{1}{2}$, use qn 1.3(i).
(x) nth term $= 1/\sqrt[n]{(1/b)}$, limit 1, by the reciprocal rule.

69 $n < 64 \Rightarrow 8 > \sqrt{n} > n/8 \Rightarrow a_n > b_n > c_n$.
$(a_n) \to 0$ like qn 35(v).
$b_n = 1/(\sqrt{(1 + 1/\sqrt{n})} + 1)$, so $(b_n) \to \frac{1}{2}$.
$c_n = \sqrt{n}(\sqrt{\frac{9}{8}} - 1)$, so $(c_n) \to +\infty$.

70 $(n^k/2^n)$ for any positive integer k. All null.
Take $\varepsilon = \frac{1}{4}$, then for sufficiently large n, $a_{n+1}/a_n < \frac{3}{4}$.

71 Take $\varepsilon = \frac{1}{2}(1 - l)$, then $0 \le a_{k+n} < (\frac{1}{2}(1 + l))^n a_k$.
$0 < l < 1 \Rightarrow 0 < \frac{1}{2}(1 + l) < 1$, so qn 39 can be applied here.

72 Take $\varepsilon = \frac{1}{2}(l - 1)$, then $(\frac{1}{2}(1 + l))^n a_k < a_{k+n}$. $1 < l \Rightarrow 1 < \frac{1}{2}(1 + l)$.

73 $(1/n)$. (n).

74 The sequences are null if and only if the sequences of absolute values are null.

(i) null when $|x| < 1$, or when $x = 1$ and $k < 0$.
(ii) null for all values of x.

75 $(a_n - a)$ is a null sequence, so by the squeeze rule (qn 34), (a_n) is null, so $a = 0$. Thus for a sequence of non-negative terms, a negative limit is impossible.

76 By the difference rule (qn 53(v)), $(b_n - a_n)$ has limit $b - a$. But $(b_n - a_n)$ is a sequence of non-negative terms, so $b - a \geq 0$.

78 Since $0 \leq a_n - A,\ 0 \leq a - A$, from qn 75.
Since $0 \leq B - a_n,\ 0 \leq B - a$, from qn 75. $0 < 1/n < 2$.

79 $a_{n+2}^2 - a_{n+1}^2 = (a_{n+1} + 2) - (a_n + 2) \Rightarrow (a_{n+2} - a_{n+1})(a_{n+2} + a_{n+1}) = a_{n+1} - a_n$. Since the terms are positive, $a_{n+2} - a_{n+1}$ has the same sign as $a_{n+1} - a_n$.

80
(a) If $A < a_{n_k} = A + \varepsilon$, then the monotonic increasing subsequence gives $n_i > n_k \Rightarrow A + \varepsilon \leq a_{n_i} \Rightarrow \varepsilon \leq a_{n_i} - A$, while convergence to A gives $|a_{n_i} - A| < \varepsilon$ for sufficiently large n_i.
(c) Convergence to A implies that $|a_{n_i} - A| < \varepsilon$ for sufficiently large n_i, or $A - \varepsilon < a_{n_i} < A + \varepsilon$, so from (a) $A - \varepsilon < a_{n_i} \leq A$.
(d) Take N as a sufficiently large n_i in (c). Then claim monotonic increasing and (b).

Theorem. If a monotonic increasing sequence has a subsequence converging to A, then the sequence tends to A.
If (a_n) is monotonic decreasing with a subsequence convergent to A, then $(-a_n)$ is monotonic increasing with a subsequence converging to $-A$. By the theorem we have proved, $(-a_n) \to -A$, and so by the scalar rule, $(a_n) \to A$. The overall theorem is that if a monotonic sequence has a convergent subsequence, then the whole sequence is convergent to the same limit.

81
(a) $p_{n+1} = p_n$.
(b) $A_{n+1} = \frac{1}{2}A_n$.
(c) (p_n) is constant, (A_n) is null.

82
(a) $3 \cdot 4^{n-1}$.
(d) from (a) and (c).
(e) note that the limit of the right-hand side as $n \to \infty$ is $\frac{9}{5}$.
(f) $A_{n+1} = A_1 + \frac{3}{9}A_1(1 + \frac{4}{9} + \ldots + (\frac{4}{9})^{n-1})$.
(g) from (b) and qn 21.

4

Completeness

What the rational numbers lack

Preliminary reading: Hemmings and Tahta, Niven, Burn (1990), Gardiner part II, Lieber, Yarnelle, Zippin.
Concurrent reading: Baylis and Haggarty, Wheeler.
Further reading: Armitage and Griffiths, part II, ch. 1, Cohen and Ehrlich chs 4 and 5, Artmann chs 1 to 3.

The first sixteen questions of this chapter are concerned with integers and their quotients.

The Fundamental Theorem of Arithmetic

	$21 = 3 \cdot 7$	41 prime	61 prime	$81 = 3^4$
2 prime	$22 = 2 \cdot 11$	$42 = 2 \cdot 3 \cdot 7$	$62 = 2 \cdot 31$	$82 = 2 \cdot 41$
3 prime	23 prime	43 prime	$63 = 3^2 \cdot 7$	83 prime
$4 = 2^2$	$24 = 2^3 \cdot 3$	$44 = 2^2 \cdot 11$	$64 = 2^6$	$84 = 2^2 \cdot 3 \cdot 7$
5 prime	$25 = 5^2$	$45 = 3^2 \cdot 5$	$65 = 5 \cdot 13$	$85 = 5 \cdot 17$
$6 = 2 \cdot 3$	$26 = 2 \cdot 13$	$46 = 2 \cdot 23$	$66 = 2 \cdot 3 \cdot 11$	$86 = 2 \cdot 43$
7 prime	$27 = 3^3$	47 prime	67 prime	$87 = 3 \cdot 29$
$8 = 2^3$	$28 = 2^2 \cdot 7$	$48 = 2^4 \cdot 3$	$68 = 2^2 \cdot 17$	$88 = 2^3 \cdot 11$
$9 = 3^2$	29 prime	$49 = 7^2$	$69 = 3 \cdot 23$	89 prime
$10 = 2 \cdot 5$	$30 = 2 \cdot 3 \cdot 5$	$50 = 2 \cdot 5^2$	$70 = 2 \cdot 5 \cdot 7$	$90 = 2 \cdot 3^2 \cdot 5$
11 prime	31 prime	$51 = 3 \cdot 17$	71 prime	$91 = 7 \cdot 13$
$12 = 2^2 \cdot 3$	$32 = 2^5$	$52 = 2^2 \cdot 13$	$72 = 2^3 \cdot 3^2$	$92 = 2^2 \cdot 23$
13 prime	$33 = 3 \cdot 11$	53 prime	73 prime	$93 = 3 \cdot 31$
$14 = 2 \cdot 7$	$34 = 2 \cdot 17$	$54 = 2 \cdot 3^3$	$74 = 2 \cdot 37$	$94 = 2 \cdot 47$
$15 = 3 \cdot 5$	$35 = 5 \cdot 7$	$55 = 5 \cdot 11$	$75 = 3 \cdot 5^2$	$95 = 5 \cdot 19$
$16 = 2^4$	$36 = 2^2 \cdot 3^2$	$56 = 2^3 \cdot 7$	$76 = 2^2 \cdot 19$	$96 = 2^5 \cdot 3$
17 prime	37 prime	$57 = 3 \cdot 19$	$77 = 7 \cdot 11$	97 prime
$18 = 2 \cdot 3^2$	$38 = 2 \cdot 19$	$58 = 2 \cdot 29$	$78 = 2 \cdot 3 \cdot 13$	$98 = 2 \cdot 7^2$
19 prime	$39 = 3 \cdot 13$	59 prime	79 prime	$99 = 3^2 \cdot 11$
$20 = 2^2 \cdot 5$	$40 = 2^3 \cdot 5$	$60 = 2^2 \cdot 3 \cdot 5$	$80 = 2^4 \cdot 5$	$100 = 2^2 \cdot 5^2$

1 Find non-negative integers a, b, c, d, e, f such that
$200\,772 = 2^a \cdot 3^b \cdot 5^c \cdot 7^d \cdot 11^e \cdot 13^f$.

The *Fundamental Theorem of Arithmetic* states that each natural number greater than 1 may be expressed in a unique way as a product of prime numbers. The proof of this theorem can be found in most undergraduate textbooks on number theory. See for example the first chapter of Burn (1997) or Hardy and Wright. For the simple enunciation of this theorem it is important that the number 1 is *not* called prime.

2 Find integers a, b, c, d, e, f such that
$2775/999\,999 = 2^a \cdot 3^b \cdot 5^c \cdot 7^d \cdot 11^e \cdot 13^f$.

3 Use the Fundamental Theorem of Arithmetic to enunciate a unique factorisation theorem for *positive rational* numbers. (The set of rational numbers, $\mathbb{Q}$, was defined at the start of chapter 2.)

Dense sets of rational numbers on the number line

4 Illustrate on a number line those portions of the sets $\{m \mid m \in \mathbb{Z}\}$, $\{m/2 \mid m \in \mathbb{Z}\}$, $\{m/4 \mid m \in \mathbb{Z}\}$, $\{m/8 \mid m \in \mathbb{Z}\}$ which lie between ± 3. Is each set contained in the set which follows in this list? What would an illustration of the set $\{m/2^n \mid m \in \mathbb{Z}\}$ look like for some large positive integer n?

5 Do you believe that if only n were large enough there would be a rational number of the form $m/2^n$ lying between 57/65 and 64/73? The difference between these two numbers is $1/(65 \cdot 73)$. Try to find such a number where m and n are positive integers.

6 To show that there is always a rational number of the form $m/2^n$ between two numbers a and b, where $a < b$, remember that $((\frac{1}{2})^n)$ is a null sequence by qn 3.39, since it is a geometric progression with common ratio less than 1. So, for sufficiently large n, $(\frac{1}{2})^n < b - a$. Only when n is that big can we expect to find an $m/2^n$ between a and b. Now let k be a positive integer such that $(\frac{1}{2})^k < b - a$. Use the integer function $\lfloor\ \rfloor$ defined just before qn 3.19, to locate two integers close to $a \cdot 2^k$, with $\lfloor a \cdot 2^k \rfloor \le a \cdot 2^k < \lfloor a \cdot 2^k \rfloor + 1$. Now show that $m = \lfloor a \cdot 2^k \rfloor + 1$ and $n = k$ satisfy the required conditions.

7 By repeatedly bisecting the interval $\{x \mid a < x < b\}$, show that between any two distinct numbers on the number line, there is an infinity of numbers of the form $m/2^n$, where m is an integer and n is a non-negative integer.

A set of numbers containing an element in every interval on the number line is said to be *dense* on the line. The argument of qn 7 can be used to show that there is an *infinity* of elements of a dense set in every interval of the number line. We have shown in qn 6 that the rational numbers are *dense* on the number line, even though the proof only used rationals with denominator a power of 2. Beware! When drawing a figure, there is no difference to be seen between a dense set of points and a continuous line segment.

8 Let $T = \{m/2^n \mid m \in \mathbb{Z},\ n \in \mathbb{Z}^+\}$. Is T dense on the number line? Is there a smallest positive number in T? Is there a smallest positive rational number?

9 Let $D = \{m/10^n \mid m \in \mathbb{Z},\ n \in \mathbb{Z}^+\}$. Prove that every element of T (in qn 8) is an element of D. Deduce that D is dense on the number line. Show that $\frac{1}{5} \in D$, and use the Fundamental Theorem of Arithmetic to show that $\frac{1}{5} \notin T$.
The elements of D are called *terminating decimals*.
Is the set T closed under

(i) addition,
(ii) subtraction,
(iii) multiplication and
(iv) division, except by 0?

Would your answers have been the same if the questions had been asked about the set D?

10 Check that every term of the sequence with terms

$$a_1 = \frac{1}{4},\ a_2 = \frac{1}{4} - \left(\frac{1}{4}\right)^2,\ a_3 = \frac{1}{4} - \left(\frac{1}{4}\right)^2 + \left(\frac{1}{4}\right)^3,\ \ldots,$$

$$a_n = -\left(\left(-\frac{1}{4}\right) + \left(-\frac{1}{4}\right)^2 + \left(-\frac{1}{4}\right)^3 + \cdots + \left(-\frac{1}{4}\right)^n\right), \ldots$$

belongs to T, as defined in qn 8.
Use qn 1.3(vi) to show that $a_n = \frac{1}{4} \cdot \dfrac{1 - (-\frac{1}{4})^n}{1 - (-\frac{1}{4})}$,
and with the help of qn 3.42, the difference rule (qn 3.54(v)) and the scalar rule (qn 3.54(i)), prove that $(a_n) \to \frac{1}{5}$. Thus a sequence in the dense set T may converge to a point outside T.

Infinite decimals

11 Use the Fundamental Theorem of Arithmetic to show that $\frac{1}{3} \notin D$ as defined in qn 9. That is to say, there is no terminating decimal equal to $\frac{1}{3}$. From qn 3.51 write down a sequence of terminating decimals which converges to $\frac{1}{3}$.

The limit of the sequence 0.3, 0.33, 0.333, 0.3333, ... (with a *single* recurring digit) is denoted by $0.\dot{3}$, or $0.3\dot{3}$, or $0.33\dot{3}$.
We have proved that $0.\dot{3} = \frac{1}{3}$.

12 Prove that the nth term in the sequence 0.9, 0.99, 0.999, 0.9999, ... is equal to $1 - 1/10^n$. Deduce that $0.\dot{9} = 1$.

13 For the sequence 0.1, 0.13, 0.131, 0.1313, ... the $2n$th term is $\frac{13}{100}(1 + 10^{-2} + 10^{-4} + \ldots + 10^{-2n+2})$.
Use qn 1.3(vi) to simplify this expression, and prove (with the help of qn 3.80) that the sequence converges to $\frac{13}{99}$.
This proves that $0.\dot{1}\dot{3}$ or $0.13\dot{1}\dot{3}$ (either of which denotes the limit of the sequence) $= \frac{13}{99}$. Here there is *a pair of* recurring digits.

14 Use qns 1.3(vi) and 3.39 to prove that the limit of the sequence with nth term

$$a + b\left(\frac{1}{10^3} + \frac{1}{10^6} + \ldots + \frac{1}{10^{3n}}\right)$$

is $a + b/999$.
Find terminating decimals a and b such that $12.45\dot{6}7\dot{8} = a + b/999$.
Here there is *a triple of* recurring digits.
Must every infinite decimal with a recurring block of digits be equal to a rational number?

The ***infinite decimal*** $d_0.d_1d_2d_3\ldots d_n\ldots$ is the limit of the infinite decimal sequence with nth term

$$d_0 + \frac{d_1}{10} + \frac{d_2}{10^2} + \ldots + \frac{d_n}{10^n},$$

where d_0 is an integer, and, when $1 \leq i$, d_i is 0, 1, 2, ... or 9. (This definition looks a bit strange when d_0 is negative, but it is convenient for proving theorems to have all infinite decimal sequences increasing.) This definition does not make the claim that every infinite decimal sequence is necessarily convergent or that every infinite decimal is necessarily a number.
In qn 3.51 we found that every number on the number line is the limit of an infinite decimal sequence.
When for all sufficiently large n, $d_n = 0$, the decimal is said to *terminate*, as in qn 9. When there is a number b such that for all sufficiently large n, $d_n = d_{n+b}$, the decimal is said to *recur* (with a recurring block of length b).

15 Show that every rational number which is not equal to a terminating decimal is the limit of a recurring decimal sequence. (*Hint*. Analyse the remainders on dividing the integer A by the integer B by long division.)

16 (*Dedekind*, 1877) Suppose two infinite decimals have the same limit, say $x = d_0.d_1d_2d_3\ldots d_n\ldots = e_0.e_1e_2e_3\ldots e_n\ldots$. If they are different as infinite decimals then they must differ at some digit. Suppose $d_0 \neq e_0$,

and in fact $d_0 < e_0$. Then $d_0 + 1 \leq e_0$ since the numbers are integers. From qn 12, $0.d_1d_2d_3 \ldots d_n \leq 1 - 1/10^n$. Since an infinite decimal sequence is monotonic increasing, $x \leq d_0 + 1$, by the inequality rule, qn 3.76. Again, since an infinite decimal sequence is monotonic increasing, $e_0 \leq x$.
Now $e_0 \leq x \leq d_0 + 1 \leq e_0$, so $x = e_0 = d_0 + 1$, and $e_i = 0, d_i = 9$ when $1 \leq i$.
Revise this argument for the case when the first digits to differ are d_j and e_j, to show that of two equal infinite decimals, one has recurring 9s and the other terminates. For example, $0.5 = 0.4\dot{9}$.

17 If the infinite decimal sequence for the infinite decimal 0.101 001 000 100 001 ..., with 1s in the $\frac{1}{2}n(n+1)$th positions and 0s elsewhere were convergent, would its limit be a rational number?

Irrational numbers

18 If p and q are non-zero integers, prove that an equation of the form $p^2 = 2q^2$ would contradict the Fundamental Theorem of Arithmetic. So there can be no rational number equal to $\sqrt{2}$. List some other square roots and cube roots which cannot equal a rational number for similar reasons.

19 If a and b are rational numbers, with $b \neq 0$, is it possible for $a + b\sqrt{2}$ to be a rational number?

20 Let a and b be rational numbers with $a < b$. Show that $(a + b\sqrt{2})/(1 + \sqrt{2})$ is not rational. Show also that $a < (a + b\sqrt{2})/(1 + \sqrt{2}) < b$.
Deduce that there is an irrational number between any two distinct rationals. This, with qn 7, shows that the irrationals are dense on the number line.

21 (Optional) Show that there can be no rational number equal to $\log_{10} 2$.

Infinity: countability

.1	2	3
.1	**7**	8
.0	5	**4**

.1	2	3	4
.1	**7**	8	9
.0	7	**9**	4
.0	3	6	**0**

.1	2	3	4	5
.9	**8**	7	6	5
.2	0	**6**	8	4
.0	1	0	**0**	2
.0	1	1	1	**1**

.1	2	3	4	5	6
.2	**2**	2	3	3	3
.2	7	**1**	8	0	9
.7	5	3	**0**	1	2
.7	7	2	2	**8**	8
.0	0	0	1	1	**1**

Figure 4.1

22 In the squares above, filled with terminating decimals between 0 and 1, think how you could fill in the last row with a new terminating decimal between 0 and 1, not in the earlier rows. Can you think of an automatic procedure that would give you a new row without further checking, for a 100×100 square or bigger?
What about looking at the diagonal entries (marked bold here) and making sure that the entries below them in your last line differed from them?

23 (*Cantor*, 1891) If you made a sequence of infinite decimals between 0 and 1, could you always construct a new infinite decimal between 0 and 1 that was not in the list you had made?

One way of describing the result of qn 23 is to say that there are too many infinite decimals between 0 and 1 to put them all in one sequence. Any sequence is of the form $a_1, a_2, a_3, \ldots, a_n, \ldots$. So if the terms are all different, the suffix shows how to match the terms of the sequence with the natural numbers, $\mathbb{N}$.

24 By starting $0, +1, -1, \ldots$ show how to put all the integers into a sequence. This will match $\mathbb{N}$ to $\mathbb{Z}$, one-to-one. Another way to construct a sequence which includes every integer is to match the positive integer n with 2^n and the negative integer $-n$ with $2^n \cdot 3$. Then running through the selected natural numbers of the form 2^n and $2^n \cdot 3$ in order of size gives a way of building a sequence which includes each of the *integers* exactly once. What are the first seven integers when this second method is used to construct a sequence running through $\mathbb{Z}$?

25 If a and b are positive integers, matching the rational number a/b (always taken in lowest terms) with the natural number $2^a \cdot 3^b$ gives a one-to-one matching of the positive rationals with a (rather small!) subset of $\mathbb{N}$. If we then run through the selected natural numbers of the form $2^a \cdot 3^b$ in order of size, we have a way of building a sequence which includes each of the *positive rationals* exactly once. What are the first four terms of this sequence? If, further, the rational number $-a/b$ is matched with $2^a \cdot 3^b \cdot 5$, show how to construct a sequence which runs through all the rational numbers once. What are the first six terms of this sequence?

26 (Optional) If a, b, c and d are positive integers, matching the number $(a/b) + (c/d)\sqrt{2}$ with the natural number $2^a \cdot 3^b \cdot 5^c \cdot 7^d$ gives a one-to-one matching of the numbers $x + y\sqrt{2}$ (where x and y are positive rationals) with a subset of $\mathbb{N}$, provided the fractions a/b and c/d are in their lowest terms. Use this matching to construct a sequence containing all the numbers of the form $x + y\sqrt{2}$, where x and y are positive rationals. List the first six numbers in your sequence.

When it is possible to put the elements of a set into a sequence without repetition, the set is said to be *countably infinite*. There is then a one-to-one correspondence between the elements of the set and the natural numbers $\mathbb{N}$.

27 Which of the following sets are countably infinite: $\mathbb{Z}$, T (of qn 8), D (of qn 9), $\mathbb{Q}$, the set of infinite decimals between 0 and 1?

When an infinite set is not *countably infinite* it is said to be *uncountable*. We will shortly claim as a principle that every infinite decimal gives a number, that is that every infinite decimal sequence is convergent. The set of limits of infinite decimals is then known as the set of real numbers and denoted by $\mathbb{R}$.

The contrast between the *countably infinite* rational numbers and the *uncountable* real numbers is remarkable. A countably infinite set of numbers can be shown to occupy only a small portion of the real line.

28 If for each $n \in \mathbb{N}$, the interval $\{x \mid n - 1/2^n \leq x \leq n + 1/2^n\}$ is deleted from the number line, is every element of $\mathbb{N}$ deleted? What is the total length deleted?

29 Let ε be a fixed positive number. If for each $n \in \mathbb{N}$, the interval $\{x \mid n - \varepsilon/2^n \leq x \leq n + \varepsilon/2^n\}$ is deleted from the number line, is every element of $\mathbb{N}$ deleted? What is the total length deleted? Can you make this length as small as you like? How much length should we say the set $\mathbb{N}$ occupies?

30 Let (a_n) be any sequence of numbers and let ε be any positive number. If for each $n \in \mathbb{N}$, the interval $\{x \mid a_n - \varepsilon/2^n \leq x \leq a_n + \varepsilon/2^n\}$ is deleted from the number line, is every element of the sequence (a_n) deleted? What is the total length deleted? Deduce that a countably infinite set of numbers does not occupy any length on the number line.

When we remember that the set of rational numbers is countably infinite (qn 25) and that the set of rational numbers is dense on the line (qn 7), it is particularly strange to realise that qn 30 implies that when every rational point is removed together with a small interval around it, most of the real number line remains untouched. Despite appearances, a real line on which only points corresponding to rational numbers can be identified is mostly empty!

In the uncountability proof for infinite decimals (qn. 23), the construction of a new infinite decimal could be done even if all the original infinite decimals were rational. Can you see why this does *not* establish that the rationals are uncountable?

Summary

Denseness

Definition	If a set of numbers has a member in every interval on the number line it is said to be *dense* on the line.
Theorem qn 7	A dense set on the number line has an infinity of members in every interval.
Theorem qn 9	The terminating decimals are dense on the number line.
Theorem qn 6	The rational numbers are dense on the number line.

Decimals and irrationals

Definition An ***infinite decimal*** $d_0.d_1d_2d_3\ldots d_n\ldots$ is the limit of the infinite decimal sequence with nth term

$$d_0 + \frac{d_1}{10} + \frac{d_2}{10^2} + \ldots + \frac{d_n}{10^n},$$

where d_0 is an integer, and, when $1 \leq i$, d_i is 0, 1, 2, ... or 9.

Definition If $d_n = 0$ for sufficiently large n, the infinite decimal is said to *terminate*.

Definition If $d_n = d_{n+b}$ for sufficiently large n and some fixed b, the infinite decimal is said to recur, with a recurring block of length b.

Theorem qn 9, 14, 15 An infinite decimal is equal to a rational number (has a rational limit) if and only if it is either terminating or recurring.

Theorem qn 16 If two distinct infinite decimals are equal, then one is terminating and the other has recurring 9s.

Theorem qn 18 There is no rational number whose square is 2.

Theorem qn 20 There is an irrational number between any two rationals.

Theorem qn 20 The irrational numbers are dense on the line.

Countability

Definition	If there is a sequence of distinct terms from a set which contains all the terms of that set, then that set is said to be *countably infinite.*
Theorem qn 25	The set of rational numbers is countably infinite.
Theorem qn 23	The set of infinite decimals between 0 and 1 is not countably infinite.

The completeness principle: infinite decimals are convergent

The properties of numbers which we have worked with so far consist of

(1) algebraic properties listed in appendix 1,

(2) properties of less than, based on the notion of positiveness, in chapter 2,

(3) the Archimedean Principle in chapter 3.

We have been able to show that in a number system satisfying these properties, every number is the limit of an infinite decimal sequence (qn 3.51). Now the rational numbers, by themselves, satisfy (1), (2) and (3), so these number properties do not ensure the existence of square roots or of limits for non-recurring infinite decimals.
To have a number system with all the properties we expect, namely, the real number system, $\mathbb{R}$, we adopt one more principle, the completeness principle.

Every infinite decimal sequence is convergent

Or, in other words, every infinite decimal is a real number. That is to say, the set of real numbers is the set of limits of infinite decimal sequences.

31 Write down the first five terms of the infinite decimal sequence with limit $0.123\,456\,789\,101\,112\ldots$.
Is the sequence monotonic increasing?
Is every term rational?
State an upper bound for this sequence.
Can you find an upper bound which is smaller than this?

32 If two numbers a and b are given as infinite decimals, explain how $a+b$, $a-b$, ab and a/b are determined when $b \neq 0$.

The remainder of this chapter will be devoted to establishing six properties of the real numbers, any one of which is in fact equivalent to the completeness principle, in the context of the number properties described as (1), (2) and (3) above. Other authors commonly select one of these six as their affirmation of completeness, but this list does not exhaust the alternatives.

I. Every bounded monotonic sequence of real numbers is convergent.
II. The intersection of a set of *nested* closed intervals is not empty (the term 'nested' is defined in qn 42).
III. Every bounded sequence of real numbers has a convergent subsequence.
IV. Every infinite bounded set of real numbers has a *cluster point* (the term is defined before qn 48).
V. Every *Cauchy sequence* of real numbers is convergent (the term 'Cauchy sequence' is defined before qn 56).
VI. Every non-empty set of real numbers which is bounded above has a least upper bound.

You will find **I, III** and **VI** used repeatedly in the rest of the book, though all six are necessary for the further study of analysis.

Bounded monotonic sequences

33 Explain why every monotonic sequence is either bounded above or bounded below.
Deduce that an increasing sequence which is bounded above is bounded, and that a decreasing sequence which is bounded below is bounded.

34 (after *du Bois-Reymond*, 1882) We suppose that (a_n) is a monotonic decreasing sequence which is bounded below, by L, and we seek to prove that (a_n) is convergent.

(i) Name a lower bound for the sequence (a_n) which is an integer.
(ii) Name an upper bound for the sequence (a_n) which is an integer.
(iii) Must there be consecutive integers $c, c+1$ for which c is a lower bound for (a_n) and $c+1$ is not? For such integers let $t_0 = d_0 = c$.
(iv) Consider the eleven numbers $t_0, t_0 + \frac{1}{10}, t_0 + \frac{2}{10}, \ldots, t_0 + \frac{10}{10}$.
Is there a $c = 0, 1, 2, \ldots$, or 9 such that $t_0 + \dfrac{c}{10}$
is a lower bound for (a_n) and $t_0 + \dfrac{c+1}{10}$ is not?
Let $d_1 = c$ and $t_1 = d_0.d_1$.
(v) Proceed inductively to define an infinite decimal sequence (t_n), in such a way that t_n is a lower bound for the sequence (a_n) while $t_n + 1/10^n$ is not a lower bound. Define $t_{n+1} = t_n + \dfrac{c}{10^{n+1}}$, with

$c = 0, 1, 2, \ldots$, or 9. where t_{n+1} is a lower bound for (a_n) while $t_{n+1} + 1/10^{n+1}$ is not.
$c = d_{n+1}$, and $t_{n+1} = d_0.d_1d_2d_3 \ldots d_nd_{n+1}$.

(vi) Why is the sequence (t_n) convergent? Let its limit be D.

(vii) For a constant k, say why $(a_k - t_n) \to a_k - D \geq 0$.

(viii) How do you know that for sufficiently large n, $D \leq a_n < D + 1/10^i$ for a given positive integer i?

(ix) Prove that $(a_n) \to D$.

35 Use qn 34 to prove that a monotonic increasing sequence which is bounded must be convergent.

Now we have established that monotonic sequences which are bounded are convergent, and it easily follows that monotonic sequences which are unbounded tend to $\pm\infty$. Questions 34 and 35 prove that the convergence of infinite decimal sequences implies the convergence of monotonic bounded sequences. The converse is obvious, since every infinite decimal sequence is bounded and monotonic. So we could have adopted the convergence of monotonic bounded sequences as our completeness principle.

36 Use qn 2.49 to prove that the sequence with nth term $(1 + 1/n)^n$ has a limit which lies between 2 and 3.

***n*th roots of positive real numbers, *n* a positive integer**

37 If a and x are positive real numbers with $a \leq x^2$, prove that $a \leq (\frac{1}{2}(x + a/x))^2 \leq x^2$. Deduce that if x_1 is positive and $a < x_1{}^2$, the sequence (x_n) defined by

$$x_{n+1} = \frac{1}{2}\left(x_n + \frac{a}{x_n}\right)$$

is convergent. Find its limit. Deduce that every positive real number has a unique positive real number as square root. Compare with qn 2.37. (This method of approximating to square roots is usually known as Horner's method, though it coincides with Newton's approximation in appendix 3, and was used centuries before by the Chinese.)

38 Prove that the sequences (a_n) and (b_n) of qn 2.38 are both convergent to the same limit. Prove the analogous result for qn 2.39.

39 Use qn 37 to prove that positive real numbers have real 4th roots, 8th roots, 16th roots, etc.

40 Let $k > 1$ be a positive integer.
If $0 < c < 1$, then $c^k < c < 1^k$, and if $1 < c$, then $1^k < c < c^k$, by qn 2.11.

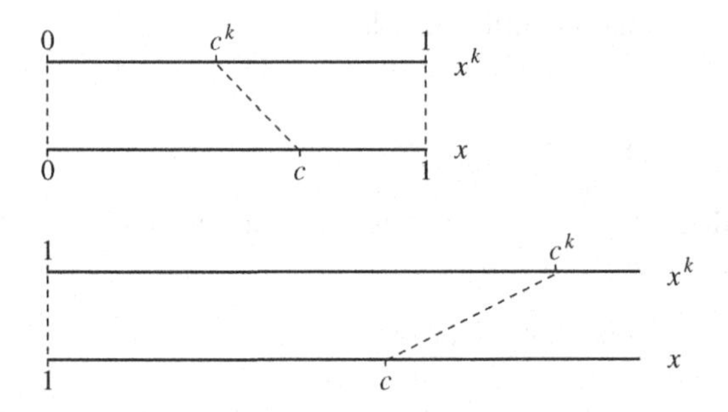

Figure 4.2

Thus for any positive number $c \neq 1$, we can choose positive numbers a_1 and b_1 such that ${a_1}^k < c < {b_1}^k$.
Now consider the number $d = (a_1 + b_1)/2$.
If $d^k = c$, we have found a kth root of c.
If $d^k < c$, let $a_2 = d$ and $b_2 = b_1$.
If $d^k > c$, let $a_2 = a_1$ and $b_2 = d$.
So $a_1 \leq a_2 < b_2 \leq b_1$, ${a_2}^k < c < {b_2}^k$, and $b_2 - a_2 = (b_1 - a_1)/2$.
Repeat this process to define sequences (a_n), (b_n) inductively such that $a_n \leq a_{n+1} < b_{n+1} \leq b_n$, ${a_n}^k < c < {b_n}^k$, and $b_{n+1} - a_{n+1} = (b_n - a_n)/2$.

(i) Why must the sequence (a_n) be convergent?
(ii) Why must the sequence (b_n) be convergent?
(iii) Why must $(b_n - a_n)$ be a null sequence, and the limits of (a_n) and (b_n) equal?
(iv) Denote the common limit by l. Why must $({a_n}^k) \to l^k$ and $({b_n}^k) \to l^k$?
(v) Why must $({b_n}^k - {a_n}^k)$ be a null sequence?
(vi) Use ${a_n}^k < c < {b_n}^k$ and the squeeze rule to prove that $(c - {a_n}^k)$ is a null sequence.
(vii) Deduce that $c = l^k$, so that c has a kth root, namely l.

41 (i) Prove that the sequence $(n(\sqrt[n]{a} - 1))$ of qn 2.50 is convergent when $1 < a$.
(ii) Prove that $\lim n(\sqrt[n]{(1/a)} - 1) = -\lim n(\sqrt[n]{a} - 1)$, using qn 3.57 and the quotient rule, qn 3.67.

(iii) Prove also that $\lim n(\sqrt[n]{(ab)} - 1) = \lim n(\sqrt[n]{a} - 1) + \lim n(\sqrt[n]{b} - 1)$, using qn 3.57 and the product rule, qn 3.54(vi).

If you consider $\lim n(\sqrt[n]{a} - 1)$ as a function of a, can you think of any other function of a with similar properties? These properties were first recognised by E. Halley (1695). This limit will play a significant role in chapter 11 in the formal definition of exponentials and logarithms.

Nested closed intervals

42 *The Chinese Box Theorem*
(a_n) and (b_n) are sequences such that $a_n \leq a_{n+1} < b_{n+1} \leq b_n$ for all values of n.
Let $[a, b]$ denote the closed interval $\{x \mid a \leq x \leq b\}$.

(i) Say why $[a_1, b_1] \cap [a_2, b_2] \cap \ldots \cap [a_n, b_n] = [a_n, b_n]$.
We describe this by saying that these intervals are *nested*.
(ii) Why must each of the two sequences be convergent?
(iii) By considering the limits of the two sequences prove that at least one number lies in all the intervals $[a_n, b_n]$.
(iv) By considering intervals of the type $\{x \mid 0 < x < 1/n\}$, show that the intersection of nested *open* intervals may be empty.

Convergent subsequences of bounded sequences

43 Find an upper bound and a lower bound for the sequences with nth term (i) $(-1)^n$, and (ii) $(-1)^n(1 + 1/n)$.
Is either sequence convergent? In each case find a convergent subsequence.

44 Look back to your proof in qn 3.62 that every convergent sequence is bounded. Is it true that every bounded sequence is convergent?

45 The ordinary decimal representation of positive integers may be used to construct a sequence of terminating decimals $(0.n)$. The first twelve terms of this sequence are

$$0.1, 0.2, 0.3, 0.4, 0.5, 0.6, 0.7, 0.8, 0.9, 0.10, 0.11, 0.12, \ldots$$

Is this sequence bounded?
Identify at least two convergent subsequences.

46 Prove that **every bounded sequence has a convergent subsequence.** Use qns 3.16 (for the existence of a monotonic subsequence) and 3.12 (on the boundedness of subsequences), and qns 34 and 35 of this chapter.

This theorem can be applied to such sequences as $(\sin n)$ and $(n\sqrt{2} - \lfloor n\sqrt{2} \rfloor)$ which appear to have no regularity at all.

Without paying proper regard to history, this theorem has sometimes been called the Bolzano–Weierstrass Theorem, the name being transferred from qn 53, to which it is closely related. The theorem will be used repeatedly in our study of real functions.

Because an infinite decimal sequence is bounded and monotonic, the limit of any subsequence is the limit of the sequence (by qn 3.80). Thus we could, equally well, have adopted 'every bounded sequence has a convergent subsequence' as our completeness principle.

47 By considering the infinite decimal sequence for $\sqrt{2}$, show that if only rational numbers are considered, a bounded sequence need not have a convergent subsequence.

Cluster points: the Bolzano–Weierstrass theorem

Because of the matching of real numbers with the points of a number line, it is sometimes helpful to refer to real numbers as 'points'. For any given positive ε, there is an infinity of $n \in \mathbb{N}$, such that $-\varepsilon < 1/n < \varepsilon$, and we know this because $(1/n)$ is a null sequence. In other words, there is an infinity of points from the set $\{1/n \mid n \in \mathbb{N}\}$ in any neighbourhood of 0. See qn 3.61 for the term 'neighbourhood'. This makes 0 a *cluster point* or an *accumulation point* for the set.

The point $\frac{1}{4}$ is *not* a cluster point for the set because we can find a positive value of ε such that $\frac{1}{4} - \varepsilon < 1/n < \frac{1}{4} + \varepsilon$ is only possible for the single natural number $n = 4$.

48 State a value of ε such that $\frac{1}{4} - \varepsilon < 1/n < \frac{1}{4} + \varepsilon$ only holds for a unique natural number n.

49 State a positive value of ε such that $\frac{2}{7} - \varepsilon < 1/n < \frac{2}{7} + \varepsilon$ does not hold for any natural number n. This means that the point $\frac{2}{7}$ is *not* a cluster point for the set $\{1/n \mid n \in \mathbb{N}\}$.

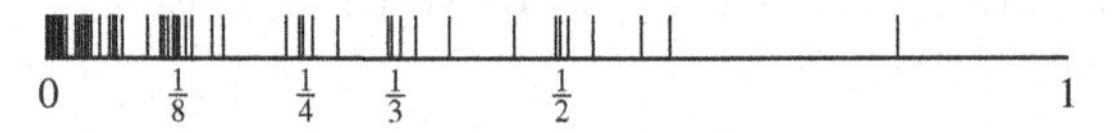

Figure 4.3

50 Name three cluster points for the set

$$\left\{ \frac{1}{2^m} + \frac{1}{3^n} \mid m, n \in \mathbb{N} \right\}.$$

Name two points between 0 and 1 which are not cluster points for the set.

51 Can a finite set of points on the number line have a cluster point?

52 Describe an infinite set of points that has no cluster point.

A set A of real numbers has an upper bound U if $a \leq U$ for all $a \in A$, and has a lower bound L if $L \leq a$ for all $a \in A$. A set is said to be *bounded* if it has an upper bound and a lower bound. The language here is like that for the upper bound and lower bound of a sequence introduced in qn 3.5.

53 *The Bolzano–Weierstrass Theorem*
Let A be an infinite set of real numbers with an upper bound U and a lower bound L.

(i) Construct a sequence of distinct points (a_n) in A.
(ii) Is the sequence (a_n) bounded?
(iii) Why must the sequence (a_n) contain a convergent subsequence?
(iv) If the convergent subsequence $(a_{n_i}) \to a$, must $a \in A$? Must $L \leq a \leq U$?
(v) Whether $a \in A$ or not, show that a is a cluster point of A.

This proves that **every infinite bounded set has a cluster point**. It was formulated by Weierstrass about 1867, and proved by a method due to Bolzano (1817) of repeated bisections of the bounded interval containing the set A, choosing always a half containing an infinity of points of A. The name, Bolzano–Weierstrass, was given by Cantor (1870) and Heine (1872). Cantor said that this theorem was the basis of all the more important mathematical truths.

To understand the significance of this theorem it is necessary to recognise that it would be false if the rational numbers were the only ones available.

54 Let A be the set of terminating decimals in the infinite decimal sequence for $\sqrt{2}$. State integer bounds for this set. Prove that no rational number can be a cluster point for A.

Cauchy sequences

The great virtue of the theorem that a bounded monotonic sequence is convergent is that it provides a method of determining the fact of convergence without knowing the value of the limit.

Is there a condition guaranteeing the convergence of a non-monotonic sequence which does not require knowledge of the value of the limit?

A reasonable conjecture would be to suggest that a sequence was convergent if the differences between consecutive terms form a null sequence. But there are counter-examples to such a conjecture. One is provided by the harmonic series which we will examine in the next chapter. Another derives from the sequence $(\sqrt{n})$.

We have shown that the differences between consecutive terms $(\sqrt{(n+1)} - \sqrt{n})$ gives a null sequence in qn 3.35(v), and that the sequence itself tends to infinity in qn 3.20. So the condition '$(a_{n+1} - a_n)$ is null' does not guarantee the convergence of (a_n).

Examining differences of the form $(a_{n+k} - a_n)$ for some constant k is no more effective, even for a large value of k, than for $k = 1$ since the sequence $(\sqrt{(n/k)})$ also tends to infinity. However, it is worth examining the hypothesis that the difference $(a_{n+k} - a_n)$ is sandwiched between two null sequences for *all* values of k. If for example

$$-1/n < a_{n+k} - a_n < 1/n, \text{ for all positive integers } k,$$

is this sufficient for the sequence (a_n) to be convergent? We first establish the converse result.

55 If $(a_n) \to a$, use the fact that

$$|a_{n+k} - a_n| \leq |a_{n+k} - a| + |a - a_n|,$$

from the triangle inequality, qn 2.61, to show that $|a_{n+k} - a_n|$ may be arbitrarily small, irrespective of the value of k, provided only that n is large enough.
Or, to use more precise language, that given $\varepsilon > 0$, there exists an N such that $|a_{n+k} - a_n| < \varepsilon$, when $n > N$.

So the condition we are examining is a necessary consequence of convergence. The argument in qn 55 does not depend upon completeness.

The condition we are investigating is called the **Cauchy criterion**. A sequence (a_n) is said to satisfy the Cauchy criterion, when

given $\varepsilon > 0$, there exists an N such that $n > N \Rightarrow |a_{n+k} - a_n| < \varepsilon$, for all positive integers k.

A sequence satisfying the Cauchy criterion is called a **Cauchy sequence**. In qn 55, we showed that every convergent sequence is a Cauchy sequence. We now explore the converse.

56 If (a_n) is an infinite decimal sequence prove that $|a_{n+k} - a_n| < 1/10^n$.
This shows that an infinite decimal sequence is necessarily a Cauchy sequence.
Thus the claim that every Cauchy sequence is convergent implies the completeness principle.

57 *The General Principle of Convergence.*
Let (a_n) be a Cauchy sequence of real numbers.
By putting $\varepsilon = 1$ in the Cauchy criterion prove that every Cauchy sequence is bounded (use qn 3.13). Use qn 46 to establish the existence of a convergent subsequence $(a_{n_i}) \to a$.

Now use the triangle inequality $|a_n - a| \le |a_n - a_{n_i}| + |a_{n_i} - a|$ to prove that $(a_n) \to a$.
So every Cauchy sequence of real numbers converges to a real limit.

In 1882, *P. du Bois-Reymond* gave the name 'The General Principle of Convergence' to the proposition that a sequence is convergent if and only if it is a Cauchy sequence.

58 Locate the continued fractions

$$1,\ 1+\frac{1}{2},\ 1+\frac{1}{2+\frac{1}{2}},\ 1+\cfrac{1}{2+\cfrac{1}{2+\frac{1}{2}}},\ 1+\cfrac{1}{2+\cfrac{1}{2+\cfrac{1}{2+\frac{1}{2}}}},$$

on a number line.
If, for all n, the terms of a sequence (a_n) satisfy the inequalities $a_{2n-1} \le a_{2n+1} \le a_{2n+2} \le a_{2n}$, and $|a_{n+1} - a_n| < 1/n$, prove that the sequence is convergent.

Least upper bounds (sup) and greatest lowest bounds (inf)

Upper bounds and greatest terms

59 Some sets of numbers have a greatest member.
What is the greatest member of $\{0, 1, 2, 4, -100, -50\}$?
Must any finite set of numbers have a greatest member?
What is the greatest member of $\{x \mid 0 \le x \le 1\}$?

60 (i) (a) Is 1 an upper bound for $\{x \mid 0 \le x \le 1\}$?
(b) Can any number less than 1 be an upper bound for $\{x \mid 0 \le x \le 1\}$?
(ii) (a) Is 3 an upper bound for $\{x \mid 0 \le x \le 1, 2 \le x \le 3\}$?
(b) Can any number less than 3 be an upper bound for $\{x \mid 0 \le x \le 1, 2 \le x \le 3\}$?

61 Check that 2 is an upper bound for each of the following sets:

(i) $\{x \mid 0 \le x \le 1\}$,
(ii) $\{x \mid 0 < x < 1\}$,
(iii) $\{x \mid 0 \le x \le \frac{1}{4}, \frac{3}{4} \le x \le 1, 1\frac{1}{4} \le x \le 1\frac{1}{2}\}$,
(iv) $\{x \mid 0 \le x \le \frac{1}{2}, 1 < x < 1\frac{1}{2}\}$,
(v) $\{1 + 1/n \mid n \in \mathbb{N}\}$,
(vi) $\{2 - 1/n \mid n \in \mathbb{N}\}$,
(vii) $\{1 + (-1)^n/n \mid n \in \mathbb{N}\}$,
(viii) $\{x \mid \frac{1}{2} < x < 2,\ x \in \mathbb{Q}\}$,
(ix) $\{q \mid q^2 < 2,\ q \in \mathbb{Q}^+\}$.

For which of these sets can you find a number less than 2 which is still an upper bound for the set?

Least upper bound (sup)

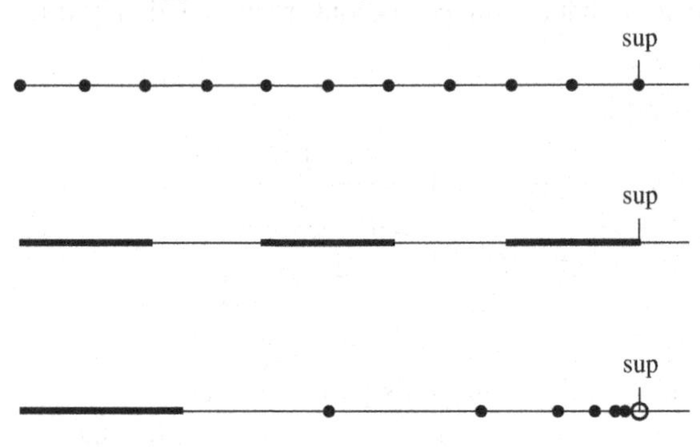

Figure 4.4

DEFINITION

When (i) u is an upper bound for a set A of real numbers, and (ii) every v less than u is not an upper bound for A, then and only then, u is called the *least upper bound* of A, and denoted by sup A.

The meanings of the terms 'least upper bound' and 'sup' lie in the words of this definition, and nowhere else!

If a set of numbers has a greatest member, the greatest member is the *least upper bound* of that set.

In qn 60 you proved that 1 is the *least upper bound* of $\{x \mid 0 \leq x \leq 1\}$, and that 3 is the *least upper bound* of $\{x \mid 0 \leq x \leq 1, 2 \leq x \leq 3\}$. A finite set of real numbers and a closed interval are examples of sets with a greatest member.

In qn 61 you should have found that 2 is the *least upper bound* of $\{2 - 1/n \mid n \in \mathbb{N}\}$ and $\{x \mid \frac{1}{2} < x < 2,\ x \in \mathbb{Q}\}$. A strictly increasing sequence and an open interval are examples of sets without a greatest member. The *least upper bound* is then a cluster point of the set. This will follow from qn 64.

62 If a set A has a least upper bound, sup A, must sup A be a member of A? Might sup A be a member of A?

63 Show that a set A cannot have two different least upper bounds.

64 If a set A has a least upper bound sup A, why must there be an $a \in A$ such that $(\sup A) - \varepsilon < a \leq \sup A$, for any positive number ε?
Use this result to construct a strictly increasing sequence of elements of A within the interval $(\sup A - 1, \sup A)$ on the assumption that $\sup A \notin A$.

65 If u is an upper bound for the set A, and for any positive number ε, you can find an $a \in A$, such that $u - \varepsilon < a \leq u$, must $u = \sup A$? (*Hint*. Draw a diagram. Suppose $v < u$ and that v is also an upper bound for A. Try putting $\varepsilon = u - v$.)

66 Let (a_n) be a monotonic increasing sequence with limit a. Prove that $a = \sup\{a_n \mid n \in \mathbb{N}\}$. Use qn 3.80.

Lower bounds and least members

67 Give some examples of sets of numbers which have a least member. Make sure some of them are finite and some are infinite.

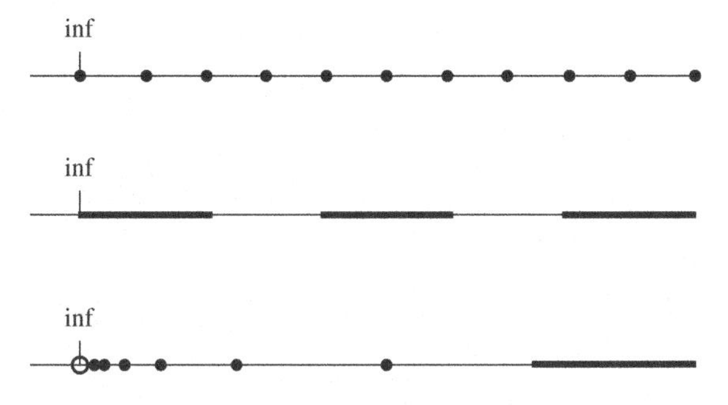

Figure 4.5

68 (a) Is 0 a lower bound for $\{x \mid 0 \leq x \leq 1\}$?
(b) Can any number greater than 0 be a lower bound for $\{x \mid 0 \leq x \leq 1\}$?

69 Give an example of a bounded set of numbers which does not have a least member.

70 Check that 0 is a lower bound for each of the following sets:

(i) $\{x \mid 0 \leq x \leq 1\}$,
(ii) $\{x \mid 0 < x < 1\}$,
(iii) $\{x \mid 0 \leq x \leq \frac{1}{4}, \frac{3}{4} \leq x \leq 1, 1\frac{1}{4} \leq x \leq 1\frac{1}{2}\}$,
(iv) $\{x \mid 0 \leq x \leq \frac{1}{2}, 1 < x < 1\frac{1}{2}\}$,
(v) $\{1 + 1/n \mid n \in \mathbb{N}\}$,
(vi) $\{2 - 1/n \mid n \in \mathbb{N}\}$,
(vii) $\{1 + (-1)^n/n \mid n \in \mathbb{N}\}$,
(viii) $\{x \mid \frac{1}{2} < x < 2,\ x \in \mathbb{Q}\}$,
(ix) $\{q \mid q^2 < 2,\ q \in \mathbb{Q}^+\}$.

For which of these sets can you find a number greater than 0 which is still a lower bound for the set?

Greatest lower bound (inf)

DEFINITION

When (i) l is a lower bound for a set A of real numbers, and (ii) every m, greater than l, is not a lower bound for A, then and only then, l is called the *greatest lower bound* of A, and denoted by inf A. The meanings of the terms 'greatest lower bound' and 'inf' lie in the words of this definition, and nowhere else!

If a set of numbers has a least member, that number is the greatest lower bound for the set.

In qn 68 you proved that 0 is the *greatest lower bound* of $\{x \mid 0 \leq x \leq 1\}$. In qn 70 you should have found that 0 is the *greatest lower bound* of $\{x \mid 0 \leq x \leq 1\}$, $\{x \mid 0 < x < 1\}$, $\{1 + (-1)^n/n \mid n \in \mathbb{N}\}$ and $\{q \mid q^2 < 2, q \in \mathbb{Q}^+\}$.

71 Show that a set A cannot have two different greatest lower bounds.

72 If a set A has a greatest lower bound inf A, and ε is any positive number, why must there be an $a \in A$ such that $\inf A \leq a < (\inf A) + \varepsilon$?

73 If l is a lower bound for the set A, and for any positive number ε, you can find an $a \in A$, such that $l \leq a < l + \varepsilon$, must $l = \inf A$? (*Hint*. Draw a diagram. Suppose $l < m$ and m is a lower bound for A. Try putting $\varepsilon = m - l$.)

74 Let A be a set of real numbers with a least upper bound sup A, and let $B = \{-x \mid x \in A\}$. Prove that $\inf B = -\sup A$.

75 (i) Prove that the set of numbers $\{x \mid x < 0\}$ has no greatest member.
(ii) Prove that the set of numbers $\{x \mid x < 10\}$ has no greatest member.

76 List some upper bounds for the set $\{x \mid x < 0\}$, and then identify the set of all upper bounds for $\{x \mid x < 0\}$. Does the set of all upper bounds have a least member?

77 List some upper bounds for the set $\{x \mid x \leq 0\}$, and then identify the set of all upper bounds for this set. Does the set of upper bounds have a least member?

78 Let (a_n) be a monotonic decreasing sequence with limit a. Prove that $a = \inf\{a_n \mid n \in \mathbb{N}\}$. Use qn 66 and qn 74.

sup, inf and completeness

79 Every infinite decimal sequence consists entirely of terminating decimals, which are rational numbers. If (a_n) is an infinite decimal sequence, what is the least upper bound of the set of terms of this sequence? Deduce that if every non-empty set of real numbers which is bounded above has a least upper bound, the completeness principle must follow.

80 (*Bolzano*, 1817) **We seek a least upper bound for a non-empty set which is bounded above.**
Let X be a non-empty set of real numbers with an upper bound u. Let $x \in X$. Then $x \leq u$.
If x is an upper bound for X, then x is the greatest member of X and so x is the least upper bound, and the search ends.

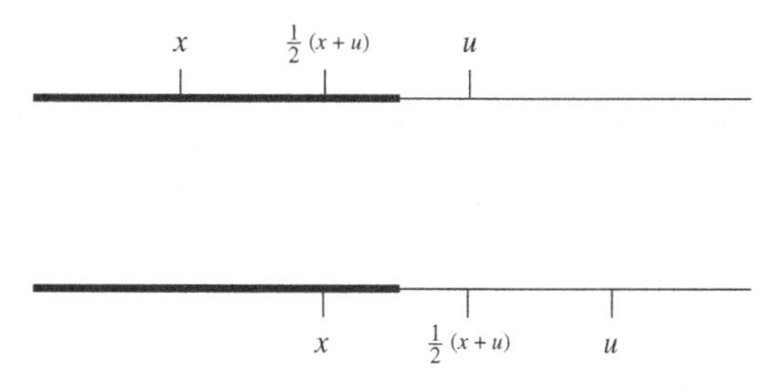

Figure 4.6

So suppose x is not an upper bound for X. Let $l_1 = x$ and $u_1 = u$.
Now consider the number $d = (l_1 + u_1)/2$.
If d is not an upper bound for X, let $l_2 = d$ and $u_2 = u_1$.
If d is an upper bound for X, let $l_2 = l_1$ and $u_2 = d$.
So $l_1 \leq l_2 < u_2 \leq u_1$, and $u_2 - l_2 = (u_1 - l_1)/2$.
Repeat this process to define sequences (l_n), (u_n) inductively such that $l_n \leq l_{n+1} < u_{n+1} \leq u_n$, where u_n is an upper bound for X, l_n is not and $u_{n+1} - l_{n+1} = (u_n - l_n)/2$.

(i) Why is the sequence (l_n) convergent?
(ii) Why is the sequence (u_n) convergent?
(iii) Why is the sequence $(u_n - l_n)$ null and why are the limits of (l_n) and (u_n) equal (to s, say)?
(iv) By considering the sequence (u_n) show that the limit, s, is an upper bound for X. (For any $x \in X$, $x \leq u_n$ for all n, so $x \leq s$ by the closed interval property, qn 3.78.)
(v) By considering the sequence (l_n) show that the limit is the least upper bound for X (from qn 65).

So any non-empty set of real numbers which is bounded above has a least upper bound.

81 Articulate a theorem for greatest lower bounds analogous to that of qn 80.

82 If A and B denote bounded sets of real numbers, how do the numbers sup A, inf A, sup B and inf B relate if $B \subseteq A$? Give examples of unequal sets for which $\sup A = \sup B$ and $\inf A = \inf B$.

lim sup and lim inf

The ideas in qns 83 and 84 will only be used in later chapters in qns 5.102 and 5.107. Question 5.107 will be used at the end of chapter 12. They have been bracketed, not because they are incidental in the development of analysis, but because they may most profitably be studied at a second reading.

(**83**) For any bounded sequence with an upper bound U and lower bound L

(i) how do you know that (a_n) contains a convergent subsequence;
(ii) if a is the limit of a convergent subsequence, how do you know that $L \leq a \leq U$;
(iii) how do you know the set of limits of convergent subsequences has a supremum and an infimum?

The supremum is called **lim sup a_n** and the infimum is called **lim inf a_n**

Question 83 shows the power of the theorems we have developed by establishing the existence of numbers about which we know very little. A concrete and constructive approach to the same concepts is given in the following question.

(**84**) (i) Illustrate on a graph the first few terms of the sequence with nth term $a_n = (-1)^n(1 + 1/n)$.
(ii) Is this sequence bounded?
(iii) Is this sequence convergent?
(iv) Identify one convergent subsequence.
(v) Find $\sup\{a_n \mid n \in \mathbb{N}\}$ and $\inf\{a_n \mid n \in \mathbb{N}\}$.
(vi) Let $u_k = \sup\{a_n \mid k \leq n\}$ and let $l_k = \inf\{a_n \mid k \leq n\}$. Find the first four terms of the sequence (u_n) and the first four terms of the sequence (l_n).
(vii) Explain why $u_{k+1} \leq u_k$ for all k, and why $l_k \leq l_{k+1}$ for all k.
(viii) Are both (u_n) and (l_n) bounded monotonic sequences?
(ix) The limits of these sequences are $\limsup a_n$ and $\liminf a_n$ respectively. Find these limits.

The method used in qn 84 applies quite generally and provides a more illuminating way of describing $\limsup a_n$ and $\liminf a_n$. This description enables us to prove, for example, that a sequence (a_n) is convergent if and only if $\limsup a_n = \liminf a_n$.

In the absence of completeness, a bounded sequence need not have any convergent subsequences (take the infinite decimal sequence of $\sqrt{2}$ for example) and therefore neither lim sup nor lim inf may be defined.

Summary: Completeness

The completeness principle	
	Every infinite decimal sequence is convergent.
Definition	The limit of every infinite decimal sequence is a *real number*.
Theorem qns 34, 35	Every bounded monotonic sequence is convergent.
Theorem qn 36	The sequence with nth term $(1 + 1/n)^n$ is convergent with limit (e) between 2 and 3.
Theorem qn 40	Every positive real number has a unique nth root.
Theorem qn 46	Every bounded sequence has a convergent subsequence.
Definition	A sequence (a_n) is said to satisfy the *Cauchy criterion*, when, given $\varepsilon > 0$, there exists an N, such that $n > N \Rightarrow |a_{n+k} - a_n| < \varepsilon$, for all positive integers k.
Definition	A sequence satisfying the Cauchy criterion is called a *Cauchy sequence*.
The General Principle of Convergence	
qns 55–57	A sequence is convergent if and only if it is a Cauchy sequence.
Definition	An upper bound for a set A of real numbers is called a least upper bound, when no lesser number is an upper bound for the set. The least upper bound of A is denoted by sup A.
Theorem qn 80	Every non-empty set of real numbers which is bounded above has a least upper bound.
Theorem	Any one of the following propositions is sufficient to imply the completeness principle:
After qn 35	1. Every bounded monotonic sequence is convergent.
After qn 46	2. Every bounded sequence has a convergent subsequence.
qn 56	3. Every Cauchy sequence is convergent.
qn 79	4. Every non-empty set of real numbers which is bounded above has a least upper bound.

Historical Note

About 500 BC, the Pythagoreans showed that the sides of a 45, 45, 90 triangle were incommensurable. This is equivalent, in modern terms, to proving that $\sqrt{2}$ is irrational. A multiplicity of irrationals is discussed in Euclid, Book X. The treatment of ratios given in Euclid, Book V (c. 300 BC) includes a careful description (in definition 5) of how to compare two incommensurable lengths. Until the nineteenth century, European mathematicians took their notions of number from a geometric view of measurements.

Eighteenth-century writers implicitly assumed that bounded monotonic sequences were convergent in their discussion of series and also assumed that lines which crossed on a graph necessarily had a point of intersection.

The description of a *least upper bound* which is not attained appears in the thesis of C. F. Gauss (1799), and the notion of lim sup and lim inf, geometrically defined, in an unpublished notebook of his about 1800. The term 'la plus grande des limites' is used by Cauchy (1821). The notation, lim sup and lim inf, was introduced by Pasch in 1887.

In 1817, B. Bolzano proved that the convergence of Cauchy sequences implied the least upper bound property, though his proof that Cauchy sequences converged was defective since he had nothing equivalent to an axiom of completeness.

In 1821, A. L. Cauchy proved that the partial sums of a convergent series (see chapter 5) satisfy the Cauchy criterion and claimed that partial sums satisfying the Cauchy criterion were those of a convergent series, but without proof. Cauchy also presumed that bounded monotonic sequences were convergent, both in his work on series and in his work on continuity. Cauchy stated that irrational numbers were limits of sequences of rational numbers, but did not argue from this proposition.

Until the 1860s there was no public discussion to clarify the status of irrational numbers. In his lectures in Berlin in 1865, K. Weierstrass gave an elaborate construction of positive irrational numbers as bounded infinite sums of rational numbers. Weierstrass insisted on the fundamental importance of the theorem that an infinite bounded set has a cluster point, which he proved in his lectures from about 1867.

In attempting to simplify the treatment of Weierstrass, both E. Heine and G. Cantor defined real numbers as the limits of rational Cauchy sequences which they called *fundamental sequences* (1872). After reading Heine's and Cantor's treatments, Dedekind decided to publish work that he had done in 1858, when he acknowledged the unprovability of the convergence of monotonic bounded sequences from a geometric standpoint. Dedekind described points on the number line (whether rational or irrational) as producing a cut in the rational numbers, separating those above the cut from those below. This description is similar to that in Euclid, Book V, but Dedekind's new definition of numbers as cuts in the rationals led to formal proofs of their algebraic properties and of completeness, in that every cut of this new system is actually a cut of the rationals.

Independently of the German tradition, C. Méray, a student of Cauchy, had proved that the convergence of bounded monotonic sequences implied the convergence of Cauchy sequences by considering their eventual upper bounds and eventual lower bounds respectively. In the same paper (1869) Méray also defined irrational numbers as fictitious limits of rational Cauchy sequences.

Defining a real number as a pair of subsets of the rationals (as Dedekind) or as a family of Cauchy sequences (as Méray and Cantor) seemed to divorce numbers which originate in counting and measurement from their roots. So in 1882, P. du Bois-Reymond proposed as a postulate that every decimal number (whether infinite or not) corresponded to a unique point on an orientable line, and it is this proposal which has inspired the choice of *completeness principle* in this book. In the same spirit, Ascoli proposed as a postulate in 1895 that a set of nested closed intervals with length tending to zero had a unique common point.

All of this work presumed that the rational numbers were well-founded and that it was only irrationals that were not precisely defined. A modern, axiomatic, approach avoids any notion that some numbers are more 'real' than others, but the fact that the real numbers can be constructed from the rational numbers (whether by Dedekind cuts, families of Cauchy sequences, infinite decimals or nested intervals) establishes that there is nothing inherently contradictory in introducing a principle (or axiom) of completeness.

Galileo noted in his *Discourses on Two New Sciences* (1638) that the mapping $n \to n^2$ matched the set $\mathbb{N}$, one-to-one, with a proper subset of $\mathbb{N}$. He also noted that the mapping $x \to \frac{1}{2}x$ matched the set $\{x \mid 0 \leq x \leq 2\}$, one-to-one, with the set $\{x \mid 0 \leq x \leq 1\}$, another one-to-one matching of a set with a proper subset. Either one of these mappings provides a counter-example to Euclid's axiom: the whole is always greater than the part. It was Bolzano (1851) who first recognised that this was always a possibility with infinite sets.

Almost at the same time that Weierstrass was successfully banishing infinitesimals and the infinite from definitions of limits (these two notions had been the main tools for discussing limits in the seventeenth and eighteenth centuries), Cantor and Dedekind were in correspondence about infinite sets and the possibility of transfinite cadinals. Cantor's first paper on the subject was published in 1874 and contained a proof that the set of algebraic numbers (solutions of polynomial equations with integer coefficients) was countably infinite while the set of real numbers was not. Modestly, he claimed that this provided a new proof that transcendental numbers (real numbers that are not algebraic) were dense, which had been shown by Liouville in 1851. The argument of qn 23 was given by Cantor in 1891. In 1879, Cantor introduced the notion of a dense set of numbers. It was in 1885 that A. Harnack showed that a countable infinity of points did not occupy length on the line.

Answers and comments

1 $a = 2, b = 3, c = 0, d = 0, e = 1, f = 2.$

2 $a = 0, b = -2, c = 2, d = -1, e = -1, f = -1.$
$2775/999\,999 = 5^2 \cdot \left(\frac{1}{3}\right)^2 \cdot \left(\frac{1}{7}\right) \cdot \left(\frac{1}{11}\right) \cdot \left(\frac{1}{13}\right).$

3 Every positive rational number, different from 1, may be expressed in a unique way as a product of prime numbers and of reciprocals of prime numbers.

4 The numbers of $\{m/2^n \mid m \in \mathbb{Z}\}$ are evenly spaced at intervals of $1/2^n$.

5 $2^{12} < 65 \cdot 73 < 2^{13}$, so $1/2^{13} < 1/(65 \cdot 73)$. $\lfloor 2^{13} \cdot \frac{64}{73} \rfloor = 7182$, so $\frac{64}{73} < 7183/2^{13} < \frac{57}{65}$.

6 $\lfloor a \cdot 2^k \rfloor \le a \cdot 2^k < \lfloor a \cdot 2^k \rfloor + 1 \Rightarrow \lfloor a \cdot 2^k \rfloor / 2^k \le a < \lfloor a \cdot 2^k \rfloor / 2^k + 1/2^k < b.$

7 Since there is an $m/2^n$ number in $\{x \mid a < x < b\}$ there must be two, one in $\{x \mid a < x < (a+b)/2\}$ and one in $\{x \mid (a+b)/2 < x < b\}$. The argument may be repeated.

8 By qn 7, T is dense on the number line. If $m/2^n$ were smallest, $m/2^{n+1}$ would be smaller. Contradiction. Likewise if a were smallest, $a/2$ would be smaller. Contradiction.

9 $m/2^n = (m \cdot 5^n)/10^n$. Since T is dense and D contains T, D is dense.
$\frac{1}{5} = \frac{2}{10} \cdot \frac{1}{5} = m/2^n \Leftrightarrow 5m = 2^n \Leftrightarrow 2$ is the only prime factor of $5m$.
Contradiction.

(i) (ii) (iii) Yes.
(ii) No, $(m/2^n)/(k/2^n) = m/k$, both m and k may be odd.

Same answers for D.

10 All terms of the sequence in T by qn 9 parts (i), (ii) and (iii).

11 $\frac{1}{3} = m/10^n \Leftrightarrow 3m = 10^n$, so 2 and 5 are the only prime factors of $3m$.
Contradiction.
$(\lfloor 10^n/3 \rfloor / 10^n) = 0.3, 0.33, 0.333, 0.3333, \ldots$
$\frac{1}{3} - 0.3 = \frac{1}{30}, \frac{1}{3} - 0.33 = \frac{1}{300}, \frac{1}{3} - 0.333 = \frac{1}{3000}, \ldots$

12 $a_n =$ decimal point followed by n 9s. So $10^n a_n = 10^n - 1$, and $a_n = 1 - 1/10^n$.
Thus $|a_n - 1| = 1/10^n$, and $(a_n - 1)$ is a null sequence.

13
$$a_{2n} = \frac{13}{100} \cdot \left(1 + \frac{1}{10^2} + \frac{1}{10^4} + \ldots + \frac{1}{10^{2n-2}}\right)$$
$$= \frac{13}{100} \cdot \frac{1 - \dfrac{1}{10^{2n}}}{1 - \dfrac{1}{10^2}} = \frac{13}{99} \cdot \left(1 - \frac{1}{10^{2n}}\right).$$
Since the sequence is monotonic increasing, the limit of the sequence is the limit of (a_{2n}).

14 $$1/10^3 + 1/10^6 + \ldots + 1/10^{3n} = \frac{1}{10^3} \cdot \frac{1 - \dfrac{1}{10^{3n}}}{1 - \dfrac{1}{10^3}} = \frac{1}{999} \cdot \left(1 - \frac{1}{10^{3n}}\right).$$

$a = 12.45,\ b = 6.78.$

With a recurring block of length l, the infinite decimal will be the limit of a sequence with nth term

$$a + b\left(\frac{1}{10^l} + \frac{1}{10^{2l}} + \ldots + \frac{1}{10^{nl}}\right) = a + \frac{b}{10^l} \cdot \frac{1 - \dfrac{1}{10^{nl}}}{1 - \dfrac{1}{10^l}}$$

$$= a + \frac{b}{10^l - 1}\left(1 - \frac{1}{10^{nl}}\right).$$

If a and b are terminating decimals, the sequence with this as nth term has a rational limit.

15 If A and B are integers and A/B is not equal to a terminating decimal, the long division of A by B will not terminate. The remainder at each stage of the division is one of the numbers $1, 2, \ldots, B - 1$. Since the process is endless, the remainders must recur and thus the dividends will recur.

16 If the first digits to differ are d_j and e_j, then we suppose $d_j < e_j$ and so $d_j + 1 \le e_j$.
Now $0.00\ldots 0d_{j+1}d_{j+2}\ldots d_{j+n} \le (1/10^j)(1 - 1/10^n)$, with equality only when each $d_{j+i} = 9$, so $x \le d_0.d_1d_2\ldots(d_j + 1)$. But also $e_0.e_1e_2\ldots e_j \le x$ since an infinite decimal sequence is monotonic increasing. Thus $e_0.e_1e_2\ldots e_j \le x \le d_0.d_1d_2\ldots(d_j + 1)$.
So $x = e_0.e_1e_2\ldots e_j = d_0.d_1d_2\ldots(d_j + 1)$ and $e_{j+i} = 0,\ d_{j+i} = 9$ for all i.

17 Every rational number is equal to a terminating or recurring decimal. The infinite decimal given here is neither terminating nor recurring, so if it converges it does not converge to a rational number.

18 In the prime factorisation of both p^2 and q^2, all primes occur to an even power. So 2 appears to an even power in p^2 and to an odd power in $2q^2$, so $p^2 = 2q^2$ contradicts the fundamental theorem of arithmetic. Likewise $\sqrt{3}$, $\sqrt{6}$ and $\sqrt[3]{2}$ cannot be rational numbers.

19 If $a + b\sqrt{2} = x$ were rational, then $(x - a)/b = \sqrt{2}$ would also have to be rational. Contradiction.

20 If $(a + b\sqrt{2})/(1 + \sqrt{2}) = x$, then $(a - x)/(x - b) = \sqrt{2}$, so x cannot be rational.

21 Suppose that there are integers p and q such that $\log_{10} 2 = p/q$, then $10^{p/q} = 2$, so $10^p = 2^q$, and 5 divides the left-hand side but not the right-hand side, contradicting the fundamental theorem of arithmetic.

22 The new decimal will be different from the others above it if each entry differs from the diagonal entry above it.

23 Constructed sequence of infinite decimals

$$\begin{array}{llllllll}
0. & a_{11} & a_{12} & a_{13} & a_{14} & a_{15} & a_{16} & \ldots \\
0. & a_{21} & a_{22} & a_{23} & a_{24} & a_{25} & a_{26} & \ldots \\
0. & a_{31} & a_{32} & a_{33} & a_{34} & a_{35} & a_{36} & \ldots \\
0. & a_{41} & a_{42} & a_{43} & a_{44} & a_{45} & a_{46} & \ldots \\
0. & a_{51} & a_{52} & a_{53} & a_{54} & a_{55} & a_{56} & \ldots \\
\ldots & \ldots & \ldots & \ldots & \ldots & \ldots & \ldots & \ldots \\
0. & a_{n1} & a_{n2} & a_{n3} & a_{n4} & a_{n5} & a_{n6} & \ldots \\
\ldots & \ldots & \ldots & \ldots & \ldots & \ldots & \ldots & \ldots
\end{array}$$

new infinite decimal

$$\begin{array}{llllllll}
0. & b_1 & b_2 & b_3 & b_4 & b_5 & b_6 & \ldots
\end{array}$$

Choose b_1 different from 0, 9 and a_{11}.
Choose b_2 different from 0, 9 and a_{22}.
Choose b_3 different from 0, 9 and a_{33}.
...
Choose b_n different from 0, 9 and a_{nn}.
By avoiding 0 and 9, unwitting duplication is avoided. By avoiding a_{nn} a new infinite decimal is guaranteed.

24 $0, +1, -1, +2, -2, +3, -3, \ldots$
$0, 1, 2, 3, 4, \ldots \rightarrow 1, 2, 4, 8, 16, \ldots$
$-1, -2, -3, \ldots \rightarrow 6, 12, 24, \ldots$
So sequence built from increasing natural numbers is
$0, 1, 2, -1, 3, -2, 4, \ldots$

25 positive a/b

$b \downarrow a \rightarrow$	1	2	3	4
1	6	12	24	48
2	18	36	72	144
3	54	108	216	432
4	162	324	648	1296

Sequence $\frac{1}{1}, \frac{2}{1}, \frac{1}{2}, \frac{3}{1}, \left[\frac{2}{2}\right], \frac{4}{1}, \frac{1}{3}, \frac{3}{2}, \ldots$

negative a/b

$b \downarrow a \rightarrow$	1	2	3	4
1	30	60	120	240
2	90	180	360	720
3	270	540	1080	2160
4	810	1620	3240	6480

Sequence of non-zero rationals:
$\frac{1}{1}, \frac{2}{1}, \frac{1}{2}, \frac{3}{1}, -\frac{1}{1}, \left[\frac{2}{2}\right], \frac{4}{1}, \frac{1}{3}, -\frac{2}{1}, \frac{3}{2}, -\frac{1}{2}, \frac{5}{1}, \ldots$

26 $1+\sqrt{2},\ 2+\sqrt{2},\ \frac{1}{2}+\sqrt{2},\ 3+\sqrt{2},\ 1+2\sqrt{2},\ 1+\frac{1}{2}\sqrt{2},\ \ldots$

27 $\mathbb{Z}$, T, D, and $\mathbb{Q}$ are countably infinite. The set of infinite decimals is not.

28 All of $\mathbb{N}$ is deleted. The total length removed is 2 units.

29 All of $\mathbb{N}$ is deleted. The total length removed is 2ε units. ε may be as small as we wish. So $\mathbb{N}$ occupies no length.

30 All of (a_n) is deleted with a total length of 2ε units. The length is arbitrarily small, so a sequence (a countably infinite set of numbers) does not occupy length on the line.
Since the rationals are countably infinite, so are the infinite decimals equal to rationals between 0 and 1. If a sequence of these infinite decimals was used the new decimal constructed would be irrational.

31 The sequence is monotonic increasing. Every term is a terminating decimal and therefore rational. 0.2, 0.13, 0.124, 0.1235 etc. are all upper bounds for the sequence.

32 Apply the sum rule (qn 3.54(iii)), the difference rule (qn 3.54(v)), the product rule (qn 3.54(vi)) and the quotient rule (qn 3.67) to the infinite decimal sequences.

33 The first term of an increasing sequence is a lower bound. The first term of a decreasing sequence is an upper bound.

34
(i) $\lfloor L \rfloor$.
(ii) $\lfloor a_1 \rfloor + 1$.
(iii) Yes. Observe the finite number of integers between (i) and (ii).
(iv) Since t_0 is a lower bound and $t_0 + 1$ is not, we can argue as in (iii).
(v) Argue as in (iii).
(vi) (t_n) is an infinite decimal sequence which converges by the completeness principle.
(vii) The limit is $a_k - D$ by the difference rule (qn 3.54(v)). Since t_n is a lower bound for (a_n), the terms of the sequence are non-negative. Use qn 3.75.
(viii) t_i is a lower bound for (a_n) while $t_i + 1/10^i$ is not, so there are terms of the sequence (a_n) between these numbers, *e.g.* $t_i \le D \le a_n < t_i + 1/10^i \le D + 1/10^i$, for some value of n and since (a_n) is decreasing, for all subsequent n.
(ix) Part (viii) implies that $0 \le a_n - D < 1/10^i$ for sufficiently large n. Given $\varepsilon > 0$, $\varepsilon > 1/10^i$ for some i, so $(a_n) \to D$ by definition.

35 If (a_n) is monotonic increasing and bounded above, then $(-a_n)$ is monotonic decreasing and bounded below. From qn 34, $(-a_n) \to A$, say, so $(a_n) \to -A$.

36 The sequence is convergent from qn 35. The limit is between 2 and 3 by the closed interval property, qn 3.78. In qn 11.32 we will see that the limit is e.

37 $a \le x^2 \Rightarrow a/x \le x \Rightarrow a/x \le \frac{1}{2}(x + a/x) \le x \Rightarrow (\frac{1}{2}(x + a/x))^2 \le x^2$ and $0 \le (x - a/x)^2 \Rightarrow a \le (\frac{1}{2}(x + a/x))^2$. (x_n) is monotonic decreasing and bounded. Let $(x_n) \to l$. Apply the sum rule (qn 3.54), the scalar rule (qn 3.54) and the reciprocal rule (qn 3.65) to obtain $l = \frac{1}{2}(l + a/l)$, and thus $l^2 = a$. Because the terms are positive, $l = \sqrt{a}$.

38 (a_n) is increasing, (b_n) is decreasing, so both are convergent by qns 34 and 35. For the sequences of qn 2.38, qn 2.38(iii) implies that $(b_n - a_n)$ is a null sequence, so by the difference rule the sequences have the same limit. In the case of the sequences of qn 2.39, the property $b_{n+1} - a_{n+1} < \frac{1}{4}(b_n - a_n)$ implies that $b_{n+1} - a_{n+1} = (\frac{1}{4})^{n-1}(b_2 - a_2)$ and so $(b_n - a_n)$ is a null sequence, and the sequences have the same limit.

39 If $\sqrt{a}$ exists for all positive real a, then $\sqrt{\sqrt{a}}$ etc. exist.

40
(i) (a_n) is increasing and bounded above.
(ii) (b_n) is decreasing and bounded below.
(iii) $b_{n+1} - a_{n+1} = (\frac{1}{2})^n(b_1 - a_1)$, so the difference rule with qn 3.39 makes the limits equal.
(iv) Repeated use of product rule.
(v) Difference rule.
(vi) $0 < c - {a_n}^k < {b_n}^k - {a_n}^k$.
(vii) By difference rule.

41
(i) The sequence is decreasing and bounded below by 0, so it is convergent by qn 34.
(ii) $n(\sqrt[n]{(1/a)} - 1) = n(1 - \sqrt[n]{a})/\sqrt[n]{a}$.
(iii) $\sqrt[n]{(ab)} - 1 = (\sqrt[n]{a} - 1) \cdot \sqrt[n]{b} + (\sqrt[n]{b} - 1)$.

In chapter 11 we will find that $\lim n(\sqrt[n]{a} - 1) = \log_e a = \ln a$.

42
(i) $a_1 \le a_2 \le \ldots \le a_n < b_n \le \ldots \le b_2 \le b_1$.
(ii) (a_n) is monotonic increasing and bounded above by b_1.
(b_n) is monotonic decreasing and bounded below by a_1.
(iii) Let $(a_n) \to a$ and $(b_n) \to b$. Then by the inequality rule (qn 3.76) $a \le b$. By qn 3.80(b) $a_n \le a$ and by analogy $b \le b_n$ for all n, so $a_n \le a \le b \le b_n$. Thus any c such that $a \le c \le b$, lies in all the $[a_n,\ b_n]$.
(iv) Suppose c was in all the intervals. Consider $n > 1/c$.

43
(i) $[-1,\ +1]$, even terms constant, so this subsequence is convergent.
(ii) $[-2,\ +2]$, subsequence of even terms $(1 + 1/2n) \to 1$.

44 Convergent $\Rightarrow$ bounded, but not conversely by qn 43.

45 All terms within $[0, 1]$.
$0.1, 0.10, 0.100$, etc. $\to 0.1$;
$0.1, 0.11, 0.111, 0.1111$, etc. $\to \frac{1}{9}$.

47 The infinite decimal sequence for $\sqrt{2}$ is a bounded sequence of terminating decimals, that is, of rational numbers. From qn 3.80, if any subsequence is convergent, it is convergent to $\sqrt{2}$. So if the completeness principle is not adopted, qn 46 can fail.

48 Choose positive $\varepsilon < \frac{1}{4} - \frac{1}{5} = \frac{1}{20}$, say $\varepsilon = \frac{1}{21}$.

49 $\frac{1}{4} < \frac{2}{7} < \frac{1}{3}$. $|\frac{2}{7} - 1/n| = |(2n-7)/7n| \geq \frac{1}{28}$. Take $\varepsilon = \frac{1}{29}$.

50 $\frac{1}{2}$, $\frac{1}{3}$ and $\frac{1}{4}$ are cluster points. $\frac{3}{5}$ and $\frac{2}{5}$ are not. $\frac{1}{2} + \frac{1}{27} < \frac{3}{5} < \frac{1}{2} + \frac{1}{9}$, $\frac{1}{4} + \frac{1}{9} < \frac{2}{5} < \frac{1}{2}$.

51 No. Can take $\varepsilon <$ shortest distance between two points in the set.

52 $\mathbb{Z}$.

53
(i) Choose any a_1 in A. Now choose a_2 in A but different from a_1. Construct the sequence by choosing a new point in A for each successive term.
(ii) $L \leq a_n \leq U$.
(iii) By qn 46.
(iv) a need not be in A, but $L \leq a \leq U$ by the closed interval property (qn 3.78).
(v) Given $\varepsilon > 0$, there is an N such that $n_i > N \Rightarrow |a_{n_i} - a| < \varepsilon$.

54 Each term in the infinite decimal sequence for $\sqrt{2}$ lies in the interval $[1, 2]$. If q is a rational number in $[1, 2]$ and we choose ε such that $0 < \varepsilon < |q - \sqrt{2}|$ then an ε-neighbourhood of q contains at most a finite number of terms of the sequence.

55 If $n > N \Rightarrow |a_n - a| < \varepsilon/2$, then also $|a_{n+k} - a| < \varepsilon/2$.

56 $d_0.d_1d_2 \ldots d_{n+k} - d_0.d_1d_2 \ldots d_n = 0.00\ldots0d_{n+1}d_{n+2}\ldots d_{n+k}$
$\leq (1/10^n)(1 - 1/10^k)$, so $|a_{n+k} - a_n| < 1/10^n$.

57 Suppose $n \geq N \Rightarrow |a_{n+k} - a_n| < 1$, then the sequence is bounded above by $\max(a_1, a_2, \ldots, a_{N-1}, a_N + 1)$ and bounded below by $\min(a_1, a_2, \ldots, a_{N-1}, a_N - 1)$. Now (a_n) is a Cauchy sequence, so given $\varepsilon > 0$, there is an N_1 and an N_2 such that $n_i > N_1 \Rightarrow |a_{n_i+k} - a_{n_i}| < \frac{1}{2}\varepsilon$, and $n_i > N_2 \Rightarrow |a_{n_i} - a| < \frac{1}{2}\varepsilon$, so for any $n_i > \max(N_1, N_2)$, $n > n_i \Rightarrow |a_n - a| < \varepsilon$.

58 With the given inequalities, all the later terms of the sequence lie between a_n and a_{n+1}, so $|a_{n+k} - a_n| < 1/n$, and the sequence is a Cauchy sequence.

59 The greatest member is 4. Any finite set of numbers, which is not empty, has a greatest member. 1.

60
(i) (a) Yes. (b) No.
(ii) (a) Yes. (b) No.

61 UB < 2 except for (v), (vi) and (viii).

62 Not for 61(vi) or (viii). Is for 61(v).

63 Suppose u and v are least upper bounds for A, and $v < u$, then v is not a least upper bound by definition (ii).

64 If not, then $(\sup A) - \varepsilon$ is an upper bound and so $\sup A$ is not least. So there is an $a_1 \in (\sup A - 1,\ \sup A]$ with $a_1 \in A$. But if $\sup A \notin A,\ a_1 \in (\sup A - 1,\ \sup A)$, and then there is an $a_2 \in (a_1,\ \sup A)$ with $a_2 \in A$, and an $a_3 \in (a_2,\ \sup A)$, etc.

65 Now suppose there is an upper bound v with $v < u$. Taking $\varepsilon = u - v$, there is an a such that $u - \varepsilon < a \leq u$, and thus $v < a \leq u$. So v is not an upper bound and $u = \sup A$.

66 Since $(a_n) \to a$, for any $\varepsilon > 0,\ |a_n - a| < \varepsilon$ for sufficiently large n, or $a - \varepsilon < a_n < a + \varepsilon$. From qn 3.80(b), $a_n \leq a$ for all n, so a is an upper bound for $\{a_n | \, n \in \mathbb{N}\}$ and $a - \varepsilon < a_n \leq a$, for sufficiently large n. By qn 65, $a = \sup\{a_n | \, n \in \mathbb{N}\}$.

68 (a) Yes. (b) No.

69 $\{x \mid 0 < x < 1\}$.

70 Can find lower bound > 0 for (v), (vi), (viii).

71 If there were two, then the greater of the two would not be a lower bound.

72 If not, $\inf A + \varepsilon$ would be a lower bound.

73 If l is not the greatest lower bound, then there is a lower bound m with $l < m$. Now let $\varepsilon = m - l$, and there is an a such that $l \leq a < l + \varepsilon = m$, so m is not a lower bound and $l = \inf A$.

74 For all $a,\ a \leq \sup A \Rightarrow -\sup A \leq -a$, and $-\sup A$ is a lower bound for B. For any $\varepsilon > 0$, there is an $a \in A$ such that $\sup A - \varepsilon < a \leq \sup A$, so that $-\sup A \leq -a < -\sup A + \varepsilon$, and by qn 73, $\inf B = -\sup A$.

75 (i) If g were greatest, then $g/2$ would be greater!
(ii) If g were greatest, $(g + 10)/2$ would be greater!

76 All positive numbers and 0 are upper bounds for the set of negative numbers. The least is 0.

77 Answer as qn 76.

78 $(-a_n)$ is monotonic increasing with limit $-a$, so $-a = \sup\{-a_n | \, n \in \mathbb{N}\}$ by qn 66. Now $a = \inf\{a_n | \, n \in \mathbb{N}\}$ by qn 74.

79 By qn 66, the least upper bound of the terms of an infinite decimal sequence is the infinite decimal. So if every non-empty set of real numbers which is bounded above has a least upper bound, then every infinite decimal sequence is convergent.

80 (i) The sequence (l_n) is monotonic increasing and bounded above by u_1.
(ii) The sequence (u_n) is monotonic decreasing and bounded below by l_1.
(iii) $u_{n+1} - l_{n+1} = (\frac{1}{2})^n(u_1 - l_1)$, so by the difference rule and qn 3.39, the limits are equal.
(v) Since l_n is not a upper bound for X and s is, there is an x with $l_n < x \le s$, for every n. Now $(l_n) \to s$, so given $\varepsilon > 0$, $|l_n - s| < \varepsilon$ for sufficiently large n, and $s - \varepsilon < l_n < x \le s$. So $s = \sup X$ by qn 65.

81 A non-empty set of real numbers which is bounded below has a greatest lower bound.

82 $\inf A \le \inf B \le \sup B \le \sup A$. Can take $A = \{x \mid 0 \le x \le 1\}$ and $B = \{x \mid 0 < x < 1\}$.

83 (i) (a_n) is bounded, qn 46.
(ii) By the closed interval property, qn 3.78.
(iii) By qns 80 and 81.

84

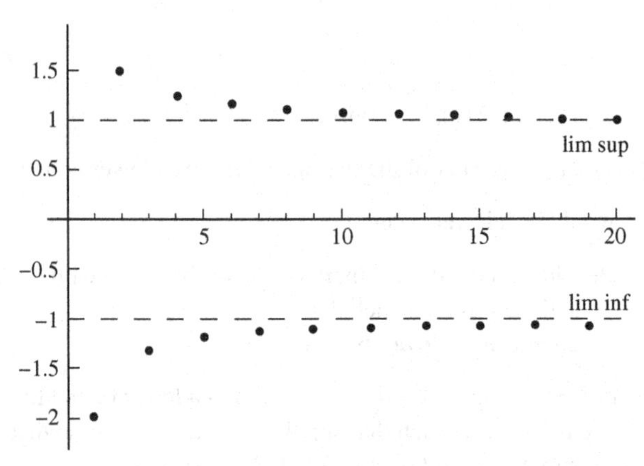

Figure 4.7

(ii) All terms between ± 2.
(iv) $a_{2n} = 1 + 1/2n$. $(a_{2n}) \to 1$.
(v) $\sup = \frac{3}{2}$. $\inf = -2$.
(vi) $u_1 = 1\frac{1}{2}, u_2 = 1\frac{1}{2}, u_3 = 1\frac{1}{4}, u_4 = 1\frac{1}{4}$.
$l_1 = -2, l_2 = -1\frac{1}{3}, l_3 = -1\frac{1}{3}, l_4 = -1\frac{1}{5}$.
(vii) $\{a_n \mid k + 1 \le n\} \subseteq \{a_n \mid k \le n\}$ and qn 82.
(ix) $\limsup a_n = 1$, $\liminf a_n = -1$.

5

Series

Infinite sums

Preliminary reading: Cohen, D., Northrop ch. 7, Nelsen, R. B.
Concurrent reading: Ferrar.
Further reading: Bonar and Khoury, ch. 3 onward, Rudin, ch. 3, Knopp.

Sequences of partial sums

1 Criticise the following argument:
If $S = 1 + x + x^2 + x^3 + \ldots$,
then $xS = x + x^2 + x^3 + x^4 + \ldots$,
so $S - xS = 1$,
and therefore $S = \dfrac{1}{1-x}$.
Try putting $x = 2$!

(*Grandi*, 1703) If the argument in qn 1 were sound we could put $x = -1$ and obtain the sum of the series

$1 - 1 + 1 - 1 + 1 - \ldots$

to be $\frac{1}{2}$. But the same series could be plausibly thought to have a sum of 0:

$(1 - 1) + (1 - 1) + (1 - 1) + \ldots$;

or a sum of 1:

$1 + (-1 + 1) + (-1 + 1) + (-1 + 1) + \ldots$.

These paradoxical conclusions show that great care must be used in arguing with infinite sums. The ordinary associative law for finite sums, $a + (b + c) = (a + b) + c$, does not lead to a clear answer.

2 For $x \neq 1$, let $s_n = 1 + x + x^2 + \ldots + x^{n-1}$, a sum of only n terms. By considering $x \cdot s_n - s_n$, prove that $s_n = (x^n - 1)/(x - 1)$.

Compare this with qn 1.3(vi). It is also conventional to write s_n, as defined in the first line, in the form

$$\sum_{r=1}^{r=n} x^{r-1} \text{ or } \sum_{r=0}^{r=n-1} x^r.$$

3 By decomposing $1/r(r+1)$ into partial fractions, or by induction, prove that

$$\frac{1}{1\cdot 2}+\frac{1}{2\cdot 3}+\ldots+\frac{1}{n(n+1)}=1-\frac{1}{n+1}.$$

Express this result using the Σ notation.

4 If $s_n = \sum_{r=1}^{r=n} \frac{1}{r(r+1)}$, prove that $(s_n) \to 1$ as $n \to \infty$.

This result is also written $\sum_{r=1}^{\infty} \frac{1}{r(r+1)} = 1.$

(*Cauchy*, 1821) When discussing whether the series
$a_1 + a_2 + a_3 + \ldots + a_n + \ldots$
has a sum, we construct the sequence (s_n) thus:

$$s_1 = a_1,$$
$$s_2 = a_1 + a_2,$$
$$s_3 = a_1 + a_2 + a_3,$$
$$\ldots$$
$$s_n = a_1 + a_2 + a_3 + \ldots + a_n,$$
$$\ldots$$

and so on.

The sequence (s_n) is called the *sequence of partial sums* of the series. When the sequence (s_n) is convergent to s we say that *the series is convergent* to s, or has the sum s. Symbolically, the series $\Sigma\, a_r$ is said to be *convergent* when *the sequence of partial sums* (s_n), defined by

$$s_n = \sum_{r=1}^{r=n} a_r,$$

is convergent. When $(s_n) \to s$ as $n \to \infty$, we write

$$\sum_{r=1}^{\infty} a_r = s,$$

and we say that the series has the sum s.

5 Write down a formula for $\Sigma_{r=1}^{r=n} \left(\frac{1}{2}\right)^{r-1}$ and find the sum $\Sigma_{r=1}^{\infty} \left(\frac{1}{2}\right)^{r-1}$. So, if you go half way, and then half of what is left, and then half again, and so on, do you ever get there?

6 Find the sum $\Sigma_{r=1}^{\infty} \left(\frac{1}{10}\right)^{r}$. What recurring decimal have you evaluated?

7 Construct an infinite decimal equal to $\frac{1}{11}$.

8 Find the sum $\Sigma_{r=0}^{\infty} \left(\frac{9}{10}\right)^{r}$.

9 Use qn 2 to determine whether the series $\Sigma\, x^n$ is convergent when

(i) $|x| < 1$,
(ii) $|x| > 1$.

Also examine the two cases $x = 1$ and $x = -1$.

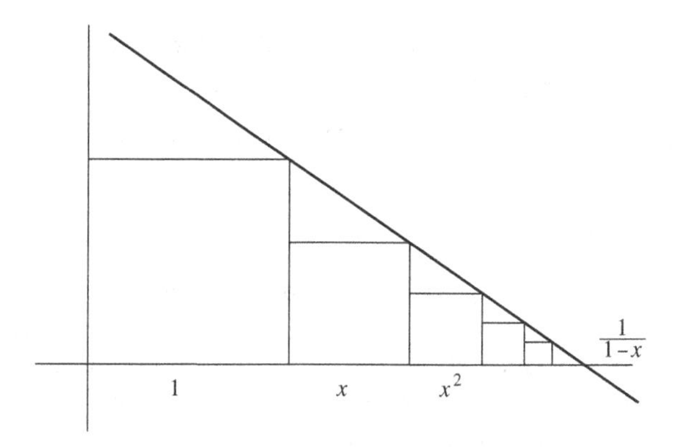

Figure 5.1

(Diagram after Gregory of St Vincent, 1647)

10 (*Oresme*, c. 1350) If $s_n = \Sigma_{r=1}^{r=n} \dfrac{r}{2^r}$ show by induction or otherwise that $s_n = 2 - (n+2)/2^n$.
Use qn 3.74 to deduce that $(s_n) \to 2$ as $n \to \infty$. Compare with qn 1.3(vii).
Oresme considered $\Sigma_1^{\infty} 1/2^r + \Sigma_2^{\infty} 1/2^r + \Sigma_3^{\infty} 1/2^r + \ldots$.
There is a geometrical description of Oresme's proof in Kopp, p. 18.

Any series which is *not convergent* is said to be *divergent*. The partial sums of a divergent series need not 'diverge' to $\pm\infty$ but may oscillate, as with $\Sigma(-1)^n$. When all the terms of a series are *positive*, the sequence of partial sums is monotonic increasing: so, if the sequence is bounded, the series is convergent; and, if it is not bounded, the series tends to $+\infty$. See qn 25.

The null sequence test

When discussing the series, there are two sequences, (a_n) and (s_n), which we will look at. The relationship between these two sequences has one straightforward aspect.

11 Suppose that the series $\Sigma\, a_n$ is convergent, and that its sequence of partial sums $(s_n) \to s$ as $n \to \infty$. Use the equation $a_{n+1} = s_{n+1} - s_n$ and the difference rule, qn 3.54(v), to prove that (a_n) is a null sequence.

So $\Sigma\, a_n$ convergent $\Rightarrow (a_n)$ is null. So what happens if (a_n) is *not* null? Clearly $\Sigma\, a_n$ cannot be convergent, if we wish to avoid a contradiction. So we can say (a_n) is not null $\Rightarrow \Sigma\, a_n$ is divergent. *This* is the *null sequence test.*

The argument here goes from $(P \Rightarrow Q)$ to (not $Q \Rightarrow$ not P). Repeating this argument gets you from (not $Q \Rightarrow$ not P) to $(P \Rightarrow Q)$. So $(P \Rightarrow Q) \Leftrightarrow$ (not $Q \Rightarrow$ not P).
(not $Q \Rightarrow$ not P) is called the *contrapositive* of $(P \Rightarrow Q)$.

12 By considering when the sequence (x^n) is *not* null, use qn 11 to check your claims of divergence in qn 9.

13 Is there a positive real number α such that $\Sigma\, n^\alpha$ is convergent?

Simple consequences of convergence

14 If the series $a_1 + a_2 + a_3 + \ldots + a_n + \ldots$ is convergent to the sum s, what can you say about the series $a_2 + a_3 + a_4 + \ldots + a_{n+1} + \ldots$? Write your claim with the Σ notation. Write a proof taking s_n as the nth partial sum of the first series and t_n as the nth partial sum of the second.
Now consider the converse problem. Suppose that $a_2 + a_3 + a_4 + \ldots + a_{n+1} + \ldots$ is convergent to the sum t; what can you say about the series $a_1 + a_2 + a_3 + \ldots + a_n + \ldots$? Write your claim with the Σ notation and prove it.
As a result of your work, you should have proved that $\Sigma_{r=1}^{\infty} a_r$ is convergent if and only if $\Sigma_{r=2}^{\infty}$ is convergent.

15 Prove that $\Sigma_{r=1}^{\infty} a_r$ is convergent if and only if $\Sigma_{r=3}^{\infty} a_r$ is convergent.

16 *The start rule*
Prove that $\Sigma_{r=1}^{\infty} a_r$ is convergent if and only if $\Sigma_{r=k}^{\infty} a_r$ is convergent.

This proves that $\Sigma\, a_n$ is not an ambiguous symbol if it is only convergence which is at stake. It *is* an ambiguous symbol if the sum is to be found.

17 If $\Sigma_{r=1}^{\infty} a_r = s$, and s_n is the nth partial sum of this series, find a value for $\Sigma_{r=n}^{\infty} a_r$ and prove that this tends to 0 as $n \to \infty$.

18 Write down the claim that a sequence of partial sums (s_n) is a Cauchy sequence. From the General Principle of Convergence (qn 4.57) this is equivalent to its convergence. Interpret this claim for the infinite series from which the partial sums were formed.

19 *The scalar rule*
For a given real number $c \neq 0$, prove that $\Sigma\, c \cdot a_n$ is convergent if and only if $\Sigma\, a_n$ is convergent.

20 Use qns 9 and 19 to determine exactly which geometric series are convergent and which are divergent.

21 *The sum rule*
If $\Sigma\, a_n$ and $\Sigma\, b_n$ are convergent, prove that $\Sigma(a_n + b_n)$ is convergent.

22 (a) If $\Sigma\, a_n$ is convergent, is $\Sigma\, a_r$ convergent?
(b) If s_n is the nth partial sum of the series
$a_1 + a_2 + a_3 + \ldots + a_n + \ldots$
which of the following is equal to s_n?

(i) $\displaystyle\sum_{1}^{n} a_r$,

(ii) $\displaystyle\sum_{i=1}^{i=n} a_i$,

(iii) Σa_n,

(iv) $\displaystyle\sum_{r=n}^{\infty} a_r$.

Summary: Convergence of series

Definition of convergence of series

For any sequence (a_n), the nth partial sum $s_n = \Sigma_{r=1}^{r=n} a_r$ may be constructed. When $(s_n) \to s$, the series $\Sigma_{n=1}^{\infty} a_n$ (or briefly $\Sigma\, a_n$) is said *to be convergent* and to converge to the sum s. A series which is not convergent is said to be *divergent*.

The start rule

qn 16 $\Sigma_{n=1}^{\infty} a_n$ is convergent if and only if $\Sigma_{n=k}^{\infty} a_n$ is convergent.

The scalar rule

qn 19 $\Sigma\, a_n$ is convergent if and only if $\Sigma\, c \cdot a_n$ is convergent, for some non-zero constant c.

The sum rule

qn 21 If $\Sigma\, a_n$ and $\Sigma\, b_n$ are convergent, then $\Sigma(a_n + b_n)$ is convergent.

The null sequence test

qn 11 If (a_n) is *not* a null sequence, then $\Sigma\, a_n$ is divergent.

Convergence of geometric series

qns 2, 9 When $|x| < 1$, $\Sigma_{n=0}^{\infty} x^n = 1/(1-x)$.

Series of positive terms

First comparison test

23 Let $e_n = 1 + 1 + 1/2! + 1/3! + \ldots + 1/n!$,
and let $s_n = 1 + 1 + \frac{1}{2} + \frac{1}{4} + \ldots + 1/2^{n-1}$, with $n \geq 1$.
Prove that $e_n \leq s_n = 3 - 1/2^{n-1}$.
Deduce that the sequence (e_n) is monotonic increasing, bounded above and so convergent. We say $(e_n) \to e$. [The relation between this e and qn 4.36 is established in qns 11.28 and 11.32.]

24 Prove that e is irrational. Suppose that e is rational $= p/q$, in lowest terms. Show that $e - e_q = k/q!$ for some integer k. Check that $q > 2$, and deduce that $e - e_q < (1/q!) \cdot \frac{1}{2}$, contradicting the previous result and showing that the hypothesis, of rational e, is false.

25 If $a_n \geq 0$ for all n, prove that the sequence of partial sums of the series $\Sigma\, a_n$ is monotonic increasing. Deduce that $\Sigma\, a_n$ is convergent if and only if its partial sums are bounded.

26 *The first comparison test*
The result of qn 25 leads to a squeeze theorem for series of positive terms.
If $0 \leq a_n \leq b_n$, and $\Sigma\, b_n$ is convergent, show that the partial sums of $\Sigma\, a_n$ are bounded and deduce that $\Sigma\, a_n$ is convergent.
Contrapositively, if $\Sigma\, a_n$ is divergent, show that the partial sums of $\Sigma\, b_n$ are unbounded and so $\Sigma\, b_n$ is divergent.

27 Use the fact that

$$0 < \frac{1}{(n+1)^2} < \frac{1}{n(n+1)},$$

and qns 4 and 26 (the first comparison test), to prove that $\Sigma\, 1/(n+1)^2$ is convergent.

Deduce from qn 14 that $\Sigma\, 1/n^2$ is convergent. (The sum of this series is $\pi^2/6$. Pólya (1954), p. 104, gives a remarkable justification of this due to Euler.)

28 Let $a_n = \sqrt{(n+1)} - \sqrt{n}$, and let $b_n = 1/\sqrt{n}$.
What is the partial sum of the first hundred terms of the series $\Sigma\, a_n$?

(i) Use qn 25 and qn 3.20 to prove that $\Sigma\, a_n$ is divergent.
(ii) Prove that $a_n = 1/(\sqrt{(n+1)} + \sqrt{n})$.
(iii) Prove that $0 < a_n < \frac{1}{2}b_n$.
(iv) Use the first comparison test, qn 26, and the scalar rule, qn 19, to prove that $\Sigma\, b_n$ is divergent.

So, although (b_n) is a null sequence, $\Sigma\, b_n$ is divergent. Rather disconcertingly, this establishes that the converse of qn 11 (the null sequence test) is false. So do not fall into the trap of thinking that $\Sigma\, b_n$ must be convergent when (b_n) is null.

29 Use the first comparison test (qn 26), and qns 27 and 28 to show that $\Sigma\, 1/n^\alpha$ is convergent when $\alpha \geq 2$ and divergent when $0 \leq \alpha \leq \frac{1}{2}$.

The harmonic series

30 For the series

$$1 + \frac{1}{2} + \frac{1}{3} + \frac{1}{4} + \ldots + \frac{1}{n} + \ldots,$$

let s_n denote the sum of the first n terms. For each n, show that

$$s_{2n} - s_n \geq \frac{n}{2n} = \frac{1}{2}.$$

Write down these inequalities for $n = 1, 2, 4, 8, \ldots, 2^{k-1}$, and add them to prove that

$$\sum_{r=1}^{r=2^n} \frac{1}{r} \geq 1 + \tfrac{1}{2}n.$$

Is $\Sigma\, 1/n$ convergent or divergent?

The convergence of $\Sigma 1/n^\alpha$

31 For what values of the positive real number α can you be sure that the series $\Sigma\, 1/n^\alpha$ is divergent? Just use the first comparison test (qn 26) and qn 30 at this stage.

32 In qn 29 we proved that the series $\Sigma\, 1/n^\alpha$ was convergent when $\alpha \geq 2$, and in qn 31 we saw that this series is divergent when $\alpha \leq 1$. We have not yet examined the convergence of $\Sigma\, 1/n^\alpha$ for values of α between 1 and 2. Let $1 < \alpha < 2$. Let s_n denote the sum of the first n terms of the series

$$1+\frac{1}{2^\alpha}+\frac{1}{3^\alpha}+\frac{1}{4^\alpha}+\ldots+\frac{1}{n^\alpha}+\ldots.$$

Verify that $s_{2n-1} - s_{n-1} < \dfrac{n}{n^\alpha}$.
Write down these inequalities for $n = 2, 4, 8, \ldots, 2^k$ and add them to show that

$$\sum_{r=1}^{r=2^{n+1}-1}\frac{1}{r^\alpha} < 1+\frac{2}{2^\alpha}+\frac{4}{4^\alpha}+\ldots+\frac{2^n}{2^{n\alpha}} = \frac{1-(1/2^{\alpha-1})^{n+1}}{1-(1/2^{\alpha-1})},$$

using qn 2 for the equality.
Show that $\alpha > 1$ implies $1/2^{\alpha-1} < 1$.
Deduce that the partial sums of $\Sigma\, 1/r^\alpha$ are bounded and so the series is convergent. This proof holds for all real $\alpha > 1$.

Cauchy's nth root test

33 Use the fact that $n/(2n+1) < \frac{1}{2}$ and the first comparison test (qn 26) to prove that the series $\Sigma(n/(2n+1))^n$ is convergent.

34 If $n > 3$, prove that

$$\frac{4n^2}{5n^2-2} < \frac{36}{43},$$

and using the start rule (qn 16) and the first comparison test (qn 26) deduce that the series $\Sigma(4n^2/(5n^2-2))^n$ is convergent.

35 *Cauchy's nth root test* (1821)
To generalise the results of qns 33 and 34, we suppose that we have a series of positive terms, $\Sigma\, a_n$, for which the sequence $(\sqrt[n]{a_n}) \to k$ as $n \to \infty$, and that $0 \le k < 1$. By choosing $\varepsilon = \frac{1}{2}(1-k)$, show that for sufficiently large n, $a_n < \rho^n$ where $\rho = k + \varepsilon = \frac{1}{2}(1+k) < 1$. Deduce from the start rule (qn 16) and the first comparison test (qn 26) that $\Sigma\, a_n$ is convergent.

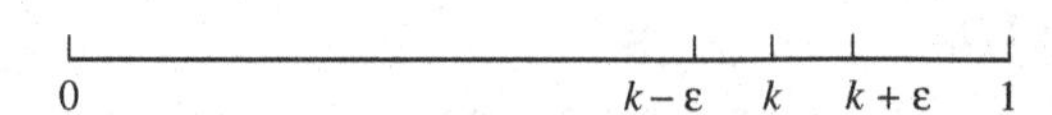

Figure 5.2

36 Apply Cauchy's nth root test to the series $\Sigma\, n/2^n$ (compare with qn 10), and to the series $\Sigma\, n^2(0.8)^n$.

37 If $\sqrt[n]{a_n} \geq 1$ for all n, prove that $\Sigma\, a_n$ is divergent.

38 A series of positive terms $\Sigma\, a_n$ is given, for which the sequence $(\sqrt[n]{a_n}) \to k$ as $n \to \infty$, where $1 < k$. Prove that the series is divergent.

39 By considering the two series $\Sigma\, 1/n$ and $\Sigma\, 1/n^2$, show that Cauchy's nth root test gives no information about the convergence of the series, when $k = 1$ in qn 38.

d'Alembert's ratio test

40 Let $a_n = n^2/2^n$. Prove that **if $n \geq 3$**, then $a_{n+1}/a_n < \frac{8}{9}$. By using this inequality when $n = 3, 4, 5, \ldots$ prove that $a_{n+3} \leq \left(\frac{8}{9}\right)^n a_3$. Use qn 20 (geometric series) and the first comparison test (qn 26) to show that $\Sigma\, a_{n+3}$ is convergent. Then use the start rule (qn 16) to prove that $\Sigma\, a_n$ is convergent.

41 Let $a_n = n^3(0.8)^n$. Prove that **if $n \geq 14$**, then $a_{n+1}/a_n < 0.99$. Use arguments like those in qn 40 to prove that $a_{n+14} < (0.99)^n a_{14}$. Deduce that $\Sigma\, a_n$ is convergent.

42 To generalise the results of qns 40 and 41, we suppose that we have a series $\Sigma\, a_n$ of positive terms for which

$$0 < a_{n+1}/a_n < \rho < 1, \text{ for all } n.$$

Show that $a_{n+1} < \rho^n \cdot a_1$. Use geometric series (qn 20) and the first comparison test to prove that $\Sigma\, a_n$ is convergent.

Figure 5.3

43 *d'Alembert's ratio test*, with qn 47
Suppose that $\Sigma\, a_n$ is a series of positive terms and the sequence of ratios $(a_{n+1}/a_n) \to k < 1$. By a suitable choice of $\varepsilon > 0$ (what ε?), show that there exists an N such that

$$n > N \Rightarrow a_{n+1}/a_n < \tfrac{1}{2}(k+1) < 1.$$

Let $\rho = \frac{1}{2}(k+1)$, and apply qn 42 and the start rule (qn 16) to prove that $\Sigma\, a_n$ is convergent.

44 Prove that $\Sigma\, 2^n/n!$ is convergent.

45 Prove that $\Sigma\, n!/n^n$ is convergent.

46 Suppose that we have a series of positive terms $\Sigma\, a_n$ for which $1 \le a_{n+1}/a_n$ for all n. Prove that $a_1 \le a_n$ for all n. Deduce that (a_n) is not a null sequence and so $\Sigma\, a_n$ is divergent.

47 *d'Alembert's ratio test*, with qn 43
Suppose that we have a series of positive terms $\Sigma\, a_n$ for which the sequence of ratios $(a_{n+1}/a_n) \to k > 1$. By a suitable choice of $\varepsilon > 0$ (what ε?), show that there exists an N such that

$$n > N \Rightarrow 1 < \tfrac{1}{2}(1+k) < a_{n+1}/a_n.$$

Apply qn 46 and the start rule (qn 16) to show that $\Sigma\, a_n$ is divergent.

48 For which positive real numbers x is the series $\Sigma\,(2x)^n n^x$ convergent and for which is it divergent? Use both d'Alembert's ratio test and Cauchy's nth root test.

It is important to note that *d'Alembert's ratio test* (qns 43 and 47) gives no information about the convergence of $\Sigma\, a_n$ when $(a_{n+1}/a_n) \to 1$. So, in particular, d'Alembert's ratio test gives no information about the convergence or divergence of $\Sigma\, 1/\sqrt{n}$, $\Sigma\, 1/n$ or $\Sigma\, 1/n^2$.

49 Give an example to show that the condition $0 < a_{n+1}/a_n < 1$ is not sufficient to ensure the convergence of the series $\Sigma\, a_n$.

50 By considering the sequence defined by $a_{2n-1} = 1/(2n)^3$ and $a_{2n} = 1/(2n)^2$, show that a series $\Sigma\, a_n$ may be convergent even when the sequence (a_{n+1}/a_n) is unbounded.

Second comparison test

51 If $a_n = \dfrac{\sqrt{n}+1}{n^2+2}$ and $b_n = \dfrac{1}{n^{3/2}}$,
prove that $(a_n/b_n) \to 1$ as $n \to \infty$.
Deduce that for some N, $n > N \Rightarrow \frac{1}{2} < a_n/b_n < \frac{3}{2}$.
Now use the convergence of $\Sigma\, b_n$ to establish the convergence of $\Sigma\, \frac{3}{2}b_n$ (with the scalar rule, qn 19) and thus the convergence of $\Sigma\, a_n$.

52 If $a_n = \dfrac{\sqrt{n}+1}{n+2}$ and $b_n = \dfrac{1}{\sqrt{n}}$,
prove that $(a_n/b_n) \to 1$ as $n \to \infty$.
Deduce that for some N, $n > N \Rightarrow \frac{1}{2} < a_n/b_n < \frac{3}{2}$. Now use the divergence of $\Sigma\, b_n$ to establish divergence of $\Sigma\, \frac{1}{2}b_n$ (with the contrapositive of the scalar rule, qn 19) and deduce the divergence of Σa_n.

53 *Second comparison test*
Let $\Sigma\, a_n$ and $\Sigma\, b_n$ be two series of positive terms for which $0 < m < a_n/b_n < M$, for all n.
If $\Sigma\, a_n$ is convergent, prove that $\Sigma\, m \cdot b_n$ is convergent and deduce that $\Sigma\, b_n$ is convergent.
If $\Sigma\, b_n$ is convergent, prove that $\Sigma\, M \cdot b_n$ is convergent and deduce that $\Sigma\, a_n$ is convergent.
Thus $\Sigma\, a_n$ and $\Sigma\, b_n$ are both convergent or both divergent.

54 If $a_n = 1/n^2$ and $b_n = 1/n$, prove that $0 < a_n/b_n < 1$.
Why does the convergence of $\Sigma\, a_n$ and the divergence of $\Sigma\, b_n$ not contradict the second comparison test?

55 *Limit version of the second comparison test*
Let $\Sigma\, a_n$ and $\Sigma\, b_n$ be two series of positive terms for which $(a_n/b_n) \to l \neq 0$ as $n \to \infty$.
By a suitable choice of $\varepsilon > 0$ (what ε?) show that there exists an N such that $n > N \Rightarrow 0 < \frac{1}{2}l < a_n/b_n < \frac{3}{2}l$. By using the second comparison test and the start rule (qn 16) prove that $\Sigma\, a_n$ and $\Sigma\, b_n$ are both convergent or both divergent.

Integral test

This section uses the notion of integral, which is discussed formally in chapter 10, and the logarithm function, which is discussed formally in chapter 11. Sixth-form notions of integrals and logarithms are sufficiently reliable to answer qns 56–61. The natural logarithm, $\int_1^x \frac{dt}{t} = \log_e x$, is denoted by $\ln x$.

56 Sketch the graph of $f(x) = 1/x$ for positive x.
Why is $\dfrac{1}{n+1} < \displaystyle\int_n^{n+1} \frac{dx}{x} < \frac{1}{n}$?

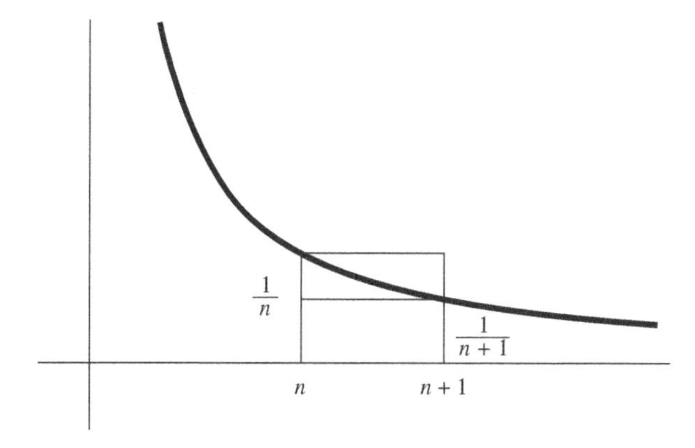

Figure 5.4

Prove that $\sum_{r=2}^{r=n} \frac{1}{r} < \int_1^n \frac{dx}{x} < \sum_{r=1}^{r=n-1} \frac{1}{r}$, for $n > 1$.

Let $s_n = \sum_{r=1}^{r=n} \frac{1}{r}$. Prove that $\frac{1}{n} < s_n - \ln n < 1$, for $n > 1$.

Let $D_n = s_n - \ln n$. Calculate some values of D_n, using a computer if possible.

Show that $D_n - D_{n+1} = \int_n^{n+1} \frac{dx}{x} - \frac{1}{n+1}$.

Deduce that (D_n) is a monotonic decreasing sequence of positive terms and so has a limit. The limit, in this case, is usually denoted by γ, and is known as Euler's constant. Its value is slightly greater than 0.577.

Let $E_n = s_n - \ln(n+1)$. Check that (E_n) is an increasing sequence, that $(D_n - E_n)$ is a null sequence (so that (D_n) and (E_n) have the same limit), and illustrate an area equal to E_n on a graph.

57 *The integral test*

Let $f(x) \geq 0$, where f is a decreasing function for $x \geq 1$. What may be said about the sequence $(f(n))$? Is it necessarily convergent? Since f is monotonic, f is integrable; see qns 10.6 and 10.7. Why is

$$f(n+1) \leq \int_n^{n+1} f(x)dx \leq f(n)?$$

Deduce that $\sum_{r=2}^{r=n} f(r) \leq \int_1^n f(x)dx \leq \sum_{r=1}^{r=n-1} f(r)$.

Let $s_n = \sum_{r=1}^{r=n} f(r)$. Prove that $0 \leq f(n) \leq s_n - \int_1^n f(x)dx \leq f(1)$.

If $\lim_{n\to\infty} \int_1^n f(x)dx$ exists, prove that $\Sigma f(n)$ is convergent.

If $\Sigma f(n)$ is convergent, prove that $\lim_{n\to\infty} \int_1^n f(x)dx$ exists.

When f is strictly decreasing, the inequalities are strict.

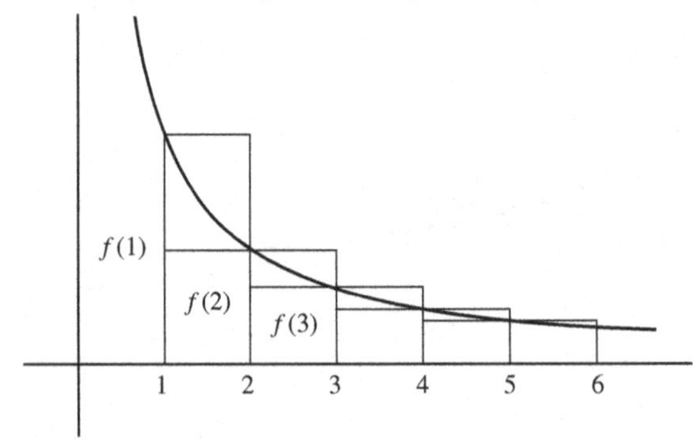

Figure 5.5

(58) Prove that $\sum_{n=1}^{\infty} \frac{1}{n^2} < 2$. Compare with qn 27.

(59) Prove that $2\sqrt{n} - 2 < \sum_{r=1}^{r=n} \frac{1}{\sqrt{r}} < 2\sqrt{n} - 1$, for $n > 1$.

60 (*Abel*, 1828) Prove that $\Sigma\, 1/(n \ln n)$ is divergent for $n \geq 2$.

61 With the notation of qn 57, let $D_n = s_n - \int_1^n f(x)dx$.

Prove that $D_n - D_{n+1} = \int_n^{n+1} f(x)dx - f(n+1) \geq 0$.

Deduce that (D_n) is convergent. Notice that this result holds whether $\Sigma f(n)$ is convergent or divergent.

Summary: Series of positive terms

First comparison test

qn 26 — If $0 \leq a_n \leq b_n$, for all n, and $\Sigma\, b_n$ is convergent, then $\Sigma\, a_n$ is convergent. If $\Sigma\, a_n$ is divergent, then $\Sigma\, b_n$ is divergent.

Theorem qns 30, 32 — $\Sigma\, 1/n^\alpha$ is convergent when $\alpha > 1$ and divergent when $\alpha \leqslant 1$.

Cauchy's nth root test

qns 35, 38 — If $0 \leq a_n$ and $(\sqrt[n]{a_n}) \to k$, then $\Sigma\, a_n$ is convergent when $k < 1$ and divergent when $k > 1$.

d'Alembert's ratio test

qns 43, 47 — If $0 < a_n$ and $(a_{n+1}/a_n) \to k$, then $\Sigma\, a_n$ is convergent when $k < 1$ and divergent when $k > 1$.

Second comparison test

qn 53 — If $0 < a_n$, $0 < b_n$ and $0 < m < a_n/b_n < M$, for all n, then $\Sigma\, a_n$ and $\Sigma\, b_n$ are both convergent or both divergent.

Limit form of second comparison test

qn 55 — If $0 < a_n$, $0 < b_n$ and $(a_n/b_n) \to l \neq 0$, then $\Sigma\, a_n$ and $\Sigma\, b_n$ are both convergent or both divergent.

Integral test

qn 57 — If, for $x \geq 1$, $f(x)$ is a decreasing positive function, then $\Sigma f(n)$ is convergent if and only if $\int_1^\infty f(x)dx$ exists.

Series with positive and negative terms

Alternating series test

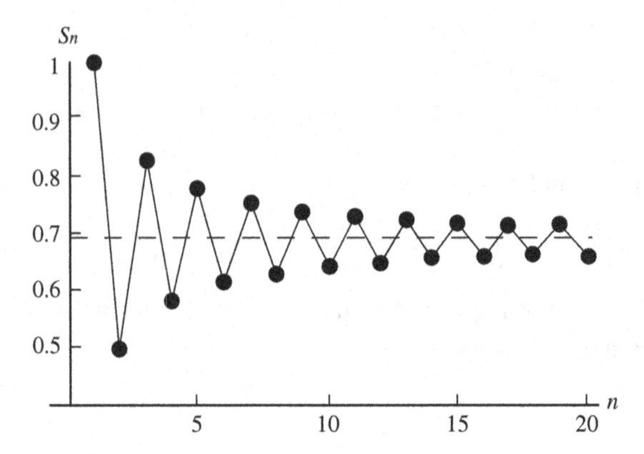

Figure 5.6

62 From our knowledge of $\Sigma\, 1/n$ we might expect the series

$$1-\frac{1}{2}+\frac{1}{3}-\frac{1}{4}+\frac{1}{5}-\ldots+\frac{(-1)^{n+1}}{n}+\ldots$$

to be divergent. However, if we let the nth partial sum $s_n = \Sigma_{r=1}^{r=n}(-1)^{r+1}/r$, and examine $s_2,\ s_4 - s_2,\ s_6 - s_4$, we can see that all these are positive. Prove that $s_{2n+2} - s_{2n}$ is positive. Deduce that the subsequence (s_{2n}) is strictly increasing.
Likewise, if we examine $s_3 - s_1,\ s_5 - s_3$, these are negative. Prove that $s_{2n+3} - s_{2n+1}$ is negative. Deduce that the subsequence (s_{2n+1}) is strictly decreasing.
The proposition

$$s_2 \leq s_{2n} + \frac{1}{2n+1} = s_{2n+1} < s_1$$

shows that $s_{2m} < s_{2n+1}$ when $m = n$. Establish this inequality when $m < n$ and when $m > n$, to show that every even partial sum is less than every odd partial sum. Deduce that the intervals $[s_{2n},\ s_{2n+1}]$ are nested, and that the intersection of all these intervals (the Chinese Box Theorem, qn 4.42) contains just one number, the sum of the series. We will find the limit in qn 65.

63 (*Leibniz*, 1682) The argument of qn 62 is easily generalised to establish the convergence of any series of the form $\Sigma\,(-1)^{n+1}a_n$ when (a_n) is a monotonic decreasing null sequence of positive terms. This is called the *alternating series test*. In this case, let $s_n = \Sigma_{r=1}^{r=n}(-1)^{r+1}a_r$.

Show that $s_{2n+2} - s_{2n} > 0$,
and that $s_{2n+3} - s_{2n+1} < 0$.
From the proposition $s_2 \leq s_{2n} + a_{2n+1} = s_{2n+1} < s_1$, show that the intervals $[s_{2n},\ s_{2n+1}]$ are nested, and their intersection contains only one number.
Deduce that $\Sigma\,(-1)^{n+1}a_n$ is convergent

64 Why is the series $1 - \frac{1}{3} + \frac{1}{5} - \frac{1}{7} + \ldots + (-1)^{n+1}\frac{1}{2n-1} + \ldots$ convergent?

65 We examine the series of qn 62, using the symbols D_n and γ from qn 56.

$$\begin{aligned} 1 - \frac{1}{2} + \frac{1}{3} - \frac{1}{4} + \ldots - \frac{1}{2n} & \\ &= 1 + \frac{1}{2} + \frac{1}{3} + \frac{1}{4} + \ldots + \frac{1}{2n} - 2\left(\frac{1}{2} + \frac{1}{4} + \ldots + \frac{1}{2n}\right) \\ &= \ln 2n + D_{2n} - (\ln n + D_n) \\ &= \ln 2 + D_{2n} - D_n. \end{aligned}$$

Now use the fact that $(D_n) \to \gamma$ as $n \to \infty$, to prove that the sum of the series in qn 62 is $\ln 2$.

A vivid illustration of $\Sigma\,(-1)^{n+1}/n = \ln 2$ was devised by Brouncker in 1668 and given by Bonar and Khoury (p. 201) and Nelsen (p. 128).

Absolute convergence

66 For the series $\Sigma\,(-1)^n/n$ both the subseries of positive terms and the subseries of negative terms, taken separately, are divergent. In the case of the series $\Sigma\,(-1)^n/n^2$, the subseries of positive terms and the subseries of negative terms, taken separately, are convergent.
For $\Sigma\,(-1)^n/n^2$, the alternating series test will establish convergence, but a more direct proof is available in such a case. We construct two new series thus.

$n =$	1	2	3	4	5	6	...
$a_n =$	-1	$+\frac{1}{4}$	$-\frac{1}{9}$	$+\frac{1}{16}$	$-\frac{1}{25}$	$+\frac{1}{36}$	...
$u_n =$	0	$\frac{1}{4}$	0	$\frac{1}{16}$	0	$\frac{1}{36}$	...
$v_n =$	1	0	$\frac{1}{9}$	0	$\frac{1}{25}$	0	...

In each case the terms of the new sequences (u_n) and (v_n) are equal either to 0 or to $|a_n|$, so clearly $0 \leq u_n \leq |a_n|$, and $0 \leq v_n \leq |a_n|$. Now $|a_n| = 1/n^2$, and $\Sigma\,1/n^2$ is convergent from qn 27, so

(i) $\Sigma\,u_n$ is convergent (why?),
(ii) $\Sigma\,v_n$ is convergent (why?),
(iii) $\Sigma\,(u_n - v_n)$ is convergent (why?),
(iv) $\Sigma\,a_n$ is convergent.

67 A series $\Sigma\, a_n$ for which $\Sigma\, |a_n|$ is convergent is said to be *absolutely convergent*. The argument used in qn 66 can be generalised to establish that *a series which is absolutely convergent is necessarily convergent.* The clue, as we have seen, is to construct the series formed by the positive terms and the series formed by the negative terms.
If we define $u_n = \frac{1}{2}(|a_n| + a_n)$, what values may u_n take?
If we define $v_n = \frac{1}{2}(|a_n| - a_n)$, what values may v_n take?
Now use the convergence of $\Sigma\, |a_n|$ to prove that $\Sigma\, u_n$ and $\Sigma\, v_n$ are convergent and hence that $\Sigma\, (u_n - v_n)$ is convergent. What is $u_n - v_n$ equal to?

68 Use d'Alembert's ratio test (qn 43) to devise a condition for a series to be absolutely convergent. When combined with qn 67, this gives a stronger form of d'Alembert's ratio test.

69 Does the condition $\left(\left|\frac{a_{n+1}}{a_n}\right|\right) \to k > 1$ as $n \to \infty$ imply that the series $\Sigma\, a_n$ is divergent?

70 Use Cauchy's nth root test (qn 35) to devise a condition for a series to be absolutely convergent. When combined with qn 67, this gives a stronger form of Cauchy's nth root test.

Conditional convergence

A series $\Sigma\, a_n$ which is convergent, but not absolutely convergent, is said to be *conditionally convergent*. The series $\Sigma\, (-1)^n/n$ is conditionally convergent.

71 With the notation of qn 67, use the equations $a_n = u_n - v_n$ and $|a_n| = u_n + v_n$ to prove that, if $\Sigma\, a_n$ is conditionally convergent, then both $\Sigma\, u_n$ and $\Sigma\, v_n$ are divergent, and the partial sums of both series tend to infinity.

Rearrangements

In qn 62 we proved that $1 - \frac{1}{2} + \frac{1}{3} - \frac{1}{4} + \frac{1}{5} - \frac{1}{6} + \ldots$ was convergent.
In qn 65 we showed that the limit was $\ln 2 = 0.693\ldots$.
Starting from the ln 2 series, adding the series to half of itself gives a rearrangement of the original.

$$\begin{array}{lcccccccccccccccccc}
S & = & 1 & - & \frac{1}{2} & + & \frac{1}{3} & - & \frac{1}{4} & + & \frac{1}{5} & - & \frac{1}{6} & + & \frac{1}{7} & - & \frac{1}{8} & + \\
\frac{1}{2}S & = & & & \frac{1}{2} & & & - & \frac{1}{4} & & & + & \frac{1}{6} & & & - & \frac{1}{8} & \\
\frac{3}{2}S & = & 1 & & & + & \frac{1}{3} & - & \frac{1}{2} & + & \frac{1}{5} & & & + & \frac{1}{7} & - & \frac{1}{4} & +
\end{array}$$

The $\frac{3}{2}S$ series is a rearrangement of the S series with two positive terms preceding each negative term.

$\frac{3}{2}S = 1 + \frac{1}{3} - \frac{1}{2} + \frac{1}{5} + \frac{1}{7} - \frac{1}{4} + \frac{1}{9} + \frac{1}{11} - \frac{1}{6} + \frac{1}{13} + \frac{1}{15} - \frac{1}{8} + \ldots$

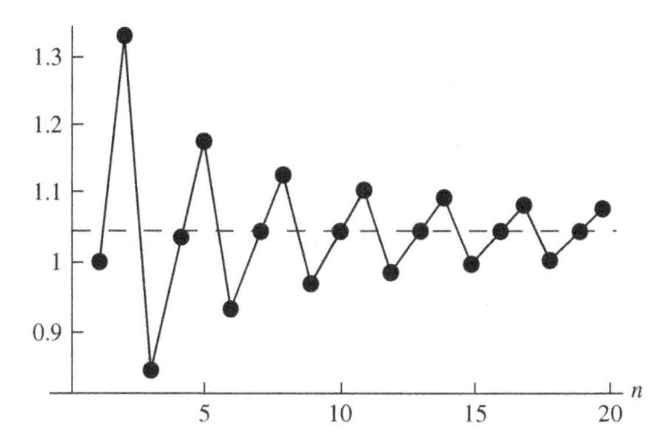

Figure 5.7

$\frac{3}{2}\ln 2 = 1.0397\ldots$

72 For the series $\Sigma\, a_n = 1 - 1 + \frac{1}{2} - \frac{1}{2} + \frac{1}{3} - \frac{1}{3} + \frac{1}{4} - \frac{1}{4} + \ldots$ find the partial sum s_{2n} of the first $2n$ terms, and the partial sum s_{2n+1} of the first $2n + 1$ terms. Deduce that the series is convergent and find its sum.

73 For the series

$$\Sigma\, b_n = 1 + \frac{1}{2} - 1 + \frac{1}{3} + \frac{1}{4} - \frac{1}{2} + \ldots$$
$$+ 1/(2n-1) + 1/2n - 1/n + \ldots$$

compare the partial sum of the first $3n$ terms with the partial sum of the first $2n$ terms of the series in qn 62. Deduce with the help of qn 65 that this series has the sum ln 2.

74 Find expressions for a_{2n-1} and a_{2n} in qn 72 and for b_{3n}, b_{3n-1} and b_{3n-2} in qn 73. Prove that the set $\{a_n \mid n \in \mathbb{N}\}$ of qn 72 is identical to the set $\{b_n \mid n \in \mathbb{N}\}$ of qn 73. Check that there is no repetition within terms of either series.
Deduce that a rearrangement of the terms of a conditionally convergent series may alter its sum.

75 (*Dirichlet*, 1837)

(i) Write down the first nine terms of the series whose $(3n-2)$th term is $1/\sqrt{(4n-3)}$, $(3n-1)$th term is $1/\sqrt{(4n-1)}$ and $3n$th term is $-1/\sqrt{(2n)}$.

(ii) Check that this series is a rearrangement of $\Sigma\,(-1)^{n+1}/\sqrt{n}$.

(iii) Is $\Sigma\,(-1)^{n+1}/\sqrt{n}$ convergent?

(iv) Prove that, if $n \geq 12$, then

$$\frac{1}{n} < \frac{1}{\sqrt{4n-3}} + \frac{1}{\sqrt{4n-1}} - \frac{1}{\sqrt{2n}}.$$

(*Hint*. Multiply through by $2\sqrt{n}$.)

(v) Prove that the series in (i) is divergent.

So a conditionally convergent series may be rearranged to diverge!

With the result of qn 71 it is possible to see how to rearrange the terms of any conditionally convergent series, to produce another series which converges to any limit we please, say, l. First take *positive* terms until l is just exceeded, then take *negative* terms until the sum is just less than l, and so on. The divergence of $\Sigma\, u_n$ and $\Sigma\, v_n$ means this is always possible, and the fact that both (u_n) and (v_n) are null sequences guarantees the convergence. (*Riemann*, 1854.)

76 Let $\Sigma\, a_n$ be a sequence of *positive* terms with sum A, and let $\Sigma\, b_n$ be a rearrangement of the same series. Let $s_n = \Sigma_{r=1}^{r=n} a_r$ and let $t_n = \Sigma_{r=1}^{r=n} b_r$. Suppose that the first N terms of the sequence of bs are included within the first M terms of the sequence of as. Prove that $t_N \leq s_M \leq A$. Deduce that $\Sigma\, b_n$ is convergent to a sum B (say) and that $B \leq A$.
Now suppose that the first N terms of the sequence of as are contained within the first L terms of the sequence of bs. Prove that $s_N \leq t_L \leq B$. Deduce that $A \leq B$, so that $A = B$.
This establishes that when a convergent series of *positive* terms is rearranged it is always convergent to the same sum.

77 Let $\Sigma\, a_n$ be an absolutely convergent series and let $\Sigma\, b_n$ be a rearrangement of the same series. As in qn 67, we let
$u_n = \frac{1}{2}(|a_n| + a_n)$, $v_n = \frac{1}{2}(|a_n| - a_n)$,
$x_n = \frac{1}{2}(|b_n| + b_n)$, $y_n = \frac{1}{2}(|b_n| - b_n)$.
Why are the two series $\Sigma\, u_n$ and $\Sigma\, v_n$ necessarily convergent?
Are they both sequences of positive or zero terms?
How does $\Sigma\, u_n$ relate to $\Sigma\, x_n$, and $\Sigma\, v_n$ to $\Sigma\, y_n$?
Prove that $\Sigma\, a_n = \Sigma\,(u_n - v_n) = \Sigma\,(x_n - y_n) = \Sigma\, b_n$.
This establishes that when an *absolutely convergent* series is rearranged, it is always convergent to the same sum.

Summary: Series of positive and negative terms

Alternating series test

qn 63 If (a_n) is a monotonic null sequence then $\Sigma\,(-1)^{n+1} a_n$ is convergent.

Definition If $\Sigma\ |a_n|$ is convergent, then $\Sigma\ a_n$ is said to be *absolutely convergent.*

Absolute convergence test

qn 67 If $\Sigma\ a_n$ is absolutely convergent then $\Sigma\ a_n$ is convergent.

Definition A series $\Sigma\ a_n$ which is convergent, but not absolutely convergent, is said to be *conditionally convergent.*

Rearrangements qns 74, 76 If the terms of a conditionally convergent series are rearranged, the sum of the series may be changed, or the rearranged series may diverge.

Theorem qn 77 If the terms of an absolutely convergent series are rearranged, the rearranged series has the same sum.

Power series

Applications of d'Alembert's ratio test and Cauchy's nth root test for absolute convergence

78 For what values of x is $\Sigma\ x^n$ convergent? divergent? See qn 9.

79 By using d'Alembert's test for absolute convergence (qns 68 and 69) determine for what values of x the series $\Sigma\ nx^n$ is convergent and for what values it is divergent.

80 By using Cauchy's nth root rest for absolute convergence (qn 70) determine for what values of x the series $\Sigma\ nx^n$ is convergent and for what values it is divergent.

81 For what values of x is the series $\Sigma\ n^2x^n$ convergent and for what values is it divergent?

82 For what values of x is the series $\Sigma\ x^n/n$ convergent and for what values is it divergent? This question is particularly illuminating because of the possibilities at the critical values when $|x| = 1$.

83 For what values of x is the series $\Sigma\ x^n/n^2$ convergent and for what values is it divergent?

84 For what values of x is the series for arctan x convergent:

$$x - \frac{x^3}{3} + \frac{x^5}{5} - \frac{x^7}{7} + \ldots + (-1)^{n+1}\frac{x^{2n-1}}{2n-1} + \ldots?$$

85 Discuss the convergence of the series $\Sigma\ n^\alpha x^n$ for various values of x and α.

86 Prove that $\Sigma\, x^n/n!$ is absolutely convergent for each value of x.

87 Prove that $\Sigma\, (-1)^{n+1}x^{2n-1}/(2n-1)!$ is absolutely convergent for each value of x. This question relates to qn 9.39.

88 Prove that $\Sigma\, n!x^n$ is only convergent when $x = 0$.

89 For what values of x is the series $\Sigma\, (2x)^n$ convergent and for what values is it divergent?

90 For what values of x is the series $\Sigma\, (2x)^n/n$ convergent and for what values is it divergent?

Radius of convergence

91 (*Abel*, 1827) If $\Sigma\, a_n y^n$ is convergent, use qn 11 (the null sequence test) to prove that, for sufficiently large n, $|a_n y^n| < 1$.
For any real number x such that $|x| < |y|$, prove that, for *these* values of n, $|a_n x^n| \le |x/y|^n$, and deduce from the first comparison test (qn 26) that the series $\Sigma\, a_n x^n$ is absolutely convergent.
It is illuminating to illustrate this result on the number line. If a power series is convergent for a particular value of the variable other than 0, then it is absolutely convergent for any value of the variable which is nearer to the origin.

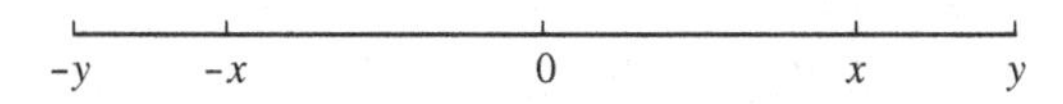

Figure 5.8

92 The result of qn 91 is available under weaker conditions. Suppose only that the terms of the series $\Sigma\, a_n y^n$ are bounded, say, $|a_n y^n| < K$. Use the argument of qn 91 to prove that the series $\Sigma\, a_n x^n$ is absolutely convergent when $|x| < |y|$.

93 Use the contrapositive of qn 91 to show that if $\Sigma\, a_n y^n$ is divergent, then $\Sigma\, a_n x^n$ is divergent when $|x| > |y|$.

94 For a given power series $\Sigma\, a_n x^n$, we consider the values of x for which the power series is convergent. Every power series is convergent for $x = 0$, so the set of such values is not empty.
Let $C = \{|x| : \Sigma\, a_n x^n \text{ is convergent}\}$.
The set C is a set of real numbers ≥ 0. Use qn 91 to show that if C is unbounded then $\Sigma\, a_n x^n$ is convergent for all values of x.
If C is bounded, let R denote its least upper bound, $R = \sup C$.

For any x with $|x| < R$, use the properties of a least upper bound (qn 4.64) to say why there exists a y with $|x| < |y| < R$ and $\Sigma\, a_n y^n$ convergent, and deduce that $\Sigma\, a_n x^n$ is absolutely convergent.
If, for some x with $|x| > R$, the series $\Sigma\, a_n x^n$ were to be convergent, say why this would contradict the definition of R as an upper bound for C.

The work of qn 94 gives the basis for the definition of the *radius of convergence*, R, of a power series $\Sigma\, a_n x^n$.

If the power series is only convergent when $x = 0$, we say that its radius of convergence $R = 0$.

If the power series is convergent for $|x| < R$ and divergent for $|x| > R$, for some positive real number R, then its radius of convergence is that real number R.

If the power series is convergent for all values of x, we say that its radius of convergence $R = \infty$.

When these results are extended to the plane of complex numbers we obtain a circle $|z| = R$ called the *circle of convergence*. On the real line, only the diameter of the 'circle of convergence' is recognisable.

Qn 94 establishes that every power series has a radius of convergence.

95 (*Cauchy*, 1821) If $(|a_{n+1}/a_n|) \to k \neq 0$, prove that the radius of convergence of the power series $\Sigma\, a_n x^n$ is $1/k$.

96 (*Cauchy*, 1821) If $(\sqrt[n]{|a_n|}) \to k \neq 0$, prove that the radius of convergence of the power series $\Sigma\, a_n x^n$ is $1/k$.

97 Find the radius of convergence of $\Sigma\, (nx)^n/n!$.

98 (*Cauchy*, 1821) If a is a real number which is not a positive integer or 0, and

$$\binom{a}{n} = \frac{a(a-1)\dots(a-n+1)}{n!}, \text{ the binomial coefficient,}$$

prove that the power series $\sum \binom{a}{n} x^n$ has radius of convergence 1.

99 If the sequence $(\sqrt[n]{|a_n|})$ is unbounded, show that, for any value of $x \neq 0$, infinitely many terms of the sequence are $> 1/|x|$, so the terms of the power series $\Sigma\, a_n x^n$ do not form a null sequence. Determine the radius of convergence.

100 If the sequence $(\sqrt[n]{|a_n|}) \to 0$ as $n \to \infty$, show that, for any value of $x \neq 0$, and for sufficiently large n, the terms of the sequence are $< \frac{1}{2}/|x|$, and hence show that the power series $\Sigma\, a_n x^n$ is absolutely convergent for all values of x. Determine the radius of convergence in this case.

101 Show that the series $\Sigma\, ((\frac{1}{2}x)^{2n-1} + (\frac{1}{3}x)^{2n})$ has radius of convergence 2. Notice that in this case, the sequence $(\sqrt[n]{|a_n|})$ is not convergent, but oscillates between $\frac{1}{2}$ and $\frac{1}{3}$.

Cauchy–Hadamard formula

Question 102 uses the notation of lim sup which was developed as an optional part of the book in qns 4.83 and 4.84 to extend the result of qn 96.

102 (*Hadamard*, 1892) If the sequence $(\sqrt[n]{|a_n|})$ is bounded but not necessarily convergent, we let $\limsup \sqrt[n]{|a_n|} = A$. We further suppose that $A \neq 0$, since the case $A = 0$ has been considered in qn 100.

(i) Show that if $|x| > 1/A$ then $\sqrt[n]{|a_n|} > 1/|x|$ for infinitely many n and so the terms of the power series $\Sigma\, a_n x^n$ do not form a null sequence and the series is divergent.

(ii) Show also that if $|x| < 1/A$, then, choosing r such that $|x| < r < 1/A$, for sufficiently large n, $\sqrt[n]{|a_n|} < 1/r$, so we have $\sqrt[n]{|a_n x^n|} < |x|/r < 1$, and so the series $\Sigma a_n x^n$ is convergent. Identify the radius of convergence.

The exploration in qns 99, 100 and 102 leads to the *Cauchy–Hadamard formula*

$$R = 1/\limsup \sqrt[n]{|a_n|},$$

which, with suitable interpretations, is a general statement of the radius of convergence of the power series $\Sigma a_n x^n$. You may wish to take advantage of qn 3.59 for the following three questions.

103 Give an example of a power series $\Sigma\, a_n x^n$, with radius of convergence 1, which is divergent at each point on the circle of convergence, that is, at the ends of its diameter.

104 Give an example of a power series $\Sigma\, a_n x^n$, with radius of convergence 1, which is convergent at one point on the circle of convergence, and divergent at another.

105 Give an example of a power series $\Sigma\, a_n x^n$, with radius of convergence 1, which is convergent at each point on the circle of convergence.

106 Compare the regions of convergence of the three power series

$$\ln(1+x) = x - \frac{x^2}{2} + \frac{x^3}{3} - \frac{x^4}{4} + \ldots,$$
$$\frac{1}{1+x} = 1 - x + x^2 - x^3 + \ldots,$$
$$\frac{-1}{(1+x)^2} = -1 + 2x - 3x^2 + 4x^3 - \ldots.$$

107 Use the Cauchy–Hadamard formula to show that the three series $\Sigma\, a_n x^n$, $\Sigma\, n a_n x^{n-1}$ and $\Sigma\, a_n x^{n+1}/(n+1)$ have the same radius of convergence. This establishes that the radius of convergence is not changed when a power series is differentiated or integrated term by term.

The Cauchy product

108 If $A_N(x) = \Sigma_{n=0}^{n=N} a_n x^n$ and $B_N(x) = \Sigma_{n=0}^{n=N} b_n x^n$, calculate the coefficients, c_n, in the product

$$A_N(x) \cdot B_N(x) = \sum_{n=0}^{n=N} c_n x^n + \sum_{n=N+1}^{n=2N} d_n x^n.$$

The series $\Sigma_{n=0}^{\infty} c_n$ is said to be the *Cauchy product* of the series $\Sigma_{n=0}^{\infty} a_n$ and $\Sigma_{n=0}^{\infty} b_n$.

109 By putting $a_n = b_n = (-1)^n/\sqrt{(n+1)}$, show that even if the series $\Sigma\, a_n$ and $\Sigma\, b_n$ are convergent, their Cauchy product need not be convergent.

110 Evaluate the Cauchy product of the series $\Sigma\, a_n$ and the series $\Sigma\, b_n$ when $a_n = x^n/n!$ and $b_n = y^n/n!$, showing that $c_n = (x+y)^n/n!$.

With the result of qn 111, qn 110 illustrates that $\exp(x) \cdot \exp(y) = \exp(x+y)$, a result which we will establish by a different route in chapter 11. See qns 11.20(i) and 11.27$^+$.

111 The series $\Sigma\, a_n$ is absolutely convergent and $\Sigma_{n=0}^{\infty} a_n = A$. The series $\Sigma\, b_n$ is absolutely convergent and $\Sigma_{n=0}^{\infty} b_n = B$. We will show that the Cauchy product of such series is convergent to the sum $A \cdot B$.

The first array below is an array of products from these two series; the second array is a renaming of the first array, position by position.

$$\begin{array}{cccccccc}
a_0b_0 & a_0b_1 & a_0b_2 & a_0b_3 & \cdots & a_0b_n & \cdots \\
a_1b_0 & a_1b_1 & a_1b_2 & a_1b_3 & \cdots & a_1b_n & \cdots \\
a_2b_0 & a_2b_1 & a_2b_2 & a_2b_3 & \cdots & a_2b_n & \cdots \\
a_3b_0 & a_3b_1 & a_3b_2 & a_3b_3 & \cdots & a_3b_n & \cdots \\
\vdots & \vdots & \vdots & \vdots & & \vdots & \\
a_nb_0 & a_nb_1 & a_nb_2 & a_nb_3 & \cdots & a_nb_n & \cdots \\
\vdots & \vdots & \vdots & \vdots & & \vdots &
\end{array}$$

$$\begin{array}{cccccccc}
d_1 & d_4 & d_9 & d_{16} & \cdots & d_{(n+1)^2} & \cdots \\
d_2 & d_3 & d_8 & d_{15} & \cdots & d_{(n+1)^2-1} & \cdots \\
d_5 & d_6 & d_7 & d_{14} & \cdots & d_{(n+1)^2-2} & \cdots \\
d_{10} & d_{11} & d_{12} & d_{13} & \cdots & d_{(n+1)^2-3} & \cdots \\
\vdots & \vdots & \vdots & \vdots & & \vdots & \\
d_{n^2+1} & d_{n^2+2} & d_{n^2+3} & d_{n^2+4} & \cdots & d_{n^2+n+1} & \cdots \\
\vdots & \vdots & \vdots & \vdots & & \vdots &
\end{array}$$

(i) Prove that

$$\sum_{n=1}^{n=(N+1)^2} d_n = \sum_{n=0}^{n=N} a_n \cdot \sum_{n=0}^{n=N} b_n.$$

(ii) Prove that

$$\sum_{n=1}^{n=(N+1)^2} |d_n| = \sum_{n=0}^{n=N} |a_n| \cdot \sum_{n=0}^{n=N} |b_n|.$$

(iii) Prove that Σd_n is absolutely convergent. Use the product rule, qn 3.54(vi), and also qn 3.80.

(iv) Show that the Cauchy product of $\Sigma_{n=0}^{\infty} a_n$ and $\Sigma_{n=0}^{\infty} b_n$ is a rearrangement of $\Sigma_{n=1}^{\infty} d_n$.

(v) Prove that the Cauchy product of Σa_n and Σb_n is convergent to the sum $A \cdot B$.

112 By calculating the Cauchy product of $\Sigma\, x^n$ with itself when $-1 < x < 1$, prove that $1/(1-x)^2 = \Sigma\,(n+1)x^n$. The series begins with $n = 0$.

113 Prove the Binomial Theorem for negative integral index by induction using a Cauchy product, or in other words, give a power series which converges to $1/(1-x)^n$ when $-1 < x < 1$.

Summary: Power series and the Cauchy product

Theorem qns 91, 92	If the terms of $\Sigma\, a_n y^n$ are bounded, then $\Sigma\, a_n x^n$ is absolutely convergent when $\|x\| < \|y\|$.
Definition	When $\{\|x\| : \Sigma\, a_n x^n$ is convergent$\}$ is bounded above, $\sup\{\|x\| : \Sigma\, a_n x^n$ is convergent$\}$ is called the *radius of convergence* of $\Sigma\, a_n x^n$. When $\{\|x\| : \Sigma\, a_n x^n$ is convergent$\}$ is not bounded above, the radius of convergence of $\Sigma\, a_n x^n$ is said to be infinite.
Radius of convergence qn 94	When $\|x\| <$ the radius of convergence, $\Sigma\, a_n x^n$ is absolutely convergent.
The Cauchy–Hadamard formula qn 102	The radius of convergence of $\Sigma\, a_n x^n$ is $1/\limsup \sqrt[n]{\|a_n\|}$.
Theorem qn 107	If a power series is differentiated or integrated term by term, its radius of convergence does not change.
Definition qn 108	If $c_n = a_0 b_n + a_1 b_{n-1} + \ldots + a_i b_{n-i} + \ldots + a_n b_0$, then the series $\Sigma_{n=0}^{\infty} c_n$ is called the *Cauchy product* of the series $\Sigma_{n=0}^{\infty} a_n$ and $\Sigma_{n=0}^{\infty} b_n$.
Theorem qn 111	If $\Sigma\, a_n$ and $\Sigma\, b_n$ are both absolutely convergent, then their Cauchy product is convergent to the product of their sums.

Historical Note

For his quadrature of the parabola, Archimedes (250 BC) did the work which Renaissance mathematicians would interpret to mean $\Sigma_{n=0}^{\infty}(\frac{1}{4})^n = \frac{4}{3}$. Oresme (1323–1382) was able to sum $\Sigma\,(\frac{3}{4})^n$ and $\Sigma\, n(\frac{1}{2})^n$ and he also showed the divergence of the harmonic series in the way we have done.

P. Mengoli found $\Sigma\,(-1)^{n+1}/n = \ln 2$ and summed the series $\Sigma\, 1/n(n+1)$ in 1650. It was the method of summing this latter series which formed the basis of Leibniz' theory of integration. I. Newton in 1667 and N. Mercator in 1668 obtained the result $\Sigma\,(-1)^{n+1}x^n = \ln(1+x)$ by integrating the power series for $(1+x)^{-1}$. James Gregory in 1671 and Leibniz in 1673 obtained the result $\Sigma\,(-1)^{n+1}/(2n-1) = \frac{1}{4}\pi$ by integrating the power series for $(1+x^2)^{-1}$. During the seventeenth and eighteenth centuries the word 'series' was equally used to describe a sequence or a series, and the word 'convergence' was used to refer to the sequence (a_n) or the series $\Sigma\, a_n$, even if it was the latter which appeared to be under discussion. In the eighteenth century, discussion of the sums of series was not confined to those cases where the sequence of partial sums was convergent. Insistence on at least the boundedness of partial sums for meaningful discussion of an overall sum is due to Gauss (1813). In 1742, C. Maclaurin used integrals to approximate to values of series and vice versa, and illustrated his method with $\Sigma\, 1/n^2$. In 1748, by factorising the power series for $\sin x$ like a polynomial, Euler showed that $\Sigma\, 1/n^2 = \pi^2/6$. In 1785, E. Waring showed that $\Sigma\, 1/n^\alpha$ was convergent/divergent according as $\alpha > / \leq 1$. His proof used a rudimentary form of the integral test. The proof we have given is a special form of Cauchy's condensation test (1821): that $\Sigma\, a_n$ is convergent if and only if $\Sigma\, (k^n)a_{k^n}$ is convergent when (a_n) is a decreasing sequence of positive terms.

The title of d'Alembert's ratio test derives from work which he published in 1768 establishing the convergence of the binomial series by comparison with geometric progressions with common ratio less than 1. The ratio test was used with modern precision in Gauss' paper on the hypergeometric series (1813). In 1815, Fourier proved that e, defined by the series $\Sigma\, 1/n!$, was irrational. The first comparison test, or sandwich theorem, was described in relation to geometric progressions by Bolzano (1816). Bolzano (1817) also explained why, for $\Sigma\, a_n$ to be convergent, it is necessary for (a_n) to be a null sequence, but not sufficient because of the harmonic series. Many of the theorems in this chapter were stated and proved for the first time by A. L. Cauchy in his *Analyse Algébrique* (1821). Cauchy considered the convergence of a series only in terms of the convergence of its sequence of partial sums. He proved that $\Sigma\, a_n$ was convergent/divergent when $\limsup \sqrt[n]{|a_n|} < / > 1$ and also when $\lim |a_{n+1}/a_n| < / > 1$. In 1827, it was proposed that $(na_n) \to 0$ was sufficient to ensure the convergence of $\Sigma\, a_n$, but Abel exhibited the counter-example $\Sigma\, 1/n \log n$ in 1828. Cauchy used his results to give the radius of convergence of a power series in some particular cases, though the key result on the radius of convergence (qn 91) is due to Abel (1827). The main result on radius of convergence (Cauchy, qn 96) was rediscovered and generalised by Hadamard (1892) in his thesis. The 'circle of convergence' was so-named by Bouquet and Briot in 1853. In his *Analyse Algébrique* Cauchy argued that absolutely convergent series were convergent and he established the convergence of $\Sigma\, (-1)^n/n$ adding that 'the same reasoning can clearly be applied to every series of this kind', that is to say the alternating series test. He proved that the Cauchy product of two absolutely convergent series converges to the product of the two limits, and showed the necessity for the requirement of absolute convergence with the counter-example we have given (qn 109). He used the Cauchy product to discuss Newton's binomial series. In 1827, Cauchy gave the integral test for convergence, though Cauchy's integrals were only for continuous functions.

In 1837, P. G. L. Dirichlet showed that a conditionally convergent series might be rearranged either to give a different sum or to diverge. The example of divergence that we have given (qn 75) is his. In the same paper, Dirichlet proved that any rearrangement of an absolutely convergent series converged to the same sum. In 1854, G. B. H. Riemann showed that a rearrangement of a conditionally convergent series could be devised to converge to any preassigned sum. The result was published posthumously in 1867.

Answers and comments

1 $a + (b + c) = (a + b) + c$ can be extended to *finite* sums. If the sum is infinite, S may have different values depending on how brackets are inserted. So an agreed system for inserting brackets is needed.

3 $1/(r(r + 1)) = 1/r - 1/(r + 1)$.

$$s_n = (1 - \tfrac{1}{2}) + (\tfrac{1}{2} - \tfrac{1}{3}) + \ldots + \left(\frac{1}{n} - \frac{1}{n+1}\right).$$

$$\sum_{r=1}^{r=n} \frac{1}{r(r+1)} = 1 - \frac{1}{n+1}.$$

5 $(1 - (\frac{1}{2})^n)/(1 - \frac{1}{2})$. Using qn 3.39 on geometric progressions, $1/(1 - \frac{1}{2}) = 2$. You do not 'get there' in a finite number of steps, but you can get as close as you like.

6 $\frac{1}{9} = 0.111\ldots$.

7 $\sum_{r=1}^{\infty} 9 \cdot (10^{-2})^r = 0.090909\ldots$.

8 10.

9 (i) Convergent, from qns 3.39 and 3.42.
(ii) Divergent from qn 3.21. Partial sums tend to infinity when $x = 1$. Partial sums $= \frac{1}{2}(1 + (-1)^{n+1})$ when $x = -1$.

10 $s_n + (n + 1)/2^{n+1} = 2 - (n + 3)/2^{n+1} = s_{n+1}$ gives the inductive step. Put $k = 1$ and $x = \frac{1}{2}$ in qn 3.74(i).

11 $\lim a_{n+1} = s - s = 0$.

13 $\alpha > 0 \Rightarrow (n^\alpha) \to +\infty \Rightarrow \Sigma\, n^\alpha$ is divergent.

14 If s_n is the partial sum of the first n terms of
$a_1 + a_2 + a_3 + \ldots + a_n + \ldots$,
then $s_n - a_1$ is the sum of the first $n - 1$ terms of
$a_2 + a_3 + a_4 + \ldots + a_{n+1} + \ldots$.

$$\lim\,(s_n - a_1) = s - a_1.\ \sum_{r=1}^{\infty} a_n = s \Rightarrow \sum_{r=2}^{\infty} a_n = s - a_1.\ t_{n-1} = s_n - a_1.$$

Conversely

$$\sum_{r=2}^{\infty} a_n = t \Rightarrow \sum_{r=1}^{\infty} a_n = t + a_1.$$

This follows from $s_n = a_1 + t_{n-1}$.
The equation $s_n = a_1 + t_{n-1}$ guarantees that (s_n) has a limit if and only if (t_n) has a limit.

15 Let u_n be the partial sum of the first n terms of $\Sigma_{r=3}^{\infty} a_n$, then $s_n = a_1 + a_2 + u_{n-2}$. Now (s_n) has a limit if and only if (u_n) has a limit.

16 Let v_n be the partial sum of the first n terms of $\Sigma_{r=k}^{\infty} a_n$, then $s_n = a_1 + a_2 + \ldots + a_{k-1} + v_{n-k}$. Now (s_n) has a limit if and only if (v_n) has a limit.

17 $s - s_{n-1}$. This tends to 0 by definition.

18 Given $\varepsilon > 0$, there exists an N such that $n > N \Rightarrow |s_{n+k} - s_n| < \varepsilon$ or

$$\left|\sum_{r=n+1}^{r=n+k} a_r\right| < \varepsilon.$$

19 If s_n is the nth partial sum of $\Sigma\, a_n$, then $c \cdot s_n$ is the nth partial sum of $\Sigma\, c \cdot a_n$. By qn 3.54(i), the scalar rule, (s_n) is convergent if and only if $(c \cdot s_n)$ is convergent, provided $c \neq 0$.

20 When $c \neq 0$, $\Sigma\, c \cdot x^n$ is convergent if and only if $|x| < 1$.

21 If the nth partial sum of $\Sigma\, a_n$ is s_n and the nth partial sum of $\Sigma\, b_n$ is t_n, and $(s_n) \to s$, $(t_n) \to t$, then $(s_n + t_n) \to s + t$, by the sum rule, qn 3.54(iii). But $s_n + t_n$ is the nth partial sum of $\Sigma\, (a_n + b_n)$.

22
(a) Yes. Unless the subscript is given to be constant, it is assumed to be a variable taking the values 1, 2, 3, ...
(b) (i) and (ii) only.

23 $n \geq 2 \Rightarrow n! \geq 2^{n-1} \Rightarrow 1/n! \leq 1/2^{n-1} \Rightarrow \mathrm{e}_n \leq s_n < 3$. Use qn 4.35 for the convergence of monotonic bounded sequences.

24 Since $2\frac{2}{3} < \mathrm{e} < 3$, $\mathrm{e} = p/q \Rightarrow q > 3$. $p/q - n/q! = k/q!$ for some integer k. $1/(q+1)! + 1/(q+2)! + \ldots < (1/q!)(1/3 + 1/3^2 + \ldots) = \frac{1}{2}/q!$ which contradicts the previous line. The proof of this result is set out very fully in Bryant, pp. 20–22.

25 $s_{n+1} = s_n + a_{n+1} \geq s_n$. Since (s_n) is increasing it is convergent if and only if it is bounded by qn 4.35.

26 Let $A_n = \displaystyle\sum_{r=1}^{r=n} a_r$ and let $B_n = \displaystyle\sum_{r=1}^{r=n} b_r$.
$a_r \leq b_r$ for all $r \Rightarrow A_n \leq B_n$ for all n.
$0 \leq a_r \leq b_r \Rightarrow$ both (A_n) and (B_n) are monotonic increasing.
(B_n) convergent $\Rightarrow (B_n)$ bounded $\Rightarrow (A_n)$ bounded $\Rightarrow (A_n)$ convergent by qn 4.35.
$\Sigma\, a_r$ divergent $\Rightarrow (A_n) \to +\infty \Rightarrow (B_n) \to +\infty \Rightarrow \Sigma\, b_r$ divergent.

27 $$\sum_{n=1}^{\infty} \frac{1}{(n+1)^2} = \sum_{n=2}^{\infty} \frac{1}{n^2}.$$

28 $\surd 101 - 1$.

(i) With the notation of qn 26, $A_n = \surd(n+1) - 1 \to +\infty$ as $n \to \infty$.
(ii) $\surd n < \surd(n+1) \Rightarrow 2\surd n < \surd n + \surd(n+1)$
$\Rightarrow 1/(2\surd n) > 1/(\surd(n+1) + \surd n)$.

29 $\alpha \geq 2 \Rightarrow 0 < 1/n^\alpha \leq 1/n^2 \Rightarrow \Sigma\, 1/n^\alpha$ is convergent, by the first comparison test.
$0 \leq \alpha \leq \frac{1}{2} \Rightarrow 0 < 1/\surd n \leq 1/n^\alpha \Rightarrow \Sigma\, 1/n^\alpha$ is divergent, by the first comparison test.

30 Partial sums are unbounded, so $\Sigma\, 1/n$ is divergent.

31 $\alpha \leq 1 \Rightarrow 0 < 1/n \leq 1/n^\alpha \Rightarrow \Sigma\, 1/n^\alpha$ is divergent.

32 $\alpha > 1 \Rightarrow \alpha - 1 > 0$. $2 > 1 \Rightarrow 2^{\alpha-1} > 1^{\alpha-1}$, though we have only proved this for rational α so far, from qn 2.20. The result for all real numbers follows from qn 11.34. Thus $1/2^{\alpha-1} < 1$. So all partial sums $< 1/(1 - 1/2^{\alpha-1})$.

33 $0 < (n/(2n+1))^n < (\frac{1}{2})^n$. Now use qn 5 and the first comparison test.

34 $n > 3 \Rightarrow 8n^2 > 72 \Leftrightarrow (180 - 172)n^2 > 72 \Leftrightarrow 36(5n^2 - 2) > 43 \cdot 4n^2$. As a geometric progression $\Sigma \left(\frac{36}{43}\right)^n$ is convergent (qn 9), so $\Sigma (4n^2/(5n^2 - 2))^n$ is convergent by the first comparison test (qn 26) for $n > 3$. From the start rule (qn 16) $\Sigma\, (4n^2/(5n^2 - 2))^n$ is convergent.

35 There is an N such that $n > N \Rightarrow |\sqrt[n]{a_n} - k| < \frac{1}{2}(1-k)$. So $n > N \Rightarrow 0 \leq \sqrt[n]{a_n} < \frac{1}{2}(1+k) = \rho < 1$. Now, as a geometric progression (qn 9) $\Sigma\, \rho^n$ is convergent. So by the first comparison test (qn 26) $\Sigma\, a_n$ is convergent for $n > N$. From the start rule (qn 16) $\Sigma\, a_n$ is convergent.

36 $(\sqrt[n]{(n/2^n)}) \to \frac{1}{2}$ and $(\sqrt[n]{(n^2(0.8)^n)}) \to 0.8$, using qn 3.59 in each case.

37 $\sqrt[n]{a_n} \geq 1 \Rightarrow a_n \geq 1 \Rightarrow (a_n)$ is not null. Null sequence test (qn 11).

38 There is an N such that $n > N \Rightarrow |\sqrt[n]{a_n} - k| < \frac{1}{2}(1-k)$. So $n > N \Rightarrow 1 < \frac{1}{2}(1+k) < \sqrt[n]{a_n}$. So $\Sigma\, a_n$ is divergent for $n > N$, by the first comparison test (qn 26). Now $\Sigma\, a_n$ is divergent by the start rule (qn 16).

39 $(\sqrt[n]{(1/n)}) \to 1$ and $\Sigma\, 1/n$ is divergent. $(\sqrt[n]{(1/n^2)}) \to 1$ and $\Sigma\, 1/n^2$ is convergent. Use qn 3.59 in each case.

40 $n \geq 3 \Leftrightarrow 1 + 1/n \leq \frac{4}{3} \Rightarrow (1 + 1/n)^2 \leq \frac{16}{9} \Leftrightarrow a_{n+1}/a_n = \frac{1}{2}(1 + 1/n)^2 \leq \frac{8}{9}$; $a_4 \leq a_3 \cdot \frac{8}{9}$; $a_5 \leq a_4 \cdot \frac{8}{9} \leq a_3 \cdot (\frac{8}{9})^2$. By qn 20, $\Sigma\, a_3(\frac{8}{9})^n$ is convergent, so by the first comparison test (qn 26) $\Sigma\, a_{n+3}$ is convergent. By the start rule (qn 16) $\Sigma\, a_n$ is convergent.

41 $n \geq 14 \Leftrightarrow 1 + 1/n \leq \frac{15}{14} \Rightarrow (1 + 1/n)^3 \leq \frac{3375}{2744} \Leftrightarrow a_{n+1}/a_n = 0.8(1 + 1/n)^3 \leq \frac{675}{686} < 0.99$; $a_{15} \leq a_{14} \cdot (0.99)$;

$a_{16} \leq a_{15} \cdot (0.99) \leq a_{14} \cdot (0.99)^2$. By qn 20, $\Sigma\, a_{14}(0.99)^n$ is convergent, so by the first comparison test (qn 26) $\Sigma\, a_{n+14}$ is convergent. By the start rule (qn 16) $\Sigma\, a_n$ is convergent.

42 $\Sigma\, a_1 \rho^n$ is a convergent geometric progression (qn 20), so $\Sigma\, a_{n+1}$ is convergent by the first comparison test (qn 26) and so $\Sigma\, a_n$ is convergent by the start rule (qn 16).

43 Choose $\varepsilon = \frac{1}{2}(1-k)$.

44 $a_{n+1}/a_n = 2/(n+1) \to 0$.

45 $a_{n+1}/a_n = 1/(1+1/n)^n \to l < \frac{1}{2}$, by qn 4.36.

46 Use the null sequence test (qn 11).

47 Choose $\varepsilon = \frac{1}{2}(k-1)$.

48 $\sqrt[n]{a_n} = 2x(\sqrt[n]{n})^x$; $a_{n+1}/a_n = 2x(1+1/n)^x$. Both sequences tend to $2x$, so $\Sigma\, a_n$ is convergent when $0 \leq x < \frac{1}{2}$, and divergent when $\frac{1}{2} < x$. If $x = \frac{1}{2}$ then $\Sigma\, a_n$ is divergent by qn 13, or the null sequence test.

49 $a_n = 1/n$.

50 $0 < a_n \leq 1/n^2$, so $\Sigma\, a_n$ is convergent by the first comparison test. But $a_{2n}/a_{2n-1} = 2n \to +\infty$.

51 $$\frac{a_n}{b_n} = \frac{1+1/\sqrt{n}}{1+2/n^2}.$$

Choose $\varepsilon = \frac{1}{2}$. $\Sigma\, b_n$ is convergent by qn 32. So $\Sigma\, \frac{3}{2}b_n$ is convergent by the scalar rule (qn 19). Now $n > N \Rightarrow 0 < a_n < \frac{3}{2}b_n$, so $\Sigma\, a_n$ is convergent by the first comparison test (qn 26) and the start rule (qn 16).

52 $$\frac{a_n}{b_n} = \frac{1+1/\sqrt{n}}{1+2/n}.$$
Choose $\varepsilon = \frac{1}{2}$. $\Sigma\, b_n$ is divergent from qn 28, so $\Sigma\, \frac{1}{2}b_n$ is divergent by the scalar rule (qn 19). Now $n > N \Rightarrow 0 < \frac{1}{2}b_n < a_n$, so $\Sigma\, a_n$ is divergent by the first comparison test (qn 26) and the start rule (qn 16).

53 Since $0 < mb_n < a_n$, $\Sigma\, a_n$ convergent $\Rightarrow \Sigma\, mb_n$ convergent by the first comparison test $\Rightarrow \Sigma\, b_n$ is convergent by the scalar rule.
Since $0 < a_n < Mb_n$, $\Sigma\, b_n$ convergent $\Rightarrow \Sigma\, Mb_n$ convergent by the scalar rule $\Rightarrow \Sigma\, a_n$ convergent by the first comparison test.

54 $a_n/b_n = 1/n$. The test requires a positive number m such that $0 < m < a_n/b_n$.

55 Choose $\varepsilon = \frac{1}{2}l$. $n > N \Rightarrow 0 < \frac{1}{2}l < a_n/b_n < \frac{3}{2}l$. Now apply the second comparison test (qn 53) and the start rule (qn 16).

56 When $n \le x \le n+1$, $1/(n+1) \le 1/x \le 1/n$, and see chapter 10. Sum the inequalities for $n = 1, 2, 3, \ldots, n-1$.
$s_n - 1 < \int_1^n \frac{dx}{x} < s_n - 1/n$.
$D_1 = 1,\ D_2 = 0.807\ldots,\ D_3 = 0.735\ldots,\ D_4 = 0.697\ldots$.

$$E_{n+1} - E_n = \frac{1}{n+1} - \int_{n+1}^{n+2} \frac{dx}{x} > 0$$

from the first inequality in the question.

$$D_n - E_n = \int_n^{n+1} \frac{dx}{x} < 1/n.$$

So $(D_n - E_n)$ is a null sequence by the squeeze rule.

57 $(f(n))$ is monotonic decreasing and bounded below, by 0, so it is convergent (qn 4.34). Since f is decreasing,
$n \le x \le n+1 \Rightarrow f(n+1) \le f(x) \le f(n)$.
Lower sum $\le$ integral $\le$ upper sum. Add the inequalities for
$n = 1, 2, 3, \ldots, n-1$. (s_n) is monotonic increasing since f is non-negative. Since

$$s_n \le f(1) + \int_1^n f(x)dx,$$

if the integral is bounded above, (s_n) is bounded above and so is convergent by qn 4.35. Conversely

$$\int_1^n f(x)dx \le s_{n-1}$$

so that, if (s_n) is convergent and so bounded, the integral is bounded above, and so convergent by qn 4.35.

58 $\sum_1^n 1/r^2 - \int_1^n dx/x^2 \le 1 \Rightarrow s_n \le 2 - 1/n$.

59 $1/\sqrt{x}$ is strictly decreasing.

$$\int_1^n \frac{dx}{\sqrt{x}} = 2\sqrt{n} - 2.$$

$\frac{1}{\sqrt{n}} < \sum_1^n \frac{1}{\sqrt{r}} - (2\sqrt{n} - 2) < 1$, from working in qn 57.

60 $\int_2^n \frac{dx}{x \cdot \ln x} = [\ln(\ln x)]_2^n = \ln(\ln n) - \ln(\ln 2)$.
Now claim the contrapositive of the integral test, qn 57.

61 The sequence (D_n) is monotonic decreasing but positive, and so bounded below. Thus it is convergent from qn 4.34.

62 $s_{2n+2} - s_{2n} = 1/(2n+1) - 1/(2n+2) = 1/(2n+1)(2n+2) > 0$, so (s_{2n}) is strictly increasing.
$s_{2n+3} - s_{2n+1} = -1/(2n+2) + 1/(2n+3) = -1/(2n+3)(2n+2) < 0$, so (s_{2n+1}) is strictly decreasing. When $m < n$, $s_{2m} < s_{2n} < s_{2n+1}$. When $m > n$, $s_{2m} < s_{2m+1} < s_{2n+1}$. The length of the interval $[s_{2n}, s_{2n+1}]$ is $1/(2n+1)$ which tends to 0, so at most one point is contained in all the intervals. But there is at least one point from the Chinese Box Theorem, qn 4.42.

63 Argument reflects qn 62.

64 $(1/(2n-1))$ is a monotonically decreasing null sequence of positive terms. Apply the alternating series test. The sum is in fact $\pi/4$.

65 $s_{2n} = \log 2 + D_{2n} - D_n$, so $s = \log 2 + \gamma - \gamma$.

66 (i) and (ii) Use first comparison test with $\Sigma|a_n|$. (iii) From the sum rule (qn 21) and the scalar rule (qn 19). (iv) $a_n = u_n - v_n$.

67 $u_n = 0$ when $a_n \leq 0$, $u_n = a_n$ when $0 \leq a_n$.
$v_n = -a_n$ when $a_n \leq 0$, $v_n = 0$ when $0 \leq a_n$.
$0 \leq u_n$, $v_n \leq |a_n|$, so Σu_n and Σv_n are convergent by the first comparison test. Now $\Sigma(u_n - v_n)$ is convergent by the scalar rule and the sum rule.
$a_n = u_n - v_n$.

68 $|a_{n+1}/a_n| \to k < 1 \Rightarrow \Sigma\, a_n$ is absolutely convergent.

69 Yes, by the null sequence test.

70 $(\sqrt[n]{|a_n|}) \to k < 1 \Rightarrow \Sigma a_n$ is absolutely convergent.

71 Since $a_n = u_n - v_n$, $u_n = a_n + v_n$ and $v_n = u_n - a_n$. Now $\Sigma\, a_n$ is convergent, so $\Sigma\, u_n$ and $\Sigma\, v_n$ are both convergent or both divergent by the sum rule and the scalar rule. If both were convergent, then so would $\Sigma\,(u_n + v_n)$ be by the sum rule, and $u_n + v_n = |a_n|$. But $\Sigma\,|a_n|$ is divergent, so both $\Sigma\, u_n$ and $\Sigma\, v_n$ are divergent and both are series of non-negative terms.

72 $s_{2n} = 0$, $s_{2n+1} = 1/(n+1)$. So $(s_n) \to 0$.

73 Let $s_n = \sum_1^n b_i$ and let t_n denote the partial sum of n terms in qn 62. Since

$$\frac{1}{2n-1} + \frac{1}{2n} - \frac{1}{n} = \frac{1}{2n-1} - \frac{1}{2n},$$

$s_{3n} = t_{2n}, s_{3n+1} = t_{2n+1}, s_{3n-1} = t_{2n} + 1/n$. So (s_n) and (t_n) have the same limit.

74 $a_{2n-1} = 1/n$, $a_{2n} = -1/n, b_{3n} = -1/n, b_{3n-1} = 1/2n, b_{3n-2} = 1/(2n-1)$.
$\Sigma a_n = 0. \Sigma b_n = \ln 2$.

75 (i) $1+\sqrt{\frac{1}{3}}-\sqrt{\frac{1}{2}}+\sqrt{\frac{1}{5}}+\sqrt{\frac{1}{7}}-\sqrt{\frac{1}{4}}+\sqrt{\frac{1}{9}}+\sqrt{\frac{1}{11}}-\sqrt{\frac{1}{6}}$.

(iii) Yes, by the alternating series test.

(iv) Apply hint. L.H.S. $= 2/\sqrt{n}$ which gives a decreasing null sequence.

$$\text{R.H.S.} = \frac{1}{\sqrt{1-3/4n}} + \frac{1}{\sqrt{1-1/4n}} - \sqrt{2},$$

which decreases but is bounded below by $2-\sqrt{2}$.
When $n = 12$, L.H.S. $< 2-\sqrt{2}$.

(v) Use first comparison test. Series is divergent by comparison with $\Sigma\, 1/n$.

77 $\Sigma\, u_n$ and $\Sigma\, v_n$ are both convergent by comparison with $\Sigma\, |a_n|$. $\Sigma\, x_n$ is a rearrangement of $\Sigma\, u_n$, and $\Sigma\, y_n$ is a rearrangement of $\Sigma\, v_n$. But $\Sigma\, u_n$ and $\Sigma\, v_n$ are series of positive or zero terms, so $\Sigma\, x_n = \Sigma\, u_n$ and $\Sigma\, y_n = \Sigma\, v_n$.

79 $|a_{n+1}/a_n| = |x|(1+1/n) \to |x|$. Series convergent when $|x| < 1$, divergent when $|x| > 1$. Divergent because terms not a null sequence when $|x| = 1$.

80 $\sqrt[n]{|a_n|} = |x|\sqrt[n]{n} \to |x|$.

81 $\sqrt[n]{|a_n|} = |x|\sqrt[n]{(n^2)} \to |x|$. Conclusion as in qn 79.

82 $\sqrt[n]{|a_n|} = |x|/\sqrt[n]{n} \to |x|$. Convergent when $|x| < 1$ and divergent when $|x| > 1$ by Cauchy's nth root test. Convergent when $x = -1$ by the alternating series test. Divergent when $x = 1$, harmonic series.

83 $\sqrt[n]{|a_n|} = |x|/(\sqrt[n]{n})^2 \to |x|$. Conv./div. as $x < / > 1$. Absolutely convergent when $|x| = 1$.

84 Conv./div. as $|x| < / > 1$ by Cauchy's nth root test. Convergent by alternating series test when $|x| = 1$.

85 $\sqrt[n]{|a_n|} = |x|\sqrt[n]{n^\alpha} \to |x|$. Conv./div. as $|x| < / > 1$. When $x = 1$, convergent when $\alpha < -1$ from qn 32. When $x = -1$, convergent when $\alpha < 0$ by alternating series test.

86 $|a_{n+1}/a_n| = |x|/(n+1) \to 0$ for all values of x.

87 $|a_{n+1}/a_n| = x^2/(2n+1)(2n) \to 0$ for all values of x.

88 $|a_{n+1}/a_n| = |x|(n+1) \to +\infty$ unless $x = 0$.

89 $\sqrt[n]{|a_n|} = |2x|$. Con./div. as $|x| < / > \frac{1}{2}$. Divergent when $|x| = \frac{1}{2}$ since terms not null.

90 $\sqrt[n]{|a_n|} = |2x|/\sqrt[n]{n} \to |2x|$. Conv./div. as $x < / > \frac{1}{2}$. Convergent when $x = -\frac{1}{2}$ by the alternating series test. Divergent when $x = \frac{1}{2}$, harmonic series.

91 $|a_n x^n| = |x/y|^n \cdot |a_n y^n|$ and $|x/y| < 1$.

92 $|a_n x^n| = |x/y|^n \cdot |a_n y^n| < K \cdot |x/y|^n$.

93 If $\Sigma\, a_n x^n$ were convergent with $|x| > |y|$, then by qn 91, $\Sigma\, a_n y^n$ would be convergent. Contradiction.

94 C unbounded $\Rightarrow$ for any given y, there is an x with $|x| > |y|$ and $\Sigma\, a_n x^n$ convergent, so $\Sigma\, a_n y^n$ must be convergent for all such y by qn 91.
If C is bounded, let $\varepsilon = R - |x|$, then by qn 4.64, there is a y with $\Sigma\, a_n y^n$ convergent with $R - \varepsilon < |y| \le R$, and then $\Sigma\, a_n x^n$ is absolutely convergent by qn 91.
$|x| \in C \Rightarrow |x| \le R$.

95 $|a_{n+1}x^{n+1}/a_n x^n| = |a_{n+1}/a_n| \cdot |x| \to |kx|$. Convergent when $|kx| < 1$, divergent when $|kx| > 1$, by d'Alembert's ratio test.

96 $\sqrt[n]{|a_n x^n|} = (\sqrt[n]{|a_n|}) \cdot |x| \to |kx|$.

97 $|a_{n+1}/a_n| = (1 + 1/n)^n \to$ e, so by qn 95, radius of convergence is 1/e, by qn 4.36.

98 Use qn 95.

$$\binom{a}{n+1} \Big/ \binom{a}{n} = \frac{a-n}{n+1} \text{ and } \left|\frac{a-n}{n+1}\right| \to 1.$$

99 $\sqrt[n]{|a_n|} > 1/|x| \Rightarrow |a_n x^n| > 1$, so $R = 0$.

100 Take $\varepsilon = \frac{1}{2}/|x|$, then $\sqrt[n]{|a_n|} < \varepsilon \Rightarrow |a_n x^n| < (\frac{1}{2})^n$. $R = \infty$.

101 $|x| < 2 \Rightarrow |\frac{1}{2}x| < 1 \Rightarrow \sum(\frac{1}{2}x)^{2n-1}$ is absolutely convergent.
Also $|x| < 2 \Rightarrow |\frac{1}{3}x| < 1$.
But $|x| > 2 \Rightarrow |\frac{1}{2}x| > 1 \Rightarrow \sum(\frac{1}{2}x)^{2n-1}$ divergent.
For n odd, $\sqrt[n]{|a_n|} = \frac{1}{2}$. For n even, $\sqrt[n]{|a_n|} = \frac{1}{3}$. Lim sup $\sqrt[n]{|a_n|} = \frac{1}{2}$.

102 A is the greatest limit of a subsequence of $(\sqrt[n]{|a_n|})$.

(i) When $|x| > 1/A$ we prove that the series is divergent.
Take $\varepsilon = A - 1/|x| > 0$; then there is a subsequence of $(\sqrt[n]{|a_n|})$ which tends to A and so has infinitely many terms such that $A - \varepsilon < \sqrt[n]{|a_n|}$, and thus infinitely many terms for which $1/|x| < \sqrt[n]{|a_n|}$, giving $1 < \sqrt[n]{|a_n x^n|}$. So the sequence of terms of the power series is not null.

(ii) When $|x| < 1/A$ we prove that the series is convergent.
Choose $r = (1/2)(|x| + 1/A)$ for example and take $\varepsilon = 1/r - A > 0$.
If $u_k = \sup\{\sqrt[n]{|a_n|} : n \ge k\}$, the sequence (u_k) is decreasing and tends to A, so $u_k < A + \varepsilon$ for sufficiently large k and hence $\sqrt[n]{|a_n|} < A + \varepsilon$ for that large n. So, eventually, all $\sqrt[n]{|a_n|} < A + \varepsilon = 1/r$.
The radius of convergence is $1/A$.

103 $a_n = 1$, or $a_n = n$, for example.

104 $a_n = 1/n$.

105 $a_n = 1/n^2$.

106 The first series is convergent when $-1 < x \leq 1$, the second and third when $-1 < x < 1$. All three have the same radius of convergence. Note the effect of differentiation on the first two functions.

107 $\limsup \sqrt[n]{|na_n|} = \limsup \sqrt[n]{n} \cdot \sqrt[n]{|a_n|} = \limsup \sqrt[n]{|a_n|}$.
$\sqrt[n]{(n+1)} \leq \sqrt[n]{(2n)} = \sqrt[n]{2} \cdot \sqrt[n]{n} \to 1$.
So $\limsup \sqrt[n]{|a_n/(n+1)|} = \limsup \sqrt[n]{|a_n|}/\sqrt[n]{(n+1)} = \limsup \sqrt[n]{|a_n|}$.

108 $c_n = a_0b_n + a_1b_{n-1} + \ldots + a_ib_{n-i} + \ldots + a_nb_0$.

109 The two series are convergent by the alternating series test.
$|c_n| \geq (n+1)/(\sqrt{(n+1)})(\sqrt{(n+1)}) = 1$. Not convergent by the null sequence test.

110 Use qn 1.5 for the proof. After completing qn 111, use qn 86 to understand the significance of the result.

111
(i) Use simple algebra.
(ii) Use simple algebra.
(iii) The partial sums of $\sum |d_n|$ form an increasing sequence, with a convergent subsequence from (ii) and the product rule, qn 3.54(vi). From qn 3.80, $\sum |d_n|$ is convergent, and so $\sum d_n$ is absolutely convergent.
(iv) Cauchy product $= d_1 + (d_4 + d_2) + (d_9 + d_3 + d_5) + \ldots$
(v) Since $\sum d_n$ is absolutely convergent, any rearrangement is absolutely convergent to the same sum by qn 77.

112 $\sum x^n$ is absolutely convergent when $|x| < 1$ from qn 9. In the Cauchy product c_n is a sum of $n+1$ terms.

113 The expansion for $n = 1$ was determined in qn 9. Suppose

$$(1-x)^{-n} = 1 + nx + \frac{n(n+1)}{1\cdot 2}x^2 + \ldots + \frac{n(n+1)\ldots(n+r-1)}{r!}x^r + \ldots.$$

This series has radius of convergence 1 by qn 98. The coefficient of x^r in the Cauchy product with $(1-x)^{-1}$ is

$$1 + n + \frac{n(n+1)}{1\cdot 2} + \ldots + \frac{n(n+1)\ldots(n+r-1)}{r!}$$
$$= (n+1)\left(1 + \frac{n}{2} + \ldots + \frac{n(n+2)\ldots(n+r-1)}{r!}\right)$$
$$= \frac{(n+1)(n+2)}{2}\left(1 + \frac{n}{3} + \ldots + \frac{n(n+3)\cdots(n+r-1)}{3\cdot 4\cdot \ldots \cdot r}\right)$$
$$\vdots$$
$$= \frac{(n+1)(n+2)\ldots(n+r)}{r!},$$

which is the same form as that assumed for n. When $n = 1$, the series takes the familiar form for a geometric progression. This, by induction, establishes the expansion for n.

PART II

Functions

6

Functions and continuity

Neighbourhoods, limits of functions

Preliminary activity: make sure that you have access to graph-drawing facilities on a computer or graphic calculator, and that you can use these facilities with confidence.
Preliminary reading: Leavitt ch. 1.
Concurrent reading: Swann and Johnson, Hart, Reade, Smith, Spivak chs 4, 5, 6.
Further reading: Mason.

Before coming to university you will have worked with polynomials, trigonometric, logarithmic and exponential functions. Now we explore properties shared by all of these functions: continuity, differentiability and integrability.

Functions

When you read or hear the phrase 'the function $f(x)$', what comes to your mind? Perhaps a formula, perhaps a graph.

1 Write down what x can stand for, and what is meant by f, in the expression $f(x)$. Compare your answer with the one in the summary at the end of this section.

We will introduce some special vocabulary in order to be clear what we mean when talking about functions.

The domain of a function

If $f(x) = x^2$ and the values of x are $0, \pm 1, \pm 2, \pm 3, \ldots, \pm n, \ldots$, then the values of $f(x)$ are $0, 1, 4, 9, \ldots, n^2, \ldots$. The set of possible values of x is called the *domain* of the function. When we say that x is a *variable*, we mean that the symbol x is being used to denote any member of the domain of a function. When the possible values of x are real numbers, the function is called a function of a *real variable*.

The range and co-domain of a function

The set of possible values of $f(x)$ is called the *range* of the function, and any set which contains the range may be declared to be the *co-domain* of the function. We have just given a function with domain $\mathbb{Z}$ and range $\mathbb{N} \cup \{0\}$, which we express symbolically by writing $f : \mathbb{Z} \to \mathbb{N} \cup \{0\}$, with the definition $f(x) = x^2$ (or $f : x \mapsto x^2$).

The terms *function* and *mapping* are synonymous, and we sometimes say that the function f maps x to $f(x)$. In fact each of the sequences in chapter 3 is a function with domain $\mathbb{N}$. When both the domain and co-domain of f are subsets of $\mathbb{R}$, the function f is called a *real function*.

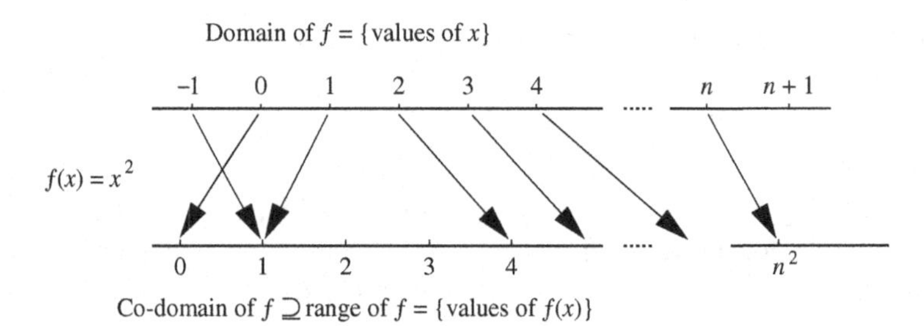

Figure 6.1 *Arrow diagram of the function f*

2 The following functions each have domain and co-domain $\mathbb{R}$. What is the range of each function?

(i) $f(x) = x^2$, (ii) $f(x) = x^3$, (iii) $f(x) = \sin x$,
(iv) $f(x) = 1/(1 + x^2)$, (v) $f(x) = e^x$

[The sine and exponential functions will be defined formally in chapter 11.]

When the co-domain and range of a function are the same, the function is said to be *onto* the co-domain and is called a *surjection*. Check that only one of the functions in qn 2 is *onto* $\mathbb{R}$.

The distinctive property of a function or mapping f is that, for a given element x of the domain, $f(x)$ is uniquely determined as a member of the co-domain. Visually, a vertical line parallel to the y-axis may cut the graph of $y = f(x)$ at most once. Thus $\pm\sqrt{(1 - x^2)}$ does not define a function of x, unless the domain is restricted to $\{\pm 1\}$, for then the range is simply $\{0\}$. But even $f(x) = \sqrt{(1 - x^2)}$ does not define a function with domain and co-domain $\mathbb{R}$, for then $f(x)$ is not defined when $|x| > 1$.

3 Each of the following functions has domain $A \subseteq \mathbb{R}$ and co-domain $\mathbb{R}$. What is the largest possible domain A for the function?

(i) $f(x) = \sqrt{x}$, (ii) $f(x) = \sqrt{(1 - x^2)}$, (iii) $f(x) = \sqrt{(x^2 - 1)}$,
(iv) $f(x) = 1/x$, (v) $f(x) = \ln x$,
(vi) $f(x) = (x^2 - 4)/(x - 2)$ (beware of $x = 2$).

It is sometimes convenient to denote the range of the function $f: A \to \mathbb{R}$ by $f(A)$ and to say that f maps A to $f(A)$.

Although for every function f, $x = y \Rightarrow f(x) = f(y)$, it is only sometimes the case that $f(x) = f(y) \Rightarrow x = y$. When this second implication holds, the function f is said to be *one-to-one* or *one–one*, and is called an *injection*.

4 Which of the functions given in qns 2 and 3 are one–one?

Bijections and inverse functions

If a function $f: A \to B$, with domain A and co-domain B, is both one–one *and* onto, not only is each element $a \in A$ matched with a unique element $f(a) \in B$, which is, of course, true for any function f, but also, for each element $b \in B$, there is a unique element $a \in A$ such that $f(a) = b$. A function which is both one–one and onto is called a *bijection*. Such a function has an inverse $g: B \to A$, defined by $g(f(a)) = a$. When this is the case, we will write $g = f^{-1}$.

5 For each of the functions described in qns 2 and 3, identify appropriate subsets $A, B \subseteq \mathbb{R}$, such that $f: A \to B$ is a bijection.

6 The points on the graph of a real function f have the form $(x, f(x))$. If $f: A \to B$ is a bijection, why do the points on the graph of f^{-1} have the form $(f(x), x)$? Sketch the graph of the function f given by $f(x) = x^2$ for positive x and sketch the graph of its inverse function $f^{-1}(x) = \sqrt{x}$. Also sketch the graph of exp (the function of qn 2(v), E in chapter 11) and its inverse, ln.

Summary: Functions

Definition qns 1, 2, 3, 4, 5

If a set A and a set B are given, then a *function* $f: A \to B$ is a pairing of each element of A with an element of B. The set A is called the *domain* of the function. The set B is called the *co-domain* of the function. The element of B which is paired with $a \in A$ is denoted by $f(a)$. The set $\{f(a)| \, a \in A\} \subseteq B$ is called the *range* of the function. When the range of a function is the whole of its co-domain, the function is said to be *onto* its co-domain and is called a *surjection*.

When $f(x) = f(y) \Rightarrow x = y$, the function is said to be *one–one* and is called an *injection*. A function which is both a surjection and an injection is called a *bijection*.

Definition A function is said to be a *real function* when both its domain and its co-domain are subsets of $\mathbb{R}$.

Theorem qns 5, 6 If $f : A \to B$ is a bijection, then there is a well-defined inverse function $f^{-1} : B \to A$ such that $f^{-1}(f(a)) = a$, for all $a \in A$.

Continuity

Throughout the eighteenth century, functions were given by algebraic formulae, with variables being added, subtracted, multiplied or divided, and combined with exponential, logarithmic or trigonometric functions. Controversy surrounded the differential equation for a plucked string, since in the ordinary starting position two straight-line parts of the string meet at an angle, and thereby require different algebraic formulae for different parts of the string. Cauchy took the example of the function defined by $f(x) = \sqrt{(x^2)}$ to show that even familiar algebraic expressions might lead to a graph with a sharp bend in it.

7 Illustrate on a graph the functions $f : \mathbb{R} \to \mathbb{R}$ given by

(i) $f(x) = 0$ when $x < 0$, and $f(x) = x$ when $x \geq 0$;
(ii) $f(x) = 0$ when $x < 0$, and $f(x) = x^2$ when $x \geq 0$;
(iii) $f(x) = -x^2$ when $x < 0$, and $f(x) = x^2$ when $x \geq 0$;
(iv) $f(x) = x \cdot |x|$.

Allowing a variety of formulae when defining a single function was a significant first step towards a more general notion of function. Further steps were taken when limiting processes were considered.

8 Examine the graphs of $y = x^2$, $y = x^3$, $y = x^4$ and $y = x^{10}$ using a computer or a graphic calculator.
For what values of x does $\lim x^n$ exist as $n \to \infty$?
What is the largest subset $A \subseteq \mathbb{R}$ for which a function $f : A \to \mathbb{R}$ may be defined by $f(x) = \lim x^n$ as $n \to \infty$?
What is the range of this function? Sketch its graph.

9 Examine the graphs of $y = \dfrac{x^n - 1}{x^n + 1}$, for $n = 2, 3, 10, 11$, using a computer or a graphic calculator.
For what values of x does $\lim \dfrac{x^n - 1}{x^n + 1}$ exist as $n \to \infty$?
What is the largest subset $A \subseteq \mathbb{R}$ for which a function $f : A \to \mathbb{R}$ may be defined by $f(x) = \lim \dfrac{x^n - 1}{x^n + 1}$ as $n \to \infty$?
What is the range of this function? Sketch its graph.

10 Examine the graphs of $y = x^2/(x^2 + 1/n)$, for $n = 2, 3, 10, 11$, using a computer or a graphic calculator.
If the function $f : \mathbb{R} \to \mathbb{R}$ is defined by $f(x) = \lim x^2/(x^2 + 1/n)$ as $n \to \infty$, what is the range of f?
Sketch the graph of f.

11 Identify the function of x defined by $\lim_{n\to\infty} \dfrac{1}{1 + n \sin^2 \pi x}$.

Questions 8, 9, 10 and 11 make it evident that the value of $f(x)$ need not be close to the value of $f(y)$, even when x and y are close to one another. The modern definition of function only requires that for each value of x in the domain there is a unique value of $f(x)$ in the co-domain, and the function is then defined by the pairing $(x, f(x))$, even when something like a fresh formula is needed for each number x.

Definition of continuity by sequences

12 Sketch the graph of the function $f : \mathbb{R} \to \mathbb{Z}$, given by $f(x) = \lfloor x \rfloor$, the integer part of x, defined before qn 3.19.

13 Find the limit of the sequence $(\lfloor 1/n \rfloor)$ as $n \to \infty$.

14 Find the limit of the sequence $(\lfloor -1/n \rfloor)$ as $n \to \infty$.

15 Find two sequences $(a_n) \to 1$, and $(b_n) \to 1$, such that $(\lfloor a_n \rfloor) \to 1$ and $(\lfloor b_n \rfloor) \to 0$.

16 For *any* sequence (a_n) which converges to $\frac{1}{2}$, prove that $(\lfloor a_n \rfloor) \to 0 = \lfloor \frac{1}{2} \rfloor$.

The results of qns 13–16 are the basis on which we say that the function f of qn 12 is *continuous* at $x = \frac{1}{2}$ and is *not continuous* at $x = 0$ or $x = 1$.

17 Is the function f of qn 12 continuous at $x = 1\frac{3}{4}$? The critical point to test is whether for *every* sequence $(a_n) \to 1\frac{3}{4}$, the sequence $(f(a_n)) \to 1 = f(1\frac{3}{4})$.

18 Describe the subset of $\mathbb{R}$ of values of x at which the integer function of qn 12 is *not* continuous. Also describe the subset of $\mathbb{R}$ of values of x at which this function *is* continuous.

The ideas in qns 12–18 lead to the following definition.

DEFINITION OF CONTINUITY

The function $f : A \to \mathbb{R}$ is continuous at $a \in A \subseteq \mathbb{R}$ when, for every sequence (a_n) converging to a, with terms in A, the sequence $(f(a_n))$ converges to $f(a)$.

This is commonly expressed by saying that '$f(x)$ tends to $f(a)$ as x tends to a', which is also expressed symbolically by writing $f(x) \to f(a)$ as $x \to a$.

The function $f : A \to \mathbb{R}$ is said to be *continuous on* A, when it is continuous at every $a \in A$.

Examples of discontinuity

Continuous functions are so familiar that to clarify the meaning of this definition we need some examples of discontinuity, illustrating the absence of continuity. The continuity of a function f at a point a is recognised by the way the function affects *every* sequence (a_n) tending to a. Discontinuity is established by finding just one sequence $(a_n) \to a$ for which $(f(a_n))$ does not tend to $f(a)$.

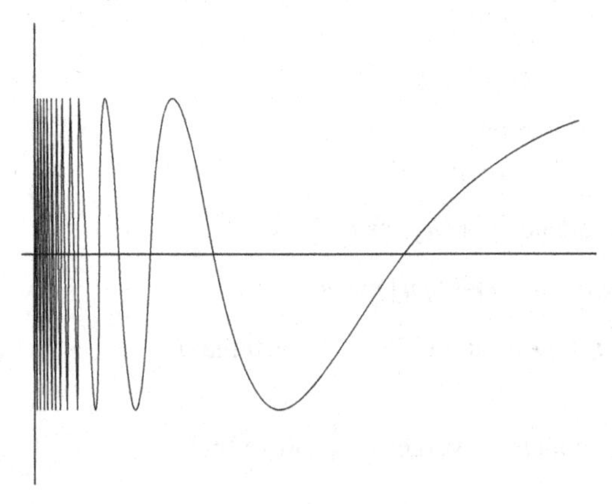

Figure 6.2

19 Examine the graph of $y = \sin 1/x$, paying particular attention to the values of y when x is small.

We define a function $f : \mathbb{R} \to \mathbb{R}$ by
$f(x) = \sin 1/x$ when $x \neq 0$, and $f(0) = 0$.
Let $a_n = 1/(2n\pi)$ and $b_n = 1/((2n + \frac{1}{2})\pi)$.

Which of the four sequences
(i)(a_n), (ii)(b_n), (iii)$(f(a_n))$, (iv)$(f(b_n))$
are null sequences?
Deduce that f is not continuous at $x = 0$.
If $f(0)$ had some value other than 0, might this redefined function f be continuous at $x = 0$?

20 (*Dirichlet*, 1829) Define a function $f : \mathbb{R} \to \mathbb{R}$ by $f(x) = 0$ when x is irrational, and $f(x) = 1$ when x is rational.
If a is a rational number, show that f is not continuous at $x = a$ by considering the sequence (a_n), where $a_n = a + \sqrt{2}/n$.
If a is an irrational number, show that f is not continuous at $x = a$ by considering the sequence $(\lfloor 10^n a \rfloor / 10^n)$, the infinite decimal sequence for a, of qn 3.51.

Question 18 gives us an example of a function with an infinity of isolated discontinuities, qn 19 an example of a function with a single discontinuity, and qn 20 an example of a function with a dense set of discontinuities. This will be enough to start with, and we now turn to the formal confirmation of continuity in the familiar context of polynomials.

Sums and products of continuous functions

21 If a real function f is defined by $f(x) = c$, that is to say, f is a constant function, prove that f is continuous at each point of its domain.

22 If a real function f is defined by $f(x) = x$, so that f is the identity function, prove that f is continuous at each point of its domain.

23 *The sum rule*
If real functions f and g have the same domain A, and are both continuous at $x = a \in A$, use qn 3.54(iii), the sum rule, to prove that the real function $f + g$ defined by

$$(f + g)(x) = f(x) + g(x)$$

is also continuous at $x = a$.

24 Use qns 21, 22 and 23 to prove that the function given by $f(x) = x + 1$ is continuous at each point of its domain.

25 If $f_1, f_2, \ldots, f_n$ are all real functions with the same domain A, and are all continuous at $x = a \in A$, prove by induction that the function $f_1 + f_2 + \ldots + f_n$, defined by

$$(f_1 + f_2 + \ldots + f_n)(x) = f_1(x) + f_2(x) + \ldots + f_n(x)$$

is continuous at $x = a$.

26 *The product rule*
If real functions f and g have the same domain A, and are both continuous at $x = a \in A$, use qn 3.54(vi), the product rule, to prove that the real function $f \cdot g$ defined by

$$(f \cdot g)(x) = f(x) \cdot g(x)$$

is also continuous at $x = a$.

27 Which of qns 21–26 are required, and in which order, to prove that the functions $f_i : \mathbb{R} \to \mathbb{R}$, given by

$$\text{(i) } f_1(x) = 2x, \quad \text{(ii) } f_2(x) = 2x + 1, \quad \text{(iii) } f_3(x) = x^2,$$
$$\text{(iv) } f_4(x) = x^3, \quad \text{(v) } f_5(x) = 2x^2 + 1,$$

are continuous on $\mathbb{R}$?

28 Prove by induction that the function $f : \mathbb{R} \to \mathbb{R}$, given by $f(x) = x^n$, where $n \in \mathbb{N}$, is continuous on $\mathbb{R}$.

29 Prove that the function $f : \mathbb{R} \to \mathbb{R}$, given by

$$f(x) = a_0 + a_1 x + a_2 x^2 + \ldots + a_n x^n,$$

where the a_i are real numbers, is continuous on $\mathbb{R}$.
This establishes that polynomial functions are continuous on $\mathbb{R}$.

30 The acid test of understanding the language we have introduced comes when deciding at what points the function given by $f(x) = 1/x$ is continuous. With reference to qn 3, determine the maximum domain of definition of this real function. Use qns 3.65 and 3.66, the reciprocal rule, to prove that this function is continuous at every point where it is defined.

31 Use qn 30 to extend the result of qn 29 to give a family of functions, each one of which is continuous on $\mathbb{R}\backslash\{0\}$.

Continuity in less familiar settings

32 Use qn 3.54(vii), the absolute value rule, to prove that the function $f : \mathbb{R} \to \mathbb{R}$ given by $f(x) = |x|$ is continuous on $\mathbb{R}$. Sketch the graph of f near the origin.

33 Sketch the graph of the real function given by $f(x) = \frac{1}{2}(x + |x|)$, and prove that it is continuous on $\mathbb{R}$.

34 Sketch the graph of the real function given by $f(x) = \frac{1}{2}(x - |x|)$, and prove that it is continuous on $\mathbb{R}$.

35 We define a function $f: \mathbb{R} \to \mathbb{R}$ by

$f(x) = 0$ when x is irrational,

$f(x) = x$ when x is rational.

Prove that, for all x,

$$\frac{1}{2}(x - |x|) \le f(x) \le \frac{1}{2}(x + |x|).$$

Locate the graph of the function f in relation to those you have drawn for qns 33 and 34.
Is there a value of x for which $\frac{1}{2}(x - |x|) = \frac{1}{2}(x + |x|)$?

A squeeze rule

36 If real functions f, g, and h have the same domain A, and

(i) $f(x) \le g(x) \le h(x)$, for all $x \in A$,
(ii) f and h are continuous at $x = a \in A$,
(iii) $f(a) = h(a)$,

use qn 3.54(viii), the squeeze rule for convergent sequences, to prove that g is continuous at $x = a$.

37 Use qn 36 to prove that the function f of qn 35 is continuous at $x = 0$, and use arguments like those of qn 20 to prove that this function is not continuous at any other point in its domain. This example highlights an important consequence of our definition of continuity since the possibility of continuity at a single isolated point does not spring from an intuitive view of the concept.

Continuity of composite functions and quotients of continuous functions

38 If $f: \mathbb{R} \to \mathbb{R}$ is defined by $f(x) = \sin x$ and $g: \mathbb{R} \to \mathbb{R}$ is defined by $g(x) = x^2$, what are $f(g(x))$ and $g(f(x))$?
Look at the graphs of these two composite functions on a computer or graphic calculator.

Generally, if f and g are two real functions and the domain of f contains the range of g, the *composite function* $f \circ g$ is defined on the domain of g by $(f \circ g)(x) = f(g(x))$.
Notice that the composite function $f \circ g$ is generally different from the composite function $g \circ f$.

39 If $\text{ABS}(x) = |x|$ and $f(x) = x + 1$, sketch the graphs of $f \circ \text{ABS}$ and $\text{ABS} \circ f$

40 *The composite rule*
If the function $g: A \to \mathbb{R}$ is continuous at $x = a$ and the function $f: g(A) \to \mathbb{R}$ is continuous at $g(a)$, prove that $f \circ g$ is continuous at $x = a$.

41 Use qn 40 to prove that, if a real function f is continuous at $x = a$, then the function ABS $\circ f$ is also continuous at a, where ABS is defined as in qn 39.

42 For any real function f, we define a function f^+ on the same domain as f by $f^+(x) = \frac{1}{2}(f(x) + |f(x)|)$. If f is continuous at a, prove that f^+ is also continuous at a.
Sketch the graph of f^+ when $f(x) = x^2 - 1$.

43 For any real function f, we define a function f^- on the same domain as f by $f^-(x) = \frac{1}{2}(f(x) - |f(x)|)$. If f is continuous at a, prove that f^- is also continuous at a.
Sketch the graph of f^- when $f(x) = x^2 - 1$.

44 For any two real numbers a and b, we define

$$\max(a,\ b) = \begin{cases} a & \text{when } b \le a, \\ b & \text{when } a < b. \end{cases}$$

Prove that $\max(a, b) = \frac{1}{2}(a + b) + \frac{1}{2}|a - b|$.
Prove that $f^+(x) = \max(f(x), 0)$.

Analogously we define the minimum of two real numbers and obtain $\min(a, b) = \frac{1}{2}(a + b) - \frac{1}{2}|a - b|$.

45 If real functions f and g have the same domain A, and are both continuous at $x = a \in A$, prove that the function $\max(f, g)$, defined by $\max(f, g)(x) = \max(f(x),\ g(x))$, is also continuous at $x = a$.

46 If $f(x) = x$ and $g(x) = -x$, identify the function $\max(f, g)$.

47 Let $r: \mathbb{R}\backslash\{0\} \to \mathbb{R}$ denote the reciprocal function $r(x) = 1/x$, and let $f(x) = 1 + x^2$. What is the maximum domain of definition of $f \circ r$ and of $r \circ f$? Are both these functions continuous throughout their domains? Use a computer or a graphic calculator to examine the graphs of $f,\ r \circ f$ and $f \circ r$.

48 Assuming that the sine function is continuous on $\mathbb{R}$ (see chapter 11), determine the largest subset of $\mathbb{R}$ on which the function f of qn 19 is continuous.

49 Use a computer or a graphic calculator to examine the graph of $y = x \sin 1/x$.

A function f is defined on $\mathbb{R}$ by

$$f(x) = x \sin 1/x \text{ when } x \neq 0,$$
$$f(0) = 0.$$

Assuming that the sine function is continuous on $\mathbb{R}$ and that $|\sin x| \leq 1$ for all values of x, prove that f is continuous on $\mathbb{R}$.
(*Hint*: use qns 48, 22 and 26 for $x \neq 0$, and qns 32, 26 and 36 for $x = 0$.)
See the figure for qn 8.21.

50 Let $f(x) = (x-2)(x-3)$ be defined on $\mathbb{R}$ and let r denote the same function as in qn 47.
What is the maximum domain of definition of $r \circ f$? Is this function continuous throughout its domain? Examine the graphs of f and $r \circ f$ on a computer or a graphics calculator.

51 Let $f(x) = a_0 + a_1 x + a_2 x^2 + \ldots + a_n x^n$, be defined on $\mathbb{R}$. Let r denote the same function as in qn 47. What is the maximum domain of definition of $r \circ f$? Is this function continuous throughout its domain?

52 Let $f : A \to \mathbb{R}$ be a real function which is continuous at $x = a$, with $f(a) \neq 0$. Let r denote the same function as in qn 47. Prove that $r \circ f$ is continuous at $x = a$.
The function $r \circ f$ is also commonly denoted by $1/f$. Give a direct proof that $1/f$ is continuous at a using qns 3.65 and 3.66, the reciprocal rule.

53 Prove that the function given by $f(x) = (x^3 - x)/(x^2 + 1)$ is continuous on $\mathbb{R}$. Examine its graph on a computer or a graphics calculator.

54 Let $f : A \to \mathbb{R}$ and $g : A \to \mathbb{R}$ be real functions which are both continuous at a. Let $G = \{x |\, g(x) = 0\}$. If $g(a) \neq 0$, prove that the function $f/g : A \backslash G \to \mathbb{R}$ defined by $(f/g)(x) = f(x)/g(x)$ is continuous at a. Either consider $f \cdot (r \circ g)$ or use qn 3.67, the quotient rule, directly.

55 (*Cauchy*, 1821) The function $f : \mathbb{R} \to \mathbb{R}$ has the property that $f(x+y) = f(x) + f(y)$ for all $x, y \in \mathbb{R}$.

(i) Prove that $f(0) = 0$.
(ii) Prove that $f(-x) = -f(x)$.
(iii) If f is continuous at $x = 0$, prove that f is continuous on $\mathbb{R}$.
(iv) If $f(1) = a$, prove that $f(n) = an$, for $n \in \mathbb{Z}$.
If $n \in \mathbb{N}$, show that $n \cdot f(x/n) = f(x)$.
(v) If $f(1) = a$, prove that $f(p/q) = ap/q$, for $p, q \in \mathbb{Z}, q \neq 0$.
(vi) If $f(1) = a$ and f is continuous on $\mathbb{R}$, prove that $f(x) = ax$.

Summary: Continuity by sequences

The sequence definition of continuity

qn 18	A function $f\colon A \to \mathbb{R}$ is said to be *continuous* at $a \in A$ when, for *every* sequence $(a_n) \to a$ with terms in A, $(f(a_n)) \to f(a)$.
Theorem	If the functions $f\colon A \to \mathbb{R}$ and $g\colon A \to \mathbb{R}$ are both continuous at $a \in A$, then
qn 41	(i) $\lvert f \rvert$ is continuous at a;
qn 52	(ii) $1/f$ is continuous at a, provided $f(a) \neq 0$;
qn 23	(iii) $f + g$ is continuous at a;
qn 26	(iv) $f \cdot g$ is continuous at a;
qn 54	(v) f/g is continuous at a, provided $g(a) \neq 0$.
Definition qn 38	If the function f is defined on the range of the function g, then the function of $f \circ g$ is defined by $f \circ g(x) = f(g(x))$
Theorem qn 40	If the function g is continuous at the point a, and the function f is continuous at the point $g(a)$, then the composite function $f \circ g$ is continuous at the point a.
A squeeze rule qn 36	If for three functions f, g and h (i) $f(x) \leq g(x) \leq h(x)$ in a neighbourhood of the point a; (ii) $f(a) = h(a)$; (iii) f and h are continuous at the point a; then the function g is continuous at a.
Theorem qn 21	A constant function is continuous at each point of its domain.
Theorem qn 22	The identity function is continuous at each point of its domain.
Theorem qn 29	A polynomial function is continuous at each point of its domain.

Neighbourhoods

We say that a function f is continuous at a point $x = a$ when '$f(x)$ tends to $f(a)$, as x tends to a', which we may think of as meaning 'as x gets near to a, $f(x)$ gets near to $f(a)$'. So far, we have always described *nearness* using sequences and convergence. Another way to describe nearness is to consider *neighbourhoods* of a point a.

56 Illustrate the sets

$$\{x\colon\ -1 < x < 3\},\ \{x\colon\ -2 < x - 1 < 2\},\ \{x\colon |x - 1| < 2\}$$

on a number line. They are, in fact, identical. Such a set is called a 2-neighbourhood of the point 1. The neighbourhood has centre 1 and radius 2. Compare this claim with qn 3.61.

57 Prove that, if $|x - 2| < 0.2$, then $|x^2 - 4| < 1$.
Illustrate this result on a graph of the function $x \mapsto x^2$.

58 Prove that, if $|x - 3| < 0.1$, then $|x^2 - 9| < 1$.

59 Prove that, if $|x - 2| < 1/3$, then $|1/x - 1/2| < 1/10$.

60 Prove that, if $|x - 2| < 0.02$, then $|\sqrt{x} - \sqrt{2}| < 0.01$.

61 By considering $x = -2$ in qn 57, $x = -3$ in qn 58, $x = 2.4$ in qn 59 and $x = 1.975$ in qn 60, show that none of the implications that you have established in qns 57–60 may be reversed.

Each of these implications takes the form:

if $|x - a| < \delta$, then $|f(x) - f(a)| < \varepsilon$.

Expressed rather crudely, each says that if x is near to a then $f(x)$ is near to $f(a)$.

If we want to pinpoint continuity this way, it matters just *how* near is *near*. We can investigate the kind of connection between δ and ε that is needed for continuity by examining a point of discontinuity.

62 Let $f(x) = \lfloor x \rfloor$, the integer function, and let $a = 2$.
If $|x - 2| < 2$, can you find an ε such that $|f(x) - f(2)| < \varepsilon$?
If $|x - 2| < 1$, can you find an ε such that $|f(x) - f(2)| < \varepsilon$?
If $|x - 2| < \frac{1}{2}$, can you find an ε such that $|f(x) - f(2)| < \varepsilon$?
If $|x - 2| < \frac{1}{4}$, can you find an ε such that $|f(x) - f(2)| < \varepsilon$?

The fact that we can find an ε every time in qn 62 tells us nothing whatever about the continuity of f.
If we turn the question round, it begins to bite.

63 Let $f(x) = \lfloor x \rfloor$ as before.
Can you find a δ such that, if $|x - 2| < \delta$, then $|f(x) - f(2)| < 2$?
Can you find a δ such that, if $|x - 2| < \delta$, then $|f(x) - f(2)| < 1$?
Can you find a δ such that, if $|x - 2| < \delta$, then $|f(x) - f(2)| < \frac{1}{2}$?
Can you find a δ such that, if $|x - 2| < \delta$, then $|f(x) - f(2)| < \frac{1}{4}$?

With our success in finding εs in qn 62 and our failure to find δs in qn 63, it looks as if we can pin down continuity at $x = a$ by saying

DEFINITION OF CONTINUITY BY NEIGHBOURHOODS
f is continuous at $x = a$,
provided that, whatever positive ε we choose,
we can find a δ such that
$|x - a| < \delta$ implies $|f(x) - f(a)| < \varepsilon$.

When we write $|x - a| < \delta$, we think 'x is near to a', and when we write $|f(x) - f(a)| < \varepsilon$, we think '$f(x)$ is near to $f(a)$'. We first of all see whether this holds in cases that we know are continuous as in qns 64(i) and (ii).

64 (i) Find a δ such that $|x - 3| < \delta$ implies $|x^2 - 9| < 0.5$.
(ii) Find a δ such that $|x - 3| < \delta$ implies $|1/x - 1/3| < 1/6$.
(iii) Find a δ such that $|x - 3| < \delta$ implies $|\sqrt{x} - \sqrt{3}| < 0.01$.

Notice that in each case here, when a satisfactory δ has been found, any lesser (but still positive) δ will also serve.

65 (i) If $|x - 3| < \delta < 1$, prove that $|\sqrt{x} - \sqrt{3}| < \delta/(\sqrt{2} + \sqrt{3})$.
(ii) Given $\varepsilon > 0$, find a δ such that $|x - 3| < \delta \Rightarrow |\sqrt{x} - \sqrt{3}| < \varepsilon$.

Question 65(ii) establishes that the neighbourhood definition of continuity holds for the function $x \to \sqrt{x}$ at the point $x = 3$. The next question uses this to show that the sequential definition of continuity holds for this function at this point. That is, the neighbourhood definition of continuity implies the sequential definition.

66 We seek to prove that for any sequence of positive terms $(a_n) \to 3$, we have $(\sqrt{a_n}) \to \sqrt{3}$.

(i) Write down what is meant by $(\sqrt{a_n}) \to \sqrt{3}$, according to the standard definition.
(ii) Write down what is guaranteed by the neighbourhood definition of continuity of $x \to \sqrt{x}$ at $x = 3$.
(iii) Now use the standard definition of $(a_n) \to 3$, and (ii), to establish (i).

So we know that the 'neighbourhood definition of continuity' fails in a case (qn 63) where we know the function is not continuous and we have found the neighbourhood definition of continuity holds in two cases (qns 64(i) and 64(ii)) where we know the function is continuous. And in qn 66 we have used the neighbourhood definition of continuity to establish the sequence definition in the case of $f(x) = \sqrt{x}$ at $x = 3$. Can we show that this neighbourhood definition of continuity is equivalent to the sequence definition that we have used in the first half of this chapter?

67 Suppose that the neighbourhood definition of continuity holds for a function f and a point $x = a$ in the domain of the function. Let $(a_n) \to a$ be a sequence in the domain of the function. Can we show that the sequence $(f(a_n)) \to f(a)$?

(i) Write down what is meant by $(f(a_n)) \to f(a)$ according to the standard definition of convergence, qn 3.60.
(ii) Use the neighbourhood definition of continuity to find a neighbourhood of a, such that $|f(a_n) - f(a)| < \varepsilon$, when a_n is in the neighbourhood.
(iii) Use the convergence of (a_n) to show that terms of the sequence (a_n) eventually lie in the neighbourhood you found in part (ii).

Question 67 establishes that neighbourhood continuity at a point implies sequence continuity at that point. Can we show that sequence continuity implies neighbourhood continuity? The sequence definition of continuity considers a collection of sets of points $\{a_n\}$, clustering about a, in the domain of the function f. Although all the sequences converge to a, the definition seems to leave in doubt the question as to whether, if f is sequentially continuous at a, given $\varepsilon > 0$, there must be a δ-neighbourhood of a such that, *whenever* x belongs to that neighbourhood, $f(x)$ is in an ε-neighbourhood of $f(a)$.

To clarify this, we suppose that sequence continuity holds and neighbourhood continuity fails, for some ε, so that every neighbourhood of a contains rogue points x, for which $|f(x) - f(a)| \geq \varepsilon$, and we obtain a contradiction.

68 Suppose that f is a real function which is sequentially continuous at a, and that, for some $\varepsilon > 0$, there is no δ-neighbourhood of a such that, for all x in the neighbourhood, $|f(x) - f(a)| < \varepsilon$.

(i) Why must there be a value of x within the neighbourhood $\{x \mid a - 1/n < x < a + 1/n\}$ such that $|f(x) - f(a)| \geq \varepsilon$?
(ii) If the value of x in (i) is called a_n, how do you know that $(a_n - a)$ is a null sequence?
(iii) With the notation of (i) and (ii), how do you know that the sequence $(f(a_n))$ does *not* tend to $f(a)$? This contradicts our starting point.

The contradiction obtained in qn 68 leads to the conclusion that, if the function f is sequentially continuous at a, then f is neighbourhood continuous at a.

Neighbourhood continuity is in fact used by many authors as the basic definition of continuity. In contrast to the sequence definition, where the infinite process is obvious, the $\varepsilon - \delta$ neighbourhood definition of continuity looks deceptively finite. The infinite process is hidden in the fact that the condition must hold for *any* ε, which must be thought to take an infinity of values, and in particular to get arbitrarily small.

There are two simple properties (that will be needed in the next chapter) which hold for a function continuous at a point, which are easily established with the neighbourhood definition of continuity.

69 (*Riemann*, 1851) If a function f is continuous at a point a, by choosing $\varepsilon = 1$, prove that $f(x)$ is bounded (above *and* below) in some neighbourhood of a.

70 (*Bolzano*, 1817) If a function f is continuous at a point a and $f(a) > 0$, by a judicious choice of ε, prove that $f(x) > 0$ in some neighbourhood of a. State and prove an analogous result in case $f(a) < 0$.

The last question in this section (qn 72) is designed to expose one of the most counter-intuitive possibilities that follows from the definition of continuity, namely that the points at which a function is continuous and the points at which a function is not continuous may be densely packed together in the domain of the function.

As a preparation for a new idea to be used in qn 72 we offer qn 71 as an option.

(71) Find a neighbourhood of $\sqrt{2}$ which contains no numbers of the form $n/2$ for any integer n.
Find a neighbourhood of $\sqrt{2}$ which contains no numbers of the form $n/3$ for any integer n.
Find a neighbourhood of $\sqrt{2}$ which contains no numbers of the form $n/4$ for any integer n.
Find a neighbourhood of $\sqrt{2}$ which contains no numbers of the form $n/5$ for any integer n.
Find a neighbourhood of $\sqrt{2}$ which contains no numbers of the form $n/2$, $n/3$, $n/4$ or $n/5$ for any integer n.

72 (*Thomae*, 1875) A function $f\colon (0, 1] \to \mathbb{R}$ is defined by

$$f(x) = \begin{cases} 0 & \text{when } x \text{ is irrational,} \\ 1/q & \text{when } x \text{ is rational and } x = p/q \text{ in lowest terms.} \end{cases}$$

This function is sometimes called the *ruler function* because of the resemblance between parts of its graph and the marks on a pre-decimal ruler. This 'ruler diagram' illustrates the graph for rational points where q is a power of 2.

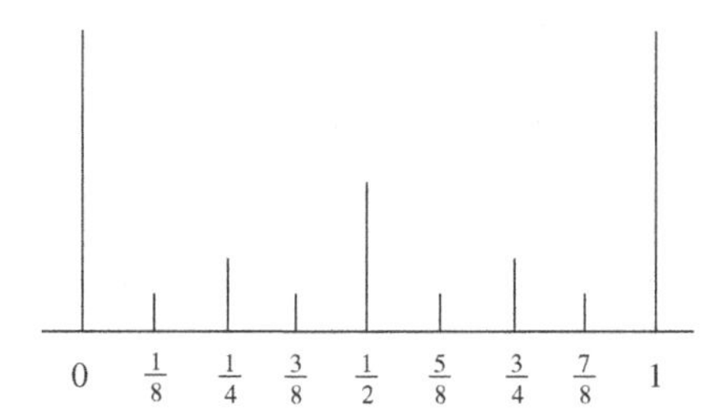

Figure 6.3

(a) By constructing a sequence of irrational numbers tending to each rational point, prove that f is not continuous at any rational point.

No surprises so far. The really astonishing property of this function is that it is *continuous* at every irrational point.

(b) Now let a be an irrational number with $0 < a < 1$.
Given $\varepsilon > 0$, we seek a δ such that
$|x - a| < \delta \Rightarrow |f(x) - f(a)| < \varepsilon$.
  (i) Is there a positive integer m such that $1/m < \varepsilon$?
  (ii) How many rational numbers p/q can there be between 0 and 1 with $q \leq m$? Might any one of them equal a?
  (iii) Is there a shortest distance $|a - p/q|$ for the rational numbers p/q in part (ii)?
  (iv) If we take the shortest distance in (iii) as δ, how big might $|f(x) - f(a)|$ get when $|x - a| < \delta$?
  (v) Does this establish continuity by the neighbourhood definition at each irrational point?

One-sided limits

Definition of one-sided limits by sequences

73 When examining the discontinuities of the integer function $f(x) = \lfloor x \rfloor$, we saw that although the sequence $(1 - 1/n)$ tends to 1, the sequence $(f(1 - 1/n))$ tends to 0 which is not equal to $f(1)$. Construct another sequence (a_n) which tends to 1, but for which the sequence $(f(a_n))$ tends to 0. What property of the sequence (a_n) is necessary for this to be the case?
If $a_n \leq 1$ and $(a_n) \to 1$, must $(f(a_n)) \to 0$?
If $a_n < 1$ and $(a_n) \to 1$, must $(f(a_n)) \to 0$?

Some kinds of discontinuity may be identified and discussed by considering the two sides of the point in question separately.

DEFINITION OF ONE-SIDED LIMIT

If $a_n < a$ and $(a_n) \to a$ implies that $(f(a_n)) \to l$, we write
$$\lim_{x \to a^-} f(x) = l,$$

which is read 'as x tends to a from below, $f(x)$ tends to l', and is also written 'as $x \to a^-$, $f(x) \to l$'.

74 What is $\lim_{x \to a^-} \lfloor x \rfloor$ for $a = 1, 2, 0$?

75 Prove that if $f: \mathbb{R} \to \mathbb{R}$ is continuous at a, then $\lim_{x \to a^-} f(x) = f(a)$.

76 Give a formal definition of what is meant by $\lim_{x \to a^+} f(x) = l$, which is read 'as x tends to a from above, $f(x)$ tends to l', and is also written 'as $x \to a^+$, $f(x) \to l$', by strict analogy with limits from below.

77 What is $\lim_{x \to a^+} \lfloor x \rfloor$ for $a = 1, 0, -1$?

78 Prove that if $f: \mathbb{R} \to \mathbb{R}$ is continuous at a, then $\lim_{x \to a^+} f(x) = f(a)$.

79 Prove that if a is a point in the domain of the real function f, and $\lim_{x \to a^+} f(x) \neq \lim_{x \to a^-} f(x)$ then f is not continuous at a.

In this case the function f is said to have a *jump discontinuity*, providing $f(a)$ exists.

Even when
$$\lim_{x \to a^+} f(x) = \lim_{x \to a^-} f(x),$$
and $f(a)$ exists, it is possible for f to be discontinuous at $x = a$, as we will see in qn 82. The discontinuity in this case is said to be *removable*.

One-sided limits for functions can be defined without knowing the value of the function at the point which seems to be in question, and this gives a way of pinpointing discontinuity at a point. The fact that the value of the function at the point in question does not come into the definition means that a one-sided limit may exist at a point even when the function itself does not.

80 *A jump discontinuity*. Find $\lim_{x \to 0^-} |x|/x$ and $\lim_{x \to 0^+} |x|/x$.
Notice that although the function $|x|/x$ is not defined at $x = 0$ the two limits have well-defined values.
If a function $f: \mathbb{R} \to \mathbb{R}$ is defined by $f(x) = |x|/x$ when $x \neq 0$, and $f(0) = k$, use qn 79 to show that f is not continuous at 0 for any choice of k.

There are important limits that cannot be reached which may be interesting in their own right quite apart from questions about continuity.

81 Let f be the function $f : \mathbb{R}\backslash\{0\} \to \mathbb{R}$ given by
$f(x) = \dfrac{\sin x}{x}$.
The function is not defined at $x = 0$, but that does not stop there being a limit from above, or a limit from below, as x tends to 0. Explore this function using a calculator. Remember to have x in radians. Investigate the limit of this function as x tends to 0 from above.
Use a calculator to find $f(x)$ when $x = 1, \frac{1}{2}, \frac{1}{4}, \frac{1}{8}$, and continue to find $f(x)$ when $x = (\frac{1}{2})^n$, for $n = 0, 1, 2, \ldots$ until no further change in value takes place, because of the limits of accuracy of your calculator.
Some calculators have a 'Fix' key, by which the number of displayed decimal places is chosen in advance.

(i) Identify the connection between the 'Fix' key and ε-neighbourhoods by deciding what value of ε makes $f(x) = 1.00$ *correct to two place of decimals* equivalent to $|f(x) - 1| < \varepsilon$.
(ii) Also decide what value of ε makes $f(x) = 1.000$ *correct to three places of decimals* equivalent to $|f(x) - 1| < \varepsilon$.
(iii) So far as you can tell using your calculator, does $0 < x < \frac{1}{2} \Rightarrow |f(x) - 1| < 0.5$?
Can you find a $\delta > \frac{1}{2}$ such that $0 < x < \delta \Rightarrow |f(x) - 1| < 0.5$?
Find a δ such that $0 < x < \delta \Rightarrow |f(x) - 1| < 0.05$.
Find a δ such that $0 < x < \delta \Rightarrow |f(x) - 1| < 0.005$.

In qn 9.27 we will prove that $\lim_{x\to 0^-} f(x) = 1 = \lim_{x\to 0^+} f(x)$, for the function of qn 81 but, despite qn 79, the function f is not continuous at $x = 0$ because it is not defined there.

82 *A removable discontinuity.* By choosing a value for $f(0)$ show that it is possible to have a function defined at $x = 0$ for which
$\lim_{x\to 0^-} f(x) = 1 = \lim_{x\to 0^+} f(x)$,
but which is still not continuous at $x = 0$.

Definition of one-sided limits by neighbourhoods

83 A real function f is defined in a neighbourhood of the point a. If, given any $\varepsilon > 0$, there exists a δ such that, when $a < x < a+\delta$, $|f(x) - l| < \varepsilon$, prove that $\lim_{x\to a^+} f(x) = l$.

The proof may be constructed like that in qn 67.

84 Prove that if

$$\lim_{x\to a^+} f(x) = l,$$

then, given $\varepsilon > 0$, there exists a δ such that, when
$a < x < a + \delta$, $|f(x) - l| < \varepsilon$.
Obtain this proof, by contradiction, by supposing that the sequence definition of limit holds, while the neighbourhood definition fails for some ε, and finding an a_n such that $a < a_n < a + 1/n$ and $|f(a_n) - l| \geq \varepsilon$, and examining the sequences (a_n) and $(f(a_n))$. The idea of this proof is used in qn 68.

In qns 83 and 84 we have established that a convergent sequence definition of limit from above and a neighbourhood definition of limit from above are equivalent.

85 State a neighbourhood definition of limit from below equivalent to the convergent sequence definition of limit from below given after qn 73.

Two-sided limits

Definition of continuity by limits

If both $\lim_{x \to a^-} f(x) = l$ and $\lim_{x \to a^+} f(x) = l$
it is customary to write

$$\lim_{x \to a} f(x) = l$$

or 'as $x \to a$, $f(x) \to l$' and to say 'as x tends to a, $f(x)$ tends to l'.

86 If $f : \mathbb{R} \to \mathbb{R}$ is continuous at a, use qns 75 and 78 to show that

$$\lim_{x \to a} f(x) = f(a).$$

87 Write down the neighbourhood definitions of

$$\lim_{x \to a^-} f(x) = l \text{ and } \lim_{x \to a^+} f(x) = l.$$

Use these definitions to show that, if $\lim_{x \to a} f(x) = l$, then, given $\varepsilon > 0$, there exists a δ such that when x is in the domain of f and $0 < |x - a| < \delta$, $|f(x) - l| < \varepsilon$.
The important point here is recognising that $0 < |x - a| < \delta$ means

$$\textit{either } a - \delta < x < a \textit{ or } a < x < a + \delta$$

and the answer to the question consists of showing how to choose the δ.

88 (i) If, given $\varepsilon > 0$, there exists a δ such that $|f(x) - l| < \varepsilon$ when $0 < |x - a| < \delta$, prove that $\lim_{x \to a} f(x) = l$.
Use neighbourhoods for this proof, not sequences.
(ii) If for every sequence $(a_n) \to a$ in the domain of the function f, with $a_n \neq a$, $(f(a_n)) \to l$, prove that

$$\lim_{x \to a} f(x) = l.$$

89 If, given $\varepsilon > 0$, there exists a δ such that when x belongs to the domain of f and $0 < |x - a| < \delta$, then $|f(x) - f(a)| < \varepsilon$, prove that when x belongs to the domain of f and $|x - a| < \delta$, then

$$|f(x) - f(a)| < \varepsilon.$$

Deduce that if $\lim_{x \to a} f(x) = f(a)$ then f is continuous at a.

Now we have established that a function f is continuous at a point a of its domain if and only if

$$\lim_{x \to a} f(x) = f(a).$$

90 Find a value of k such that the function $f : \mathbb{R} \to \mathbb{R}$ defined by

$$f(x) = \begin{cases} 2x \text{ when } x < 1, \\ k \text{ when } x \geq 1, \end{cases}$$

is continuous at $x = 1$.

91 If the function $f : [a, b] \to \mathbb{R}$ is continuous at each point of its domain, give the value of $\lim_{x \to a^+} f(x)$, $\lim_{x \to b^-} f(x)$ and $\lim_{x \to c} f(x)$ for $c \in (a, b)$.

92 If $\lim_{x \to a^+} f(x) = f(a)$, $\lim_{x \to b^-} f(x) = f(b)$ and $\lim_{x \to c} f(x) = f(c)$, for every $c \in (a, b)$, prove that f is continuous at every point of $[a, b]$. Use qn 89, and consider sequences for the one-sided limits.

Theorems on limits

93 *The algebra of limits*
Use theorems about sequences to prove that if

$$\lim_{x \to a} f(x) = l \text{ and } \lim_{x \to a} g(x) = m,$$

then

$$\lim_{x \to a} |f(x)| = |l|,$$

$$\lim_{x \to a} f(x) + g(x) = l + m,$$

$$\lim_{x \to a} f(x) \cdot g(x) = l \cdot m$$

and, provided $l \neq 0$, $\lim_{x \to a} 1/f(x) = 1/l$.

94 Prove that $\lim_{x \to a} c = c$ and $\lim_{x \to a} x = a$, and use these results to develop theorems about limits of polynomials and rational functions.

95 A function $f : \mathbb{R} \to \mathbb{R}$ is defined by

$$f(x) = x^2, \text{ when } x < 2,$$

$f(2) = k,$

$f(x) = 3x^2 - 18x + 28, \text{ when } x > 2.$

Is there a value of k which will make f continuous at every point?

96 A real function f is continuous on $[a, b]$. A real function g is continuous on $[b, c]$. $a < b < c$ and $f(b) = g(b)$. A function F is constructed by

$$F(x) = \begin{cases} f(x), & a \le x \le b \\ g(x), & b < x \le c. \end{cases}$$

Prove that F is continuous on $[a, c]$.

Thus if two continuous functions are 'contiguous' (coincide at an end point of their respective domains), the function formed by joining them is continuous.

97 A function $f : \mathbb{R} \to \mathbb{R}$ is defined by

$$f(x) = \frac{x^n - a^n}{x - a}, \text{ for } x \neq a, \text{ where } n \text{ is a positive integer;}$$

$f(a) = k.$

Is there a value of k which would make the function f continuous at every point?

98 *A squeeze rule*
If $f(x) \le g(x) \le h(x)$ for all $x \neq a$, prove that

$$|g(x) - l| \le \max(|f(x) - l|, |h(x) - l|), \text{ when } x \neq a.$$

Deduce that if $\lim_{x \to a} f(x) = l$ and $\lim_{x \to a} h(x) = l$, then $\lim_{x \to a} g(x) = l$.

There is one important theorem about continuous functions which does not carry over into a theorem about limits. If the function g is continuous at the point a and the function f is continuous at the point $g(a)$, then the function $f \circ g$ defined by $(f \circ g)(x) = f(g(x))$ is continuous at the point a, the composite rule for continuous functions.

This is equivalent to the proposition about limits that if

$$\lim_{x \to a} g(x) = g(a) \text{ and } \lim_{y \to g(a)} f(y) = f(g(a))$$

then

$$\lim_{x \to a} f(g(x)) = f(g(a)).$$

However, if

$$\lim_{x \to a} g(x) = m \text{ and } \lim_{y \to m} f(y) = l,$$

it does not necessarily follow that

$$\lim_{x \to a} (f \circ g)(x) = l.$$

99 Consider the functions f and g defined by $g(x) = 0$ for all x, and $f(x) = 1$ when $x \neq 0$ and $f(0) = 0$.
Let $a = 0$, then $m = 0$ and $l = 1$ with the notation of the previous paragraph.
Determine the function $f \circ g$ and illustrate the difficulty of applying theorems about limits to composite functions.

Limits as $x \to \infty$ and when $f(x) \to \infty$

100 Use a computer or a graphics calculator to examine the graph of

$$f(x) = \frac{x^2}{x^2 + 1}.$$

Construct a formal definition for

$$\lim_{x \to +\infty} f(x) = l \text{ and } \lim_{x \to -\infty} f(x) = l.$$

Give alternative definitions using sequences or neighbourhoods.

101 Use a computer or a graphics calculator to examine the graph of

$$f(x) = \frac{1}{x^2 - 4}.$$

Construct formal definitions for

$$\lim_{x \to a^+} f(x) = +\infty, \ \lim_{x \to a^+} f(x) = -\infty,$$
$$\lim_{x \to a^-} f(x) = +\infty, \ \lim_{x \to a^-} f(x) = -\infty.$$

Summary: Continuity by neighbourhoods and limits

Theorem	*The neighbourhood definition of continuity*
qns 67, 68	A function $f: A \to \mathbb{R}$ is continuous at $a \in A$ if and only if, given $\varepsilon > 0$, there exists a δ such that when $x \in A$ and $\lvert x - a \rvert < \delta$, then $\lvert f(x) - f(a) \rvert < \epsilon$.
Theorem qn 69	If a function f is continuous at a point a, then it is bounded in some neighbourhood of a.
Theorem qn 70	If a function is continuous at a point a and $f(a) > 0$, then f is positive in some neighbourhood of a.

Definition qns 73, 76 — *One-sided limits*
If $(f(a_n)) \to l$ whenever $(a_n) \to a$ with $a_n < a$, then we write $\lim_{x \to a^-} f(x) = l$, and say that $f(x)$ tends to l as x tends to a from below.
If $(f(a_n)) \to l$ whenever $(a_n) \to a$ with $a_n > a$, then we write $\lim\limits_{x \to a^+} f(x) = l$, and say that $f(x)$ tends to l as x tends to a from above.

Theorem qns 83, 84, 85 — $\lim_{x \to a^-} f(x) = l$ if and only if, given $\varepsilon > 0$, there exists a δ such that when x is in the domain of f and $a - \delta < x < a$, then $|f(x) - l| < \varepsilon$.
$\lim_{x \to a^+} f(x) = l$ if and only if, given $\varepsilon > 0$, there exists a δ such that when x is in the domain of f and $a < x < a + \delta$, then $|f(x) - l| < \varepsilon$.

Definition — *Two-sided limits*
$\lim_{x \to a} f(x) = l$
when both $\lim_{x \to a^-} f(x) = l$ and $\lim_{x \to a^+} f(x) = l$.

Theorem qn 88 — $\lim_{x \to a} f(x) = l$ if and only if, given $\varepsilon > 0$, there exists a δ such that when x is in the domain of f and $0 < |x - a| < \delta$, then $|f(x) - l| < \varepsilon$.

Theorem qns 75, 78, 89 — *The definition of continuity by limit*
The function f is continuous at a if and only if $\lim_{x \to a} f(x) = f(a)$.

Theorem qn 93 — If $\lim_{x \to a} f(x) = l$ and $\lim_{x \to a} g(x) = m$, then

(i) $\lim\limits_{x \to a} |f(x)| = |l|$;
(ii) $\lim\limits_{x \to a} f(x) + g(x) = l + m$,
(iii) $\lim\limits_{x \to a} f(x) \cdot g(x) = l \cdot m$,
(iv) $\lim\limits_{x \to a} 1/f(x) = 1/l$, provided $l \neq 0$.

Theorem qn 98 — If for three functions f, g and h
(i) $f(x) \leq g(x) \leq h(x)$ except possibly when $x = a$,
(ii) $\lim\limits_{x \to a} f(x) = l$ and $\lim\limits_{x \to a} h(x) = l$,
then $\lim\limits_{x \to a} g(x) = l$.

The completeness principle has not been used in this chapter, so all the definitions and theorems here are valid just with Archimedean order.

Historical Note

Throughout the seventeenth and eighteenth centuries, and for many mathematicians well on into the nineteenth century, the only functions which were under consideration were polynomials, rational functions, trigonometric and exponential functions, and combinations of these. Although the verbal definitions of a function used during the eighteenth century sound like modern definitions, only functions defined by a formula were under consideration. Such functions are continuous and infinitely differentiable except possibly at isolated points (e.g. at $x = 0$ for the function $x \mapsto 1/x$). Even continuous functions given by different formulae on different segments of their domain, such as the equation of a plucked string, were only admitted with discomfort.

It was the consideration of limiting processes by Fourier and Cauchy, among others, which began to widen the field of functions worthy of consideration. Following his study of series of trigonometric functions, Fourier insisted (1822) that the values of a function need not be smoothly connected. In 1829 Dirichlet proposed a function like that of qn 20 in his discussion of Fourier series and integration, and it was in 1837 that Dirichlet gave what amounts to the modern definition of function in which each value of x gives a unique value of $f(x)$, which need not depend in any way on the value given for a different x, and the function is defined by the pairing $(x, f(x))$.

The notation $f(x)$ or $\phi(x)$ dates from Euler (1734) though separating the variable in brackets only became standard practice after Cauchy (1821).

Until the early years of the twentieth century a real function might take the value ∞, so that, for example, if $f(x) = 1/x$, $f(0) = \infty$. In 1820, Herschel used the symbol f^{-1} for the inverse function, but this did not become standard practice until the twentieth century. Describing functions in terms of their domains and co-domains followed the set-theoretic formulation of mathematics in the early years of the twentieth century. The terms injection, surjection and bijection were devised by the twentieth-century corporate French mathematician Bourbaki.

It has always been accepted that an equation like $x^2 + y^2 = 1$ determines a functional relationship between y and x. But a given value of x between ± 1 gives two possible values of y, since $y = \pm\sqrt{(1 - x^2)}$. Such *many-valued* functions appear in the literature right through the nineteenth century, and functions were taken to be *single-valued* only when they were said to be so. However, for the purposes of defining continuity, limits, and inverse functions, single-valued functions are essential and so during the twentieth century *functions* have come to mean *single-valued* functions, though this convention is still not universal.

In order to formulate the Intermediate Value Theorem both Bolzano (1817) and Cauchy (1821) needed a definition of continuity, and said that f was continuous when $f(x + h) - f(x)$ became small with h. Their definitions are recognisable as our neighbourhood definition of continuity though they only considered functions which were continuous at least on an interval, not just at an isolated point. For his proof of the Intermediate Value Theorem, Bolzano needed the result that a function which is continuous and takes a non-zero value at a point must be non-zero in a neighbourhood of that point. The first $\varepsilon - \delta$ description of a limit appears in Cauchy's description of derivative (1823) without symbolic use of absolute values or inequalities. The

$\varepsilon - \delta$ definition of limit was formulated by Weierstrass using the modern notation for absolute value, which he introduced, in his first lectures in Berlin (1859). The equivalence of the neighbourhood definition and the sequential definition was established by Cantor in 1871 and the sequential definition of continuity was proposed by Cantor and Heine in 1872. In 1872, Heine published the definition of continuity *at a point*. It was Weierstrass who really paved the way for the modern study of continuous functions by his construction of an everywhere continuous and nowhere differentiable function, which appeared in print in 1872. In 1872, Weierstrass said that Riemann had proposed in 1861 that the function $\sum \frac{\sin n^2 x}{n^2}$ was continuous everywhere, but differentiable nowhere. It is in fact differentiable at some isolated points. Until Riemann and Weierstrass, considerations of continuity had not been clearly distinguished from those of differentiability. In fact, Bolzano had found a continuous function which was not differentiable on a dense set by 1834, but this was not published until the twentieth century. In 1851 Riemann proved that a continuous function was bounded in a neighbourhood of one of its points.

The use of 'lim' when referring to limits dates back to L'Huilier in 1786. Cauchy, in speaking of limits as x tends to 0, said that lim sin x had one value, lim $1/x$ had two values, and lim sin $1/x$ had an infinity of values. The study of limits from above and from below had been stimulated by considerations of Fourier series, and in 1837 Dirichlet proposed the notation $f(a+0)$ and $f(a-0)$ for $\lim_{x \to a^+} f(x)$ and $\lim_{x \to a^-} f(x)$, and his notation was used by many authors through the nineteenth century. Weierstrass wrote $\lim_{x=a} f(x)$, and the replacing of the equals sign by an arrow, which is now universal, comes from a suggestion of the British mathematician J. G. Leathem in 1905.

Answers

2 (i) $\mathbb{R}^+ \cup \{0\}$, (ii) $\mathbb{R}$, (iii) $[-1, 1]$, (iv) $(0, 1]$, (v) $\mathbb{R}^+$.

3 (i) $\mathbb{R}^+ \cup \{0\}$, (ii) $[-1, 1]$, (iii) $|x| \geq 1$, (iv) $\mathbb{R}\backslash\{0\}$, (v) $\mathbb{R}^+$, (vi) $\mathbb{R}\backslash\{2\}$.

4 2 (ii) and (v). 3 (i), (iv), (v) and (vi).

5 2 (i) $\mathbb{R}^+\cup\{0\} \to \mathbb{R}^+\cup\{0\}$,
(ii) $\mathbb{R} \to \mathbb{R}$,
(iii) $[-\frac{1}{2}\pi, \frac{1}{2}\pi] \to [-1, 1]$,
(iv) $\mathbb{R}^+ \cup \{0\} \to (0, 1]$,
(v) $\mathbb{R} \to \mathbb{R}^+$.

3 (i) $\mathbb{R}^+ \cup \{0\} \to \mathbb{R}^+ \cup \{0\}$,
(ii) $[0, 1] \to [0, 1]$,
(iii) $[1, \infty) \to [0, \infty)$,
(iv) $\mathbb{R}\backslash\{0\} \to \mathbb{R}\backslash\{0\}$,
(v) $\mathbb{R}^+ \to \mathbb{R}$,
(vi) $\mathbb{R}\backslash\{2\} \to \mathbb{R}\backslash\{4\}$.

6 Points on the graph of f^{-1} have the form $(b, f^{-1}(b))$, which if $b = f(a)$ have the form $(f(a), a)$. See Fig. 6.4.

7 (i) See Fig. 6.10. (iii) and (iv) are identical.
See Fig. 6.5.

8 $A = (-1, 1]$, $f(A) = \{0, 1\}$. See Fig. 6.6.

9 $A = \mathbb{R}\backslash\{-1\}$, $f(A) = \{-1, 0, 1\}$. See Fig. 6.7.

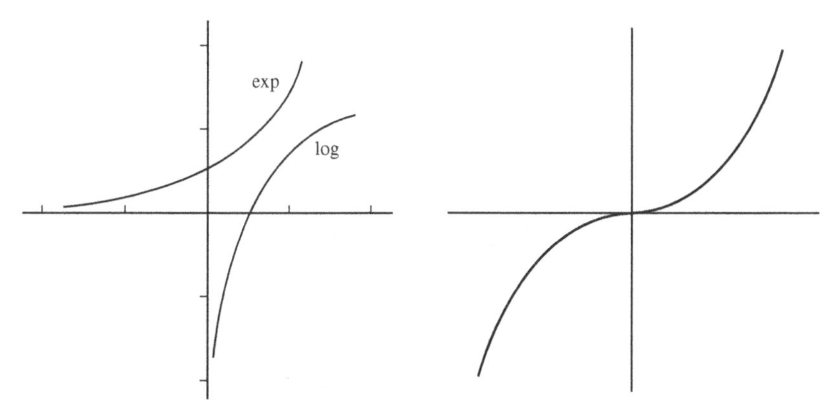

Figure 6.4

Figure 6.5

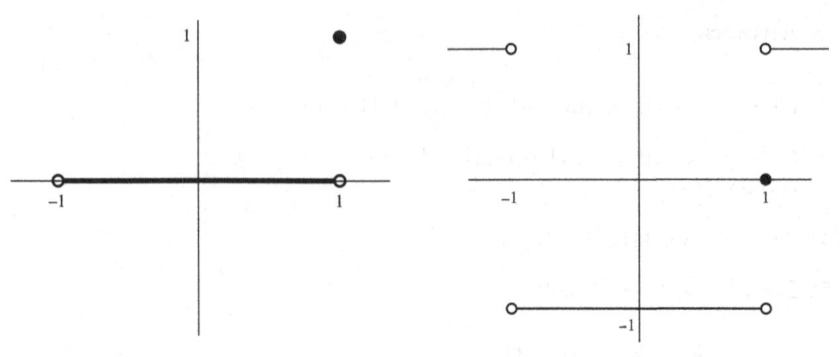

Figure 6.6

Figure 6.7

10 $f(\mathbb{R}) = \{0, 1\}$. See Fig. 6.8.

11 $f(\mathbb{Z}) = \{1\}$, $f(\mathbb{R}\backslash\mathbb{Z}\} = \{0\}$.

13 0.

14 -1.

15 $a_n = 1 + 1/n$, $b_n = 1 - 1/n$.

16 If $(a_n) \to \frac{1}{2}$, taking $\varepsilon = \frac{1}{2}$, $0 < a_n < 1$ eventually, so $\lfloor a_n \rfloor = 0$ eventually.

17 If $(a_n) \to 1\frac{3}{4}$, taking $\varepsilon = \frac{1}{4}$, $1\frac{1}{2} < a_n < 2$ eventually, so $\lfloor a_n \rfloor = 1$ eventually.

18 Not continuous on $\mathbb{Z}$. Continuous on $\mathbb{R}\backslash\mathbb{Z}$.

19 (i), (ii) and (iii) are null. (iv) $\to$ 1. For all null sequences (a_n) there is no single value of $f(0)$ for which $(f(a_n)) \to f(0)$ for continuity.

20 Let a be rational: $(a_n) \to a$, but $f(a_n) = 0$, so $(f(a_n)) \to 0 \neq 1 = f(a)$.
Let a be irrational: $(\lfloor 10^n a \rfloor / 10^n) \to a$, but $f(\lfloor 10^n a \rfloor / 10^n) = 1$ for all n, so $(f(\lfloor 10^n a \rfloor / 10^n)) \to 1 \neq 0 = f(a)$.

21 Let $(a_n) \to a$. $f(a_n) = c$, so $(f(a_n)) \to c = f(a)$.

22 Let $(a_n) \to a$. $f(a_n) = a_n$, so $(f(a_n)) = (a_n) \to a = f(a)$.

24 By qn 21, $g: x \mapsto 1$ is continuous at every point.
By qn 22, $f: x \mapsto x$ is continuous at every point.
So, by qn 23, $f + g: x \mapsto x + 1$ is continuous at every point.

27 (i) 21, 22, 26. (ii) 21, (i), 23. (iii) 22, 26. (iv) 22, 26, 26. (v) 21, 22, 26, 26, 21, 23.

28 Qn 22 gives basis for induction. Qn 26 gives the inductive step.

29 Repeated use of qns 21, 28, 26 and 25.

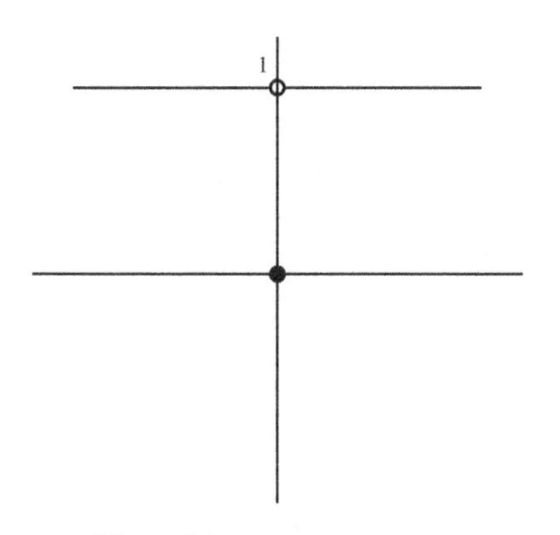

Figure 6.8

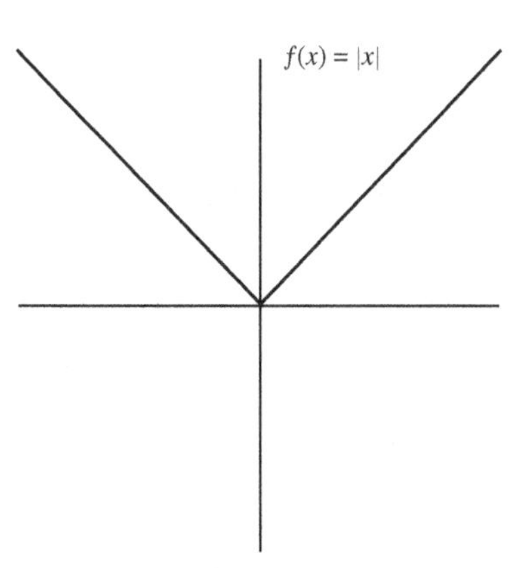

Figure 6.9

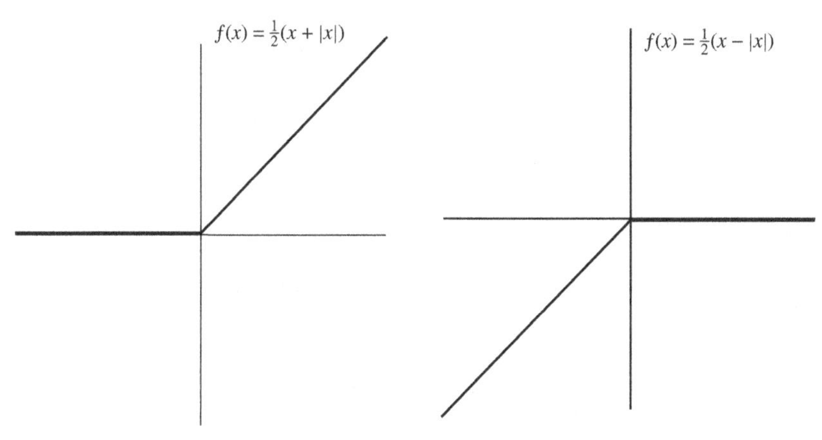

Figure 6.10

Figure 6.11

30 $\mathbb{R}\backslash\{0\}$. Let $(a_n) \to a \neq 0$, then, by qns 3.65 and 3.66, the reciprocal rule, $(f(a_n)) = (1/a_n) \to 1/a = f(a)$.

31 $f(x) = a_0 + a_1 x + a_2 x^2 + \ldots + a_n x^n + b_1/x + b_2/x^2 + \ldots + b_m/x^m$.

32 See Fig. 6.9.

33 See Fig. 6.10. Use qns 22, 32, 23 and 26.

34 See Fig. 6.11. Proof as qn 33.

35 $-|x| \leq x \leq |x| \Rightarrow -\frac{1}{2}|x| \leq \frac{1}{2}x \leq \frac{1}{2}|x| \Rightarrow \frac{1}{2}x - \frac{1}{2}|x| \leq x \leq \frac{1}{2}x + \frac{1}{2}|x|$.
Also $-\frac{1}{2}|x| \leq \frac{1}{2}x \leq \frac{1}{2}|x| \Rightarrow \frac{1}{2}x - \frac{1}{2}|x| \leq 0 \leq \frac{1}{2}x + \frac{1}{2}|x|$.
$\frac{1}{2}(x - |x|) = \frac{1}{2}(x + |x|) \Rightarrow x = 0$.

36 Let $(a_n) \to a$, then $(f(a_n)) \to f(a)$ and $(h(a_n)) \to h(a) = f(a)$.
So $(g(a_n)) \to f(a) = g(a) = h(a)$ from qn 3.54(viii), the squeeze rule.

37 In qn 36, take $f(x) = \frac{1}{2}(x - |x|)$, $g = f$ in qn 35, and $h(x) = \frac{1}{2}(x + |x|)$. By qns 33 and 34, f and h are continuous. By qns 35 and 36, g is continuous at 0.
Let $a \neq 0$ be a rational number. If $a_n = a + \sqrt{2}/n$, $g(a_n) = a_n$, so $(g(a_n)) = (a_n) \to a \neq 0 = g(a)$, so g is not continuous at a.
Let a be an irrational number, then $(\lfloor 10^n a \rfloor / 10^n) \to a$. But $g(\lfloor 10^n a \rfloor / 10^n) = 0$, so $(g(\lfloor 10^n a \rfloor / 10^n)) \to 0 \neq a = g(a)$. So g is not continuous at a.

38 $x \mapsto \sin(x^2)$. $x \mapsto (\sin x)^2$.

39 See Fig. 6.12.

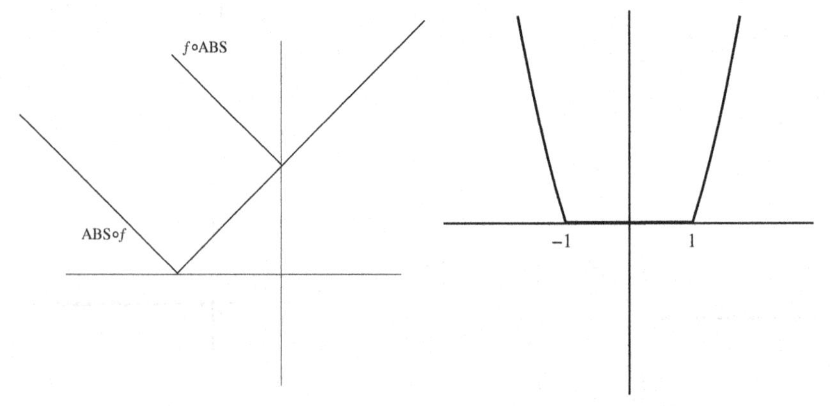

Figure 6.12

Figure 6.13

40 Let $(a_n) \to a$, then by the continuity of g at a, $(g(a_n)) \to g(a)$, so by the continuity of f at $g(a)$, $(f(g(a_n))) \to f(g(a))$.

41 ABS is continuous from qn 32.

42 f^+ is continuous by qns 41, 23, 26. See Fig. 6.13.

43 f^- is continuous by qns 41, 26, 23, 26. See Fig. 6.14.

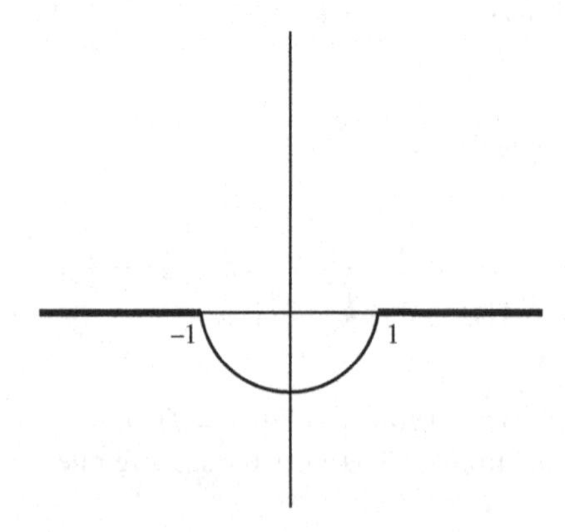

Figure 6.14

44 Consider the cases $b \leq a$ and $a < b$ separately.

45 $\max(f, g)(x) = \frac{1}{2}(f(x) + g(x)) + \frac{1}{2}|f(x) - g(x)|$. Use qns 21, 23, 26, 32, 41.

46 ABS.

47 Maximum domain for $f \circ r$ is $\mathbb{R}\backslash\{0\}$; for $r \circ f$ is $\mathbb{R}$. Both composite functions are continuous at every point by qns 29, 30 and 40.

48 Qn 19 shows discontinuity at 0. When $x \neq 0$, the function is sine $\circ\, r$ which is continuous on $\mathbb{R}\backslash\{0\}$.

50 Maximum domain for $r \circ f$ is $\mathbb{R}\backslash\{2, 3\}$. The composite function is continuous at every point of its domain, by qns 29, 30, 40.

51 Maximum domain for $r \circ f$ is $\mathbb{R}\backslash A$, where $A = \{x | f(x) = 0\}$. Continuous, by qns 29, 30, 40.

52 Since r is continuous at $f(a)$, $r \circ f$ is continuous at a, by qn 40.
If $(a_n) \to a$ then $(f(a_n)) \to f(a)$ since f is continuous at a. Now, by qns 3.65 and 3.66, the reciprocal rule, $(1/f(a_n)) \to 1/f(a)$ since $f(a) \neq 0$.
Thus $1/f$ is continuous at a.

53 If $g(x) = x^3 - x$ and $h(x) = x^2 + 1$, the function is $g \cdot (r \circ h)$.

54 If $(a_n) \to a$ is a sequence in $A\backslash G$ then $(a_n) \to a$ is a sequence in A so
$(f(a_n)) \to f(a)$ since f is continuous at a, and
$(g(a_n)) \to g(a)$ since g is continuous at a.
Also $(f(a_n)/g(a_n)) \to f(a)/g(a)$
since $g(a) \neq 0$ by qn 3.67, the quotient rule. So f/g is continuous at a.

55
(i) $f(x) = f(x + 0) = f(x) + f(0) \Rightarrow f(0) = 0$.
(ii) $0 = f(0) = f(x - x) = f(x) + f(-x) \Rightarrow f(-x) = -f(x)$.
(iii) Let $(a_n) \to a$. Then $(a_n - a)$ is a null sequence, so
$(f(a_n - a)) \to f(0)$, since f is continuous at 0.
So $(f(a_n) - f(a)) \to 0$, and so $(f(a_n)) \to f(a)$, which establishes continuity on $\mathbb{R}$.
(iv) $f(1) = a \Rightarrow f(1 + 1) = a + a$, so by induction $f(n) = na$ for $n \in \mathbb{N}$.
Now use (i) and (ii). $2 \cdot f(x/n) = f(2x/n)$ etc.
(v) $q \cdot f(p/q) = f(q(p/q)) = f(p) = pa \Rightarrow f(p/q) = (p/q)a$.
(vi) Already proved that $f(x) = ax$ for rational x. Suppose x is irrational.
$(\lfloor 10^n x \rfloor/10^n) \to x$. But f is continuous so
$(f(\lfloor 10^n x \rfloor/10^n)) \to f(x)$. Now $f(\lfloor 10^n x \rfloor/10^n) = a\lfloor 10^n x \rfloor/10^n$ and
$(a\lfloor 10^n x \rfloor/10^n) \to ax$. So $f(x) = ax$.

57 $1.8 < x < 2.2 \Rightarrow 3.24 < x^2 < 4.84 \Rightarrow 3 < x^2 < 5$.
Neither of these implications may be reversed.

58 $2.9 < x < 3.1 \Rightarrow 8.41 < x^2 < 9.61 \Rightarrow 8 < x^2 < 10$.

59 $5/3 < x < 7/3 \Rightarrow 3/7 < 1/x < 3/5 \Rightarrow 4/10 < 1/x < 6/10$.

60 $1.98 < x < 2.02 \Rightarrow 1.407 < \sqrt{x} < 1.422 \Rightarrow -0.008 < \sqrt{x} - \sqrt{2} < 0.008$.

62 Any $\varepsilon > 2$ for the first question.
Any $\varepsilon > 1$ for the second, third and fourth.

63 $|x - 2| < 1 \Rightarrow |\lfloor x \rfloor - 2| < 2$.
There is no δ such that $|x - 2| < \delta \Rightarrow |\lfloor x \rfloor - 2| < 1$ or $\frac{1}{2}$ or $\frac{1}{4}$.

64 (i) Any $\delta \leq \sqrt{9.5} - 3$, for example 0.08.
(ii) Any $\delta \leq 1$.
(iii) Any $\delta \leq (\sqrt{3} - 0.01)^2$, for example 0.03.

65 (i) $|x - 3| = |\sqrt{x} - \sqrt{3}| \cdot |\sqrt{x} + \sqrt{3}| < \delta$
$\Rightarrow |\sqrt{x} - \sqrt{3}| < \delta / |\sqrt{x} + \sqrt{3}| < \delta / (\sqrt{2} + \sqrt{3})$,
since $\sqrt{2} < \sqrt{x}$.
(ii) Take $\delta \leq \min\{\varepsilon(\sqrt{2} + \sqrt{3}),\ 1\}$.

66 (i) Given $\varepsilon > 0$, there exists an N such that
$n > N \Rightarrow |\sqrt{a_n} - \sqrt{3}| < \varepsilon$.
(ii) Given $\varepsilon > 0$, there exists a δ such that
$|x - 3| < \delta \Rightarrow |\sqrt{x} - \sqrt{3}| < \varepsilon$.
(iii) Given $\delta > 0$, there exists an N such that $n > N \Rightarrow |a_n - 3| < \delta$,
which by (ii) $\Rightarrow |\sqrt{a_n} - \sqrt{3}| < \varepsilon$.

67 (i) Given $\varepsilon > 0$, there is an N such that $n > N \Rightarrow |f(a_n) - f(a)| < \varepsilon$.
(ii) Given $\varepsilon > 0$, there is a δ such that
$|a_n - a| < \delta \Rightarrow |f(a_n) - f(a)| < \varepsilon$.
(iii) Given $\delta > 0$, there is an N such that $n > N \Rightarrow |a_n - a| < \delta$.

68 (i) $\delta = 1/n$ does not ensure that $|f(x) - f(a)| < \varepsilon$ for all x satisfying $|x - a| < \delta$. So for at least one x in this δ-neighbourhood of a, $|f(x) - f(a)| \geq \varepsilon$.
(ii) $|a_n - a| < 1/n$ ensures that $(a_n - a)$ is a null sequence by qns 3.34, the squeeze rule, and 3.33, the absolute value rule.
(iii) $(a_n) \to a$. But $|f(a_n) - f(a)| \geq \varepsilon$ means $(f(a_n))$ does not tend to $f(a)$.

69 There is a δ such that $|x - a| < \delta \Rightarrow |f(x) - f(a)| < 1$

$$\Rightarrow f(a) - 1 < f(x) < f(a) + 1.$$

70 Choose $\varepsilon = \frac{1}{2} f(a)$.
There is a δ such that $|x - a| < \delta \Rightarrow |f(x) - f(a)| < \frac{1}{2} f(a)$

$$\Rightarrow \tfrac{1}{2} f(a) < f(x) < 1\tfrac{1}{2} f(a).$$

When $f(a) < 0$, choose $\varepsilon = -\frac{1}{2} f(a)$.

71 $|x - \sqrt{2}| < \delta$, where $\delta_2 = |3/2 - \sqrt{2}|$, $\delta_3 = |4/3 - \sqrt{2}|$, $\delta_4 = |6/4 - \sqrt{2}|$, $\delta_5 = |7/5 - \sqrt{2}|$, $\delta = \min(\delta_2, \delta_3, \delta_4, \delta_5)$.

72 (a) If $a \in (0, 1]$ is rational and $a_n = a - \sqrt{2}/n$, then (a_n) is eventually in the domain of the function and $f(a_n) = 0$. So $(a_n) \to a$ and $(f(a_n)) \to 0 \neq f(a)$.

(b) (i) By Archimedean order there is an integer $m > 1/\varepsilon$.

(ii) For each q, there are q possible values for p and so not more than $1 + 2 + \ldots + m = \frac{1}{2}m(m+1)$ rational numbers between 0 and 1 with denominator $\leq m$.

(iii) Of the $\frac{1}{2}m(m+1)$ or less numbers, $|a - p/q|$, none of which are zero, there is a least.

(iv) If x is irrational, then $|f(x) - f(a)| = 0$.
If x is rational and $|x - a| < \delta$, then $x = p/q$ with $q > m$.
So $|f(x) - f(a)| = f(x) = 1/q < 1/m < \varepsilon$.

73 For example $a_n = 1 - (\frac{1}{2})^n$. Eventually $a_n < 1$.
If $a_n = 1$, then $(a_n) \to 1$, but $(f(a_n)) \to 1$.

74 $0, 1, -1$.

75 If for *every* sequence $(a_n) \to a$, $(f(a_n)) \to f(a)$, then for every sequence $(a_n) \to a$ with $a_n < a$, $(f(a_n)) \to f(a)$.

76 Answer in summary.

77 $1, 0, -1$.

78 See answer to qn 75.

79 If f were continuous at a then both limits would equal $f(a)$ from qns 75 and 78.

80 $-1, 1$.

81 (i) $\varepsilon = 0.005$.
(ii) $\varepsilon = 0.0005$.
(iii) Yes. $\delta \leq 1.895$, $\delta \leq 0.551$, $\delta \leq 0.173$.

82 Let $f(x) = 1$ when $x \neq 0$, and let $f(0) = 2$.

83 Let $(a_n) \to a$ where $a < a_n$. Given $\varepsilon > 0$, eventually $a < a_n < a + \delta$, so $|f(a_n) - l| < \varepsilon$ and this proves that $(f(a_n)) \to l$.
So the neighbourhood definition of limit implies the sequence definition of limit.

84 From $a < a_n < a + 1/n$, $(a_n) \to a$ by qn 3.54(viii), the squeeze rule. But $|f(a_n) - l| \geq \varepsilon$ implies that $(f(a_n))$ does not tend to l, which contradicts our hypothesis. So the sequence definition of limit implies the neighbourhood definition of limit.

85 Answer in summary.

87 Given $\varepsilon > 0$, the limit from below says that, for some δ_1,
$a - \delta_1 < x < a \Rightarrow |f(x) - l| < \varepsilon$,
and the limit from above says that, for some δ_2,
$a < x < a + \delta_2 \Rightarrow |f(x) - l| < \varepsilon$.
So if $0 < |x - a| < \min(\delta_1, \delta_2)$, it follows that $|f(x) - l| < \varepsilon$.

88 (i) $a - \delta < x < a \Rightarrow 0 < |x - a| < \delta \Rightarrow |f(x) - l| < \varepsilon$.
$a < x < a + \delta \Rightarrow 0 < |x - a| < \delta \Rightarrow |f(x) - l| < \varepsilon$.
(ii) The two conditions for one-sided limits are obviously satisfied.

89 $0 < |x - a| < \delta$ or $x = a \Leftrightarrow |x - a| < \delta$.
$|f(x) - f(a)| < \varepsilon$ is trivial when $x = a$.
First proposition is equivalent to the limit (qn 88).
Second proposition implies continuity (qn 67).

90 $k = 2$. Calculate limits from above and below $x = 1$ and use qns 79 and 89.

91 $f(a)$ from qn 78, $f(b)$ from qn 75 and $f(c)$ from qn 86.

92 Continuous at c from qn 89.
If (a_n) is any sequence in $[a, c]$ which tends to a, it has either a finite or an infinite number of terms different from a. If it only has a finite number of terms different from a, $(f(a_n))$ is eventually constant with terms equal to $f(a)$, so $(f(a_n)) \to f(a)$. If it has an infinite number of terms different from a, and (a_{n_i}) is the subsequence of such terms, then $(f(a_{n_i})) \to f(a)$ for the resulting subsequence by the one-sided limit. In either case,
$(f(a_n)) \to f(a)$.
Continuity at b follows similarly.

93 First result from qn 3.54(vii), the absolute value rule, second from qn 3.54(iii), the sum rule, third from qn 3.54(vi), the product rule, and the fourth from qns 3.65 and 3.66, the reciprocal rule.

94 The development is like that for continuous functions in qns 21, 22, 23, 25, 26, 28, 29, and then uses qn 93 parts three and four for rational functions. Composite functions must not be used; see qn 98.

95 Use limits from above and below $x = 2$. Take $k = 4$.

96 Since $f(b) = g(b)$, $\lim_{x \to b^-} F(x) = f(b) = g(b) = \lim_{x \to b^+} F(x)$, so F is continuous at b, by qns 87 and 89. For $x \neq b$, any sequence tending to $x \in [a, c]$ is eventually in $[a, b]$ or $[b, c]$ and so F is continuous at b from the continuity of f or g.

97 $x^n - a^n = (x - a)(x^{n-1} + ax^{n-2} + \ldots + a^{n-2}x + a^{n-1})$.
So $\lim_{x \to a} f(x) = na^{n-1} = k$.

98 $f(x) - l \leq g(x) - l \leq h(x) - l$, for $x \neq a$.
If $0 \leq g(x) - l$, then $|g(x) - l| \leq |h(x) - l|$.
If $g(x) - l \leq 0$, then $|g(x) - l| \leq |f(x) - l|$.
Given $\varepsilon > 0$, there is a δ_1 such that
$0 < |x - a| < \delta_1 \Rightarrow |f(x) - l| < \varepsilon$,
and there is a δ_2 such that
$0 < |x - a| < \delta_2 \Rightarrow |h(x) - l| < \varepsilon$.
So, if $0 < |x - a| < \min(\delta_1, \delta_2)$,
$|g(x) - l| \leq \max(|f(x) - l|, |h(x) - l|) < \varepsilon$.
The result follows from qn 87.

99 $f \circ g(x) = f(g(x)) = f(0) = 0$, so $\lim_{x\to 0} f \circ g(x) = 0$.
However, $\lim_{x\to 0} g(x) = 0$ and $\lim_{y\to 0} f(y) = 1$.

100 $\lim_{x\to +\infty} f(x) = l$ means, for any sequence $(a_n) \to +\infty$, $(f(a_n)) \to l$; or, given $\varepsilon > 0$, there exists a number M such that $|f(x) - l| < \varepsilon$ when $x > M$.
$\lim_{x\to -\infty} f(x) = l$ means, for any sequence $(a_n) \to -\infty$, $(f(a_n)) \to l$, or, given $\varepsilon > 0$, there exists a number M such that $|f(x) - l| < \varepsilon$ when $x < M$.

101 $\lim_{x\to a^+} f(x) = +\infty$ means, for all sequences $(a_n) \to a$ with $a < a_n$, $(f(a_n)) \to +\infty$; or, given any number M, there exists a δ such that $M < f(x)$ when $a < x < a + \delta$.
$\lim_{x\to a^+} f(x) = -\infty$ means, for all sequences $(a_n) \to a$ with $a < a_n$, $(f(a_n)) \to -\infty$; or given any number M, there exists a δ such that $f(x) < M$ when $a < x < a + \delta$.
$\lim_{x\to a^-} f(x) = +\infty$ means for all $(a_n) \to a$ with $a_n < a$, $(f(a_n)) \to +\infty$; or, given any number M, there exists a δ such that $M < f(x)$
when $a - \delta < x < a$.
$\lim_{x\to a^-} f(x) = -\infty$ means for all sequences $(a_n) \to a$ with $a_n < a$, $(f(a_n)) \to -\infty$; or, given any number M, there exists a δ such that $f(x) < M$ when $a - \delta < x < a$.

7

Continuity and completeness

Functions on intervals

Concurrent reading: Hart, Spivak ch. 7.
Further reading: Mason, Rudin ch. 4.

Monotonic functions: one-sided limits

Increasing and decreasing functions, like increasing and decreasing sequences, have interesting properties and it is useful to be able to name such functions.

If $f(x) \leq f(y)$ when $x < y$, the function f is said to be *increasing*, or *monotonic increasing*.

If $f(x) < f(y)$ when $x < y$, the function f is said to be *strictly increasing*, or *strictly monotonic increasing*.

1 Which of the following functions with domain $\mathbb{R}$ are monotonic increasing?
(i) $x \mapsto x^2$; (ii) $x \mapsto x^3$; (iii) $x \mapsto \lfloor x \rfloor$; (iv) $x \mapsto x - \lfloor x \rfloor$;
(v) $x \mapsto \tan x (x \neq (n + \frac{1}{2})\pi)$; (vi) $x \mapsto e^x$.
For each one that is *not* monotonic increasing, name a subset of the domain on which it *is* monotonic increasing.

2 Define what is meant by saying that a real function is *decreasing* or *monotonic decreasing*.
Also define what is meant by saying that a real function is *strictly decreasing* or *strictly monotonic decreasing*.

3 Which of the following functions with domain $\mathbb{R}$ are monotonic decreasing?
(i) $x \mapsto -x^2$; (ii) $x \mapsto -x^3$; (iii) $x \mapsto 1/x$ *and* $0 \mapsto 0$;
(iv) $x \mapsto e^{-x}$.
For each one that is not monotonic decreasing, name a subset of the domain on which it is monotonic decreasing.

4 If $f : \mathbb{R} \to \mathbb{R}$ is monotonic increasing throughout its domain and the set $V = \{f(x) | x \in \mathbb{R}\}$ is bounded, prove that $\lim_{x\to\infty} f(x) = \sup V$. In this case the line $y = \sup V$ is called an *asymptote* to the graph of the function.

5 Let $f : \mathbb{R} \to \mathbb{R}$ be a function which is monotonic increasing throughout its domain, and let a be a point of the domain.
Let $L = \{f(x) | x < a\}$ and let $U = \{f(x) | a < x\}$.
Find an upper bound for L and a lower bound for U.
Deduce that $\sup L$ and $\inf U$ exist.
Prove that $\sup L = \lim_{x\to a^-} f(x)$ and $\inf U = \lim_{x\to a^+} f(x)$.
Thus one-sided limits exist for monotonic increasing functions whether they are continuous or not. What are the values of these limits for the function $x \mapsto \lfloor x \rfloor$ at integer points of the domain?

6 Formulate an analogue of qn 5 for monotonic decreasing functions, and give an example of a discontinuous decreasing function and of its one-sided limits at points of discontinuity.

7 If, in qn 5, f is continuous at a, show that $\sup L = \inf U$. If, conversely, $\sup L = \inf U$, show that f is continuous at a. If $\sup L \neq \inf U$, explain why $\sup L < \inf U$, and deduce that a rational number must lie between these two numbers. Deduce that a monotonic function can have at most a countable number of discontinuities.

Intervals

8 Find the range of each of the functions defined below. The domain in each case is $\mathbb{R}$.

(i) $f_1(x) = 1$.
(ii) $f_2(x) = \sin x$.
(iii) $f_3(x) = \arctan x$.
(iv) $f_4(x) = \dfrac{1}{1+x^2}$.
(v) $f_5(x) = e^x$.
(vi) $f_6(x) = x^2$.
(vii) $f_7(x) = x$.

The ranges of these seven functions exhibit the seven kinds of *interval* on the real line.

Intervals, or connected sets, on the real line are subsets of the real line which contain all the real numbers lying between any two points of the subset. The set I is an *interval* if when r and $s \in I$, and $r < s$, every x such that $r < x < s$ also belongs to I.

The seven types of interval are distinguished by their boundedness and their boundaries.

Bounded above and below

(i)	singleton point	$\{a\}$
(ii)	closed interval	$[a, b] = \{x \mid a \le x \le b\}$
(iii)	open interval	$(a, b) = \{x \mid a < x < b\}$
(iv)	half-open interval	$[a, b) = \{x \mid a \le x < b\}$
		$(a, b] = \{x \mid a < x \le b\}$

Bounded above **or** *below, but not both*

(v)	open half-ray	$(a, +\infty) = \{x \mid a < x\}$
		$(-\infty, a) = \{x \mid x < a\}$
(vi)	closed half-ray	$[a, +\infty) = \{x \mid a \le x\}$
		$(-\infty, a] = \{x \mid x \le a\}$

Unbounded

(vii)	the whole real line	$\mathbb{R}$

The word *open* as we have used it here indicates that there is space *within* each open set around each point of the set. For example, if $c \in (a, b)$, then $a < \frac{1}{2}(a + c) < c < \frac{1}{2}(c + b) < b$. This is also described by saying that an open set contains a neighbourhood of each of its points.

The word *closed* as we have used it here indicates that every convergent sequence within each closed set converges to a point of the set. (See qn 3.78.)

The whole real line is both open and closed!

9 Sketch graphs of continuous functions $f\colon (a, b) \to \mathbb{R}$ for which the range is

(i) a singleton point,
(ii) a closed interval,
(iii) an open interval,
(iv) a half-open interval,
(v) an open half-ray,
(vi) a closed half-ray,
(vii) $\mathbb{R}$.

10 Sketch the graph of a continuous function $f\colon A \to \mathbb{R}$ for which the range of the function is *not* an interval. What would you conjecture about the domain A of the function in such a case?

If we wish to try to prove that continuous functions always map intervals onto intervals, it is worth noticing first that this result fails if we are working with only the rational numbers.

11 Let the function $f: \mathbb{Q} \to \mathbb{Q}$ be defined by $f(x) = x^2 - 2$. This is certainly a continuous function by qn 6.29. Now restrict the domain to the set $[1, 2] \cap \mathbb{Q}$. The numbers -1 and 2 are in the range, but what rational number between them is not?

This result suggests that it will be necessary to use the completeness of the real numbers in a proof that continuous functions map intervals onto intervals.

Intermediate Value Theorem

A simple form of the problem which we are facing is this: Suppose that A is an interval and that $f: A \to \mathbb{R}$ is a continuous function.

Suppose further that a and b are points in A and that $f(a) < 0$ and $f(b) > 0$. Can we be sure that there is a point c in the interval (a, b) such that $f(c) = 0$?

12 (*Bolzano*, 1817) Let $f: [a, b] \to \mathbb{R}$ be a continuous function with $f(a) < 0$ and $f(b) > 0$. Sketch the graph of such a function. Try different sketches which illustrate the same conditions. Notice that neither of the sets $\{x \mid f(x) < 0\}$ or $\{x \mid f(x) > 0\}$ need be an interval. We construct sequences in the domain of the function which either reach a point x where $f(x) = 0$ or converge to such a point by locating smaller and smaller intervals on each of which the function f changes sign. Let $a_1 = a$, $b_1 = b$ and $d = \frac{1}{2}(a_1 + b_1)$. If $f(d) = 0$ we have found the point we want; if $f(d) \neq 0$, explain how to choose a_2 and b_2 so that $f(a_2) < 0$, $f(b_2) > 0$ and $b_2 - a_2 = \frac{1}{2}(b_1 - a_1)$. Extend this process to an inductive definition of a_n, b_n where $f(a_n) < 0 < f(b_n)$, and $a_n \le a_{n+1} < b_{n+1} \le b_n$, with $b_{n+1} - a_{n+1} = \frac{1}{2}(b_n - a_n)$.

13 Let $f: [a, b] \to \mathbb{R}$ be a continuous function with $f(a) < 0$ and $f(b) > 0$. With the notation of qn 12, suppose that the process of repeated bisection of the interval $[a, b]$ has not located a point at which the function is 0.

(i) How do you know the sequences (a_n) and (b_n) are both convergent?

(ii) If $(a_n) \to A$ and $(b_n) \to B$, why must $f(A) \le 0 \le f(B)$?

(iii) Why must $A = B$ and $f(A) = 0$?

We summarise the result of qn 13 by saying that if $f: [a, b] \to \mathbb{R}$ is a continuous function, $f(a) < 0$, $f(b) > 0$, there exists a $c \in (a, b)$ such that $f(c) = 0$.

14 By applying the result of qn 13 to a suitably chosen function, prove that if $f: [a, b] \to \mathbb{R}$ is a continuous function and $f(a) < k < f(b)$, then there exists a $c \in (a, b)$ such that $f(c) = k$.

15 By applying the result of qn 14 to a suitably chosen function, prove that if $f: [a, b] \to \mathbb{R}$ is a continuous function and $f(a) > k > f(b)$, then there exists a $c \in (a, b)$ such that $f(c) = k$.

Questions 14 and 15, together, give the general form of the *Intermediate Value Theorem*, that a continuous function $f: [a, b] \to \mathbb{R}$ takes every value between $f(a)$ and $f(b)$.

16 Give an example of a discontinuous function $f: [a, b] \to \mathbb{R}$ which takes every value between $f(a)$ and $f(b)$. The existence of such a function shows that the converse of the Intermediate Value Theorem is false.

17 Prove that a continuous function $f: \mathbb{R} \to \mathbb{Z}$ is necessarily constant.

18 Prove that a continuous function $f: \mathbb{R} \to \mathbb{Q}$ is necessarily constant.

19 (i) For any real numbers a and b show that the function defined on $\mathbb{R}$ by $f(x) = x^3 + ax + b$ has a real root. To what class of polynomial functions may this result be extended?

(ii) For any positive real number a and integer $n > 1$, show that the function f defined by $f(x) = x^n - a$ has a positive root. Use this to provide an alternative proof of the existence of nth roots for positive numbers to that given in qn 4.40.

20 If $f: [0, 1] \to [0, 1]$ is a continuous function, by applying the result of qn 13 to an appropriately chosen function, prove that there is at least one $c \in [0, 1]$ such that $f(c) = c$. Give an example to show that this result might fail if the domain of f were $(0, 1)$. Generalise this result for a continuous function $f: [a, b] \to [a, b]$.

21 Can the Intermediate Value Theorem be used to prove that the image of an interval under a continuous function is necessarily an interval? Or, in other words, under a continuous function, is a connected set always mapped onto a connected set?

Inverses of continuous functions

22 If $f: [a, b] \to \mathbb{R}$ is a continuous and one–one function with $f(a) < f(b)$, prove that f is strictly monotonic increasing. Deduce that the range of f is $[f(a), f(b)]$.
What is the analogous result if $f(a) > f(b)$?

23 If $f : [a, b] \to \mathbb{R}$ is strictly monotonic, must f be

(i) one–one;
(ii) invertible;
(iii) continuous?

24 If $f : [a, b] \to \mathbb{R}$ is continuous and invertible (see qn 6.6) must f be strictly monotonic?

25 Suppose that $f : [a, b] \to \mathbb{R}$ is a continuous and strictly monotonic function.

(i) How do you know that the range of f is a closed interval $[c, d]$, say?
(ii) How do you know that f has an inverse function $g : [c, d] \to [a, b]$?
(iii) How do you know that g must be strictly monotonic?
(iv) Let x be any point in the domain of f and let $y = f(x)$, so that $a \le x \le b$ and $c \le y \le d$. How do you know that $\lim_{t\to y^-} g(t)$ and $\lim_{t\to y^+} g(t)$ exist?
(v) Let $(x_n) \to x$ be an increasing sequence in $[a, b]$ and let $(s_n) \to x$ be a decreasing sequence in $[a, b]$. Why must $(f(x_n)) \to f(x)$ and $(f(s_n)) \to f(x)$? Of these two sequences why must one be increasing and one decreasing?
(vi) Deduce that there is a sequence $(t_n) \to y$ from below and one tending to y from above for each of which $(g(t_n)) \to g(y)$.
(vii) Why does this establish that g is continuous at y?

We have now proved that a function which is continuous and strictly monotonic on an interval has a continuous inverse. The use of completeness in qn 25 is unavoidable. Consider $f : \mathbb{Q} \to \mathbb{Q}$ given by $f(x) = x$ when $x < \sqrt{2}$ and $f(x) = x + 1$ when $\sqrt{2} < x$. An example of a one–one continuous function $\mathbb{Q} \to \mathbb{Q}$ whose inverse is discontinuous everywhere is given in the book by Dieudonné cited in the bibliography. See also Körner (1991).

26 Does the function $f : \mathbb{R} \to \mathbb{R}$ defined by $f(x) = x^2$ have an inverse? Find a maximal interval $A \subseteq \mathbb{R}$, with $1 \in A$, for which the function $f : A \to \mathbb{R}$ defined by $f(x) = x^2$ is monotonic. With this A, is the function $f : A \to A$ a bijection? Is the function $f^{-1} : A \to A$ continuous? How is it usually denoted?

27 Why does the function $f : \mathbb{R}^+ \cup \{0\} \to \mathbb{R}$ defined by $f(x) = x^n$ have an inverse whether the positive integer n is odd or even?
Say why there is a continuous inverse $f^{-1} : \mathbb{R}^+ \cup \{0\} \to \mathbb{R}^+ \cup \{0\}$. This inverse function is normally denoted by $x \mapsto \sqrt[n]{x}$ or $x^{1/n}$.

We established the existence of nth roots for positive real numbers in qn 4.40. The argument here establishes that the function $x \mapsto x^{1/n}$ is continuous on $\mathbb{R}^+ \cup \{0\}$.

28 Find the limit

$$\lim_{x \to 1} \frac{1 - x}{1 - x^{m/n}}$$

when m and n are positive integers with $m \neq n$. (Another method is given in qn 9.26.)

Continuous functions on a closed interval

29 What does the Intermediate Value Theorem allow us to claim about the possible ranges of a continuous function $f : [a, b] \to \mathbb{R}$?

30 Attempt to draw graphs of continuous functions $f : [a, b] \to \mathbb{R}$ with the seven different ranges as in qn 9. Use pencil and paper, not a computer.

We now investigate whether the fact that the ranges found in qn 30 were all bounded above and bounded below is a necessary consequence of the conditions, or an indication of our lack of imagination. We suppose that there exists a continuous function $f : [a, b] \to \mathbb{R}$ whose range is unbounded above and see whether any contradiction arises.

31 We suppose that the range of the continuous function $f : [a, b] \to \mathbb{R}$ is unbounded above. The unboundedness of the range of f means that, whatever integer n we choose, we can find an x in $[a, b]$ such that $f(x) > n$. If such an x can be found we call it x_n. This gives us a sequence (x_n).

(i) Is there any reason why the sequence (x_n) should be convergent?
(ii) Is the sequence (x_n) bounded?
(iii) Must the sequence (x_n) contain a convergent subsequence?
(iv) If the subsequence (x_{n_i}) converges to c, why must $c \in [a, b]$?
(v) What does the continuity of f allow us to say about the sequence $(f(x_{n_i}))$?
(vi) For sufficiently large n_i can we be sure that $|f(x_{n_i}) - f(c)| < 1$?
(vii) How does this lead to a contradiction, which therefore undermines our hypothesis of the unboundedness of the sequence $(f(x_n))$?

32 If the range of a continuous function $f : [a, b] \to \mathbb{R}$ were presumed to be unbounded below, outline how you would establish a contradiction.

33 *A* function $f: \mathbb{Q} \to \mathbb{Q}$ is defined by $f(x) = 1/(x^2 - 2)$.

(i) Show that f is continuous throughout its domain.
(ii) Explain why the behaviour of f on the interval $[1, 2]$ does not contradict the result established in qn 31.

So the range of a continuous real function $f: [a, b] \to \mathbb{R}$ is bounded.

Questions 31 and 32 eliminate $\mathbb{R}$, $(a, +\infty)$, $[a, +\infty)$, $(-\infty, a)$ and $(-\infty, a]$ as possible ranges for a continuous function $f: [a, b] \to \mathbb{R}$. This leaves open intervals, closed intervals, half-open intervals and singleton points as the only possibilities. We know that closed intervals and singleton points are real possibilities. Can we now eliminate the open intervals and the half-open intervals as our investigation in qn 30 would suggest we might?

34 (*Weierstrass*, 1860) Let $f: [a, b] \to \mathbb{R}$ be a continuous function and let $V = \{f(x) | \, a \le x \le b\}$.

(i) How do we know that V is an interval?
(ii) How do we know that V is bounded above and below?
(iii) Can you be sure that $\sup V$ and $\inf V$ exist?

If we let $\sup V = M$, then $M \ge f(x)$ for all $x \in [a, b]$. As a result of our graph drawing, we conjecture that there has to be a value of x at which $M = f(x)$.

(iv) Why, given $\varepsilon > 0$, must there be an $x \in [a, b]$ such that $M - \varepsilon < f(x) \le M$?
(v) Define $x_n \in [a, b]$ so that $M - 1/n < f(x_n) \le M$. Why must the sequence (x_n) contain a convergent subsequence?
(vi) Suppose $(x_{n_i}) \to c$. Why must $c \in [a, b]$?
(vii) Why must $(f(x_{n_i}))$ be convergent?
(viii) By considering the limit of the sequence $(f(x_{n_i}))$ prove that $M = f(c)$.
(ix) Having proved that the function f attains its supremum, indicate how to show that the function must also attain its infimum.

35 Explain why function $f: \mathbb{Q} \to \mathbb{R}$ defined by $f(x) = \sin x$ does not attain its bounds when examined on the domain $[0, 6]$. Does this contradict the result of the previous question?

The result of qns 31 and 32 is usually described by saying that a continuous function on a closed interval is *bounded*, and the result of qn 34 by saying that a continuous function on a closed interval *attains its bounds*. Questions 31 and 34(i)–(viii) give the *Maximum Theorem*. Questions 32 and 34(ix) give the

Minimum Theorem. Although both results depend on completeness it is not necessary for the Intermediate Value Theorem to precede their proof. Taken with the Intermediate Value Theorem, they imply, for continuous real functions, that a closed interval is mapped to a closed interval or a point.

36 Give examples to show that continuous functions on unbounded intervals or open intervals or half-open invervals need not be bounded and, even if bounded, need not attain their bounds.

Uniform continuity

Both the sequence definition of continuity and the neighbourhood definition of continuity define the continuity of a function at a *point*, and we only claim continuity on an *interval* when the function is continuous at every point of that interval. Moreover, when determining continuity at a point with the neighbourhood definition of continuity, given an ε, the necessary δ may be different for continuity at a point a from what is needed to establish continuity at a point b. We examine now, when, with a given ε, a choice of δ may be made which will establish continuity throughout an interval. The rôle of closed intervals again turns out to be critical.

37 If $f\colon [-10, 10] \to \mathbb{R}$ is defined by $f(x) = x^2$, prove that if $|x - y| < 1/20$ then $|f(x) - f(y)| < 1$. Extend this result to show that, for any positive ε, if $|x - y| < \varepsilon/20$ then $|f(x) - f(y)| < \varepsilon$.

38 If $f\colon (0, 1) \to \mathbb{R}$ is defined by $f(x) = 1/x$, $f(x)$ is unbounded.
Find a δ_1 such that if $|x - \frac{1}{2}| < \delta_1$, then $|f(x) - f(\frac{1}{2})| < 1$.
Find a δ_2 such that if $|x - \frac{1}{4}| < \delta_2$, then $|f(x) - f(\frac{1}{4})| < 1$.
Show that, if $|x - a| < \delta$ implies $|f(x) - f(a)| < 1$, then $\delta \le a^2/(1 + a)$.
Deduce that there is no constant δ such that $|x - a| < \delta$ implies $|f(x) - f(a)| < 1$ for all x and a in the domain of the function f.

DEFINITION OF UNIFORM CONTINUITY
A function $f\colon A \to \mathbb{R}$ is said to be *uniformly continuous on* A when, given $\varepsilon > 0$, there exists a δ such that $|x - y| < \delta \Rightarrow |f(x) - f(y)| < \varepsilon$.

39 Can you be sure that a function which is uniformly continuous on A is necessarily continuous at each point of A?

In qn 37, we have proved that the function $x \mapsto x^2$ *is* uniformly continuous on $[-10, 10]$. In qn 38, we have proved that the function $x \mapsto 1/x$ *is not* uniformly continuous on $(0, 1)$.

40 By considering $x = n + 1/n$ and $y = n$, and taking $\varepsilon = 1$, prove that the function defined by $f(x) = x^2$ *is not* uniformly continuous on $\mathbb{R}$. Notice the effect of changing the domain from qn 37.

41 By considering $x = 1/2n\pi$ and $y = 1/(2n + \frac{1}{2})\pi$, and taking $\varepsilon = \frac{1}{2}$, prove that the function defined by $f(x) = \sin(1/x)$ *is not* uniformly continuous on the interval (0, 1). This shows that continuous functions (see qn 6.48) may fail to be uniformly continuous even when they are bounded.

42 If a function $f \colon A \to \mathbb{R}$ satisfies a *Lipschitz condition*, namely that there is a constant real number L such that

$$|f(x) - f(y)| \leq L \cdot |x - y|$$

for any $x, y \in A$, show that f is uniformly continuous on A, by finding an appropriate δ for a given ε. Verify that the function of qn 37 satisfies a Lipschitz condition with $L = 20$.

After studying the Mean Value Theorem in chapter 9 it will be clear that any function with bounded derivatives necessarily satisfies a Lipschitz condition, with an upper bound on the absolute value of the derivatives as the Lipschitz constant.

43 (*Heine*, 1872) We seek to prove that a continuous function $f \colon [a, b] \to \mathbb{R}$ is necessarily uniformly continuous, and we do so by supposing that it is not and establishing a contradiction.
When a function *is not* uniformly continuous, it means that, for some $\varepsilon > 0$, no δ can be found for which $|x - y| < \delta$ implies that $|f(x) - f(y)| < \varepsilon$. That is, for each δ there are points x and y in the domain such that $|x - y| < \delta$ but $|f(x) - f(y)| \geq \varepsilon$.
Suppose that f *is not* uniformly continuous and construct two sequences (x_n) and (y_n) for which the difference $|x_n - y_n| < 1/n$ but for which $|f(x_n) - f(y_n)| \geq \varepsilon$.

(i) Is there any reason why the sequence (x_n) should be convergent?
(ii) Is the sequence (x_n) bounded?
(iii) Must the sequence (x_n) contain a convergent subsequence?
(iv) If the subsequence (x_{n_i}) converges to c, must the subsequence $(y_{n_i}) \to c$?
(v) Why must $c \in [a, b]$?
(vi) What can be said about the sequences $(f(x_{n_i}))$ and $(f(y_{n_i}))$, because of the continuity of f?
(vii) Use the inequality

$$|f(x) - f(y)| \leq |f(x) - f(c)| + |f(c) - f(y)|$$

to obtain the contradiction we seek, by showing that the right-hand side may be made less than the left-hand side.

44 In qns 38, 40 and 41 we found continuous functions which were not uniformly continuous. Each of these functions was, however, either unbounded or else had unbounded slope. The function $f: [0, \infty) \to \mathbb{R}$ defined by $f(x) = \sqrt{x}$ is unbounded and has unbounded slope near the origin. Prove that this function is uniformly continuous by appealing to qns 26 and 43 on $[0, 1]$ and by noting that, if either x or $y > 1$, then $|\sqrt{x} - \sqrt{y}| < |x - y|$.

45 If $f: (a, b) \to \mathbb{R}$ is continuous and $\lim_{x \to a^+} f(x)$ and $\lim_{x \to b^-} f(x)$ both exist and are finite, prove that f is uniformly continuous by constructing a function g which is continuous on $[a, b]$ and for which $g(x) = f(x)$ on (a, b).

Extension of functions on $\mathbb{Q}$ to functions on $\mathbb{R}$

46 (i) Sketch the graph of $y = \dfrac{1}{2x^2 - 1}$, for $0 \le x \le 1$.
Use computer graphics if possible. You will need a large range for y, say $[-10, 10]$. What is the equation of the asymptote?

(ii) Is a real function defined by $f(x) = 1/(2x^2 - 1)$ for all points of the domain $[0, 1] \cap \mathbb{Q}$?

(iii) Is the function given in part (ii) continuous at each point of the given domain?

(iv) As the curve is steep near the asymptote there is a possibility that the function may not be uniformly continuous there.
Let m and n be positive integers with $m < n$. If

$$\tfrac{1}{2} + \tfrac{1}{n} < x^2 < \tfrac{1}{2} + \tfrac{1}{m},$$

prove that $m/2 < f(x) < n/2$.
Deduce that if

$$\tfrac{1}{2} < x^2 < \tfrac{1}{2} + \tfrac{1}{4n} \text{ and } \tfrac{1}{2} + \tfrac{1}{2n} < y^2 < \tfrac{1}{2} + \tfrac{1}{n},$$

then $0 < y^2 - x^2 < 1/n$ and $f(x) - f(y) > n$. So the function is not uniformly continuous.

(v) If you attempted to extend f to a function $g: [0, 1] \to \mathbb{R}$ such that $f(x) = g(x)$ for rational x, is there any value that you could take for $g(1/\sqrt{2})$ which would make the function g continuous at the point $1/\sqrt{2}$?

47 The function $f: [0, 1] \cap \mathbb{Q} \to \mathbb{R}$ is uniformly continuous.

(i) If (a_n) is a Cauchy sequence of rational numbers in $[0, 1]$, show that $(f(a_n))$ is a Cauchy sequence.

(ii) If (a_n) and (b_n) are rational Cauchy sequences in $[0, 1]$ with the same real limit, prove that $(f(a_n))$ and $(f(b_n))$ have the same limit.

(iii) Is there a well-defined function $g : [0, 1] \to \mathbb{R}$ such that $g(x) = f(x)$ for rational x and $(f(a_n)) \to g(a)$, when $(a_n) \to a$ is a Cauchy sequence of rational numbers with an irrational limit a?

(iv) Finally, show that g is uniformly continuous on $[0, 1]$. Given $\varepsilon > 0$, choose δ such that $|x - y| < \delta$ implies $|f(x) - f(y)| < \frac{1}{3}\varepsilon$, for rational x and y. Let a and b be real numbers in $[0, 1]$ such that $|a - b| < \delta$, and let $(a_n) \to a$ and $(b_n) \to b$ be Cauchy sequences of rational numbers wholly within the interval $[a, b]$. Use the inequality

$$|g(a) - g(b)| \leq |g(a) - f(a_n)| + |f(a_n) - f(b_n)| + |f(b_n) - g(b)|$$

to complete the proof that the function f may be extended to a continuous function on the whole interval $[0, 1]$.

The Intermediate Value Theorem and the Maximum–Minimum Theorem, like the theorem that a continuous function on a closed interval is uniformly continuous, can only be proved using the completeness of the real numbers. However, uniform continuity, like continuity, can be defined whether the domain of the function is complete or not. What we have shown in qns 46 and 47 is that uniform continuity is the necessary and sufficient condition that a continuous function defined on a dense subset of a closed interval may be extended to a continuous function on the whole interval. As a final optional exercise we indicate how this theorem can be used to define exponents.

(48) For positive $a \neq 1$ and positive rational x we can construct a^x by qn 4.40, and we can define $a^0 = 1$ and $a^{-x} = 1/a^x$.

(i) Prove that $f : \mathbb{Q} \to \mathbb{R}^+$ defined by $f(x) = a^x$ is strictly monotonic. (In fact it is monotonic increasing when $1 < a$, and monotonic decreasing when $0 < a < 1$.)

(ii) For $x \neq 0$, and $a > 1$, let $g(x) = (a^x - 1)/x$. Use qn 2.50(iii) to show that $g(n) < g(m)$ for two positive integers $n < m$. By putting $b = a^{qs}$ prove that $g(p/q) < g(r/s)$ for two rational numbers $0 < p/q < r/s$, so that $g : \mathbb{Q}^+ \to \mathbb{R}^+$ is monotonic increasing. Use qn 4.41 to extend g to a monotonic increasing function on $\mathbb{Q}^+ \cup \{0\}$.

(iii) Prove that f satisfies a Lipschitz condition on any closed interval $[0, B] \cap \mathbb{Q}$, where B is a positive rational and so f is uniformly continuous on that interval by qn 42.

(iv) Deduce from qn 47 that f may be extended to a continuous function on $[0, B]$ in a unique way.

(v) Use the definition of a^{-x} to extend f to a continuous function on $[-B, B]$.

Summary

Definition qns 1, 2 — *Monotonic functions*
A real function f is said to be monotonic increasing when $x < y \Rightarrow f(x) \leq f(y)$.
A real function f is said to be strictly monotonic increasing when $x < y \Rightarrow f(x) < f(y)$.
A real function f is said to be monotonic decreasing when $x < y \Rightarrow f(x) \geq f(y)$.
A real function f is said to be strictly monotonic decreasing when $x < y \Rightarrow f(x) > f(y)$.

Theorem qns 5, 6, 7 — A monotonic function has one-sided limits at each point of its domain. It is continuous where the two one-sided limits are equal.

Definition qns 8, 9 — *Intervals*
A subset I of $\mathbb{R}$ is said to be *connected* if when $a, b \in I$, and $a < b$, then $x \in I$ for every x satisfying $a < x < b$.
A connected set is called an interval.
Bounded intervals are classified as closed intervals $[a, b]$, open intervals (a, b), half-open intervals $(a, b]$ and singletons $\{a\}$.
Unbounded intervals are classified as closed half-rays $[a, +\infty)$ or $(-\infty, a]$, open half-rays $(a, +\infty)$ or $(-\infty, a)$, or the whole line $\mathbb{R}$.

The Intermediate Value Theorem
qns 14, 15 — A continuous function $f : [a, b] \to \mathbb{R}$ takes every value between $f(a)$ and $f(b)$.

Theorem qn 21 — The range of a continuous function defined on an interval is always an interval.

Theorem qns 22, 23, 24, 25 — If a continuous function f is defined on an interval, f has an inverse function if and only if f is strictly monotonic. In this case the inverse function is continuous and strictly monotonic.

The Maximum–Minimum Theorem

qns 31, 32, 34	A continuous real function defined on a closed interval is bounded, and attains its bounds.

Uniform continuity

Definition qns 34–42	A function $f: A \to \mathbb{R}$ is said to be uniformly continuous on A when, given $\varepsilon > 0$, there exists a δ such that $\lvert x - y\rvert < \delta \Rightarrow \lvert f(x) - f(y)\rvert < \varepsilon$.
Theorem qn 43	If a real function is continuous on a closed interval then it is uniformly continuous on that interval.
Theorem qn 47	If a function is uniformly continuous on a dense subset of a closed interval, including the end points, then it may be extended to a continuous function on the whole interval.

Historical Note

The Intermediate Value Theorem had been assumed from geometrical perceptions during the eighteenth century as the basis of work on approximations to roots of equations. There were mathematicians who regarded it as the essential characterisation of continuity. (We saw how wrong that was in qn 16. The inadmissibility of defining continuous functions by the Intermediate Value Theorem was pointed out by Darboux in 1875.) In 1817 Bolzano insisted that this was a theorem which required analytical proof. He provided the first formal definition of a continuous function, showed that if a continuous function was positive at a point it must be positive in a neighbourhood of that point and likewise if it is negative at a point it must be negative in a neighbourhood of that point. Taking the function to be positive at the upper end of the domain and negative at the lower end he claimed that the set of points of the domain at which the function was negative was bounded above and so must have a least upper bound. At this least upper bound only a zero value for the function is not contradictory. Bolzano proved the least upper bound property using a Cauchy sequence, though not by that name (!), derived from repeated bisections. His proof that Cauchy sequences converge was incomplete because he thought it followed from the impossibility of the convergence of such a sequence to two distinct limits. Perhaps independently of Bolzano, Cauchy also produced a formal definition of continuity in 1821 and the proof of the Intermediate Value Theorem that we constructed in this chapter was essentially his. He repeatedly divided his domain into m equal parts and then selected one part on which a change of sign had taken place. We took $m = 2$. Cauchy assumed, without proof, that monotonic bounded sequences were convergent.

The theorem that continuous functions on a closed interval were bounded and attained their bounds had been familiar from the seventeenth century but under much stronger differentiability conditions. The integral of Riemann's discontinuous function (invented 1854, but published 1867) provided an example that distinguished clearly between continuous and differentiable functions for the first time. It was Weierstrass in his lectures in Berlin in 1861 who first proved that a continuous function on a closed interval was bounded and attained its bounds. Weierstrass called this his 'Principal theorem'. Weierstrass affirmed the importance of considering continuous but not necessarily differentiable functions by his approximation theorem that every continuous function is the uniform limit of a sequence of polynomials.

The distinction between continuity at a point and continuity on an interval was not clear in Cauchy's work. In his proof of the integrability of continuous functions (1823), Cauchy assumed continuity but used uniform continuity. Dirichlet proved that a function which was continuous on a closed interval was uniformly continuous on that interval in his lectures in Berlin in 1854. The first proof of this result in a Weierstrassian context was published by E. Heine in 1872.

Answers

1 (i) $\mathbb{R}^+ \cup \{0\}$. (ii) Yes. (iii) Yes. (iv) $[0, 1)$ for example. (v) $(-\frac{1}{2}\pi, \frac{1}{2}\pi)$ for example. (vi) Yes.

2 If $f(x) \geq f(y)$ when $x < y$, the function f is said to be decreasing or monotonic decreasing.
If $f(x) > f(y)$ when $x < y$, the function is said to be strictly decreasing or strictly monotonic decreasing.

3 (i) $\mathbb{R}^+ \cup \{0\}$. (ii) Yes. (iii) $\mathbb{R}^+$. (iv) Yes.

4 $\sup V$ exists, by qn 4.80. By 4.64, given $\varepsilon > 0$, for some x_1, $\sup V - \varepsilon < f(x_1) \leq \sup V$, but f is monotonic increasing so, for $x \geq x_1$, $\sup V - \varepsilon < f(x_1) \leq f(x) \leq \sup V$, so $|f(x) - \sup V| < \varepsilon$ and $\lim_{x \to +\infty} f(x) = \sup V$.

5 $f(a)$ is an upper bound for L and a lower bound for U, so $\sup L$ and $\inf U$ exist by qns 4.80 and 4.81. Given $\varepsilon > 0$, for some $x_1 < a$, $\sup L - \varepsilon < f(x_1) \leq \sup L$. But f is monotonic increasing so, if $x_1 < x < a$, $\sup L - \varepsilon < f(x_1) \leq f(x) \leq \sup L$, and $|f(x) - \sup L| < \varepsilon$. Similarly, for some $x_2 > a$, $\inf U \leq f(x_2) < \inf U + \varepsilon$. But f is monotonic increasing so, if $a < x < x_2$, $\inf U \leq f(x) \leq f(x_2) < \inf U + \varepsilon$, and $|f(x) - \inf U| < \varepsilon$.
$\lim_{x \to n^-} \lfloor x \rfloor = n - 1$ and $\lim_{x \to n^+} \lfloor x \rfloor = n$.

6 If f is monotonic decreasing then, with the notation of qn 5, $f(a)$ is a lower bound for L and an upper bound for U. Proceed as in qn 5 with appropriate inequalities reversed. An example: $x \mapsto -\lfloor x \rfloor$.

7 If f is continuous at a, then limits from above and below at a are equal from qn 6.86. $\sup L \leq f(a) \leq \inf U$ so, if $\sup L = \inf U$, then $\lim_{x \to a^-} f(x) = f(a) = \lim_{x \to a^+} f(x)$, and so f is continuous by qn 6.89.
If $\sup L < \inf U$, then there is a rational number between these two by qn 4.6. Because the function is monotonic this locates a distinct rational number in each discontinuity. The open intervals $(\sup L, \inf U)$, at the points of discontinuity, are disjoint because the function is monotonic. A set of rationals is countable, so the set of discontinuities of a monotonic function is countable.

8 (i) $\{1\}$, (ii) $[-1, 1]$, (iii) $(-\frac{1}{2}\pi, \frac{1}{2}\pi)$, (iv) $(0, 1]$, (v) $\mathbb{R}^+$, (vi) $\mathbb{R}^+ \cup \{0\}$, (vii) $\mathbb{R}$.

9 Make your own sketches without considering possible formulae.

10 Take A as the union of (two) disjoint and separated intervals.

11 The number 0 is not in the range.

12 If $f(d) \neq 0$, then either $f(d) > 0$ or $f(d) < 0$. If $f(d) > 0$, then take $a_2 = a_1$ and $b_2 = d$. If $f(d) < 0$, take $a_2 = d$ and $b_2 = b_1$. Now suppose $f(a_n) < 0$ and $f(b_n) > 0$ and let $d = \frac{1}{2}(a_n + b_n)$. If $f(d) = 0$ we have finished. If $f(d) > 0$, take $a_{n+1} = a_n$ and $b_{n+1} = d$. If $f(d) < 0$, take $a_{n+1} = d$ and $b_{n+1} = b_n$.

13 (i) (a_n) is monotonic increasing and bounded above by b, and thus is convergent, by qn 4.35. (b_n) is monotonic decreasing and bounded below by a, and thus is convergent, by qn 4.34.

(ii) $(f(a_n)) \to f(A)$ by the continuity of f and $f(A) \leq 0$ by qn 3.75.

(iii) From qns 12 and 3.39, $(b_n - a_n)$ is a null sequence, and so from qn 3.54(v), $B = A$.
$f(A) \leq 0 \leq f(A) \Rightarrow f(A) = 0$.

14 Define $g(x) = f(x) - k$, then $g(a) < 0$ and $g(b) > 0$ and g is continuous by qn 6.23. So $g(c) = 0$ for some $c \in (a, b)$, and $f(c) - k = 0$, so $f(c) = k$.

15 Define $g(x) = -f(x)$, then $g(a) < -k < g(b)$ and g is continuous, by qn 6.26. So $g(c) = -k$ for some $c \in (a, b)$ by qn 14, and $f(c) = k$.

16 The function of qn 6.19 reaches every value between $f(0)$ and $f(x)$, but is discontinuous at 0.

The following function reaches every value between 0 and 1, but is discontinuous everywhere. $f: [0, 1] \to [0, 1]$.
$f(0) = \frac{1}{2}$; $f(\frac{1}{2}) = 0$; $f(x) = x$ when x is rational and $\neq 0, \frac{1}{2}$;
$f(x) = 1 - x$ when x is irrational.

17 If $f(a) < f(b)$ but both are integers then, by qn 14, for some $c \in (a, b)$, $f(c) = f(a) + \frac{1}{2}$ which is not an integer. This contradicts the possibility of distinct values for the function.

18 If $f(a) < f(b)$ but both are rational then, by qn 14, for some $c \in (a, b)$,

$$f(c) = \frac{f(a) + \sqrt{2}f(b)}{1 + \sqrt{2}}$$

which is irrational. See qn 4.20. This contradicts the possibility of distinct values for the function.

19 (i) If $x = |a| + |b| + 1$, then $f(x) > 0$. If $x = -|a| - |b| - 1$, then $f(x) < 0$. So there is a real root of the equation $x^3 + ax + b = 0$ by the Intermediate Value Theorem. This result may be extended to any polynomial of odd degree.

(ii) $f(0) < 0$, $f(a) > 0$, when $a > 1$, so by the Intermediate Value Theorem there is a root α in $[0, a]$. $\alpha^n - a = 0$, so $\alpha = \sqrt[n]{a}$. If $a < 1$, apply the Intermediate Value Theorem to $[0, 1]$.

20 Choose $g(x) = x - f(x)$ which is continuous, by qn 6.23. Then $g(0) = -f(0) \leq 0$ and $g(1) = 1 - f(1) \geq 0$. So $g(c) = 0$ for some

$c \in [0, 1]$, and $f(c) = c$. If $f : [a, b] \to [a, b]$ then $f(a) \geq a$ and $f(b) \leq b$, so if $g(x) = x - f(x)$, $g(a) \leq 0$ and $g(b) \geq 0$. So for some c, $a \leq c \leq b$, $g(c) = 0$, by the Intermediate Value Theorem.

21 If a and b are any two points in an interval domain of f, then $f(a)$ and $f(b)$ are two points in the range of f. The Intermediate Value Theorem says that, for any k, $f(a) \leq k \leq f(b)$ or $f(a) \geq k \geq f(b)$, there is a $c \in [a, b]$ such that $f(c) = k$. So the range is connected and is therefore an interval.

22 If $a < x < b$, then $f(a) < f(x) < f(b)$: because if $f(a) \geq f(x)$ then, for some $c \in [x, b)$, $f(c) = f(a)$ and f is not one–one; and if $f(x) \geq f(b)$ then, for some $c \in (a, x]$, $f(c) = f(b)$ and, again, f is not one–one. Applying this result to the interval $[x, b]$ we have that $a < x < y < b$ implies $f(a) < f(x) < f(y) < f(b)$, and f is strictly monotonic increasing. If $f(a) > f(b)$ and f is continuous and one–one, then f is strictly monotonic decreasing.

23 (i) Yes, immediate consequence of being strictly monotone.
(ii) Yes, consequence of (i) from qns 6.5 and 6.6.
(iii) No. For example, consider $f\colon [0, 2] \to [0, 3]$ given by $f(x) = x$ on $[0, 1]$ and $f(x) = x + 1$ on $(1, 2]$.

24 Yes, from qn 22.

25 (i) The range is an interval from qn 21. But, for $x \in [a,\ b]$, $f(a) \leq f(x) \leq f(b)$, taking f as increasing. So $c = f(a)$ and $d = f(b)$. If f is decreasing then $c = f(b)$ and $d = f(a)$.
(ii) Strictly monotonic functions are one–one and so invertible.
(iii) $x < y \Leftrightarrow f(x) < f(y) \Rightarrow g(f(x)) < g(f(y)) \Leftrightarrow f(x) < f(y)$.
(iv) From qn 5, using completeness.
(v) The sequences are convergent from the continuity of f. They are increasing and decreasing because f is strictly monotone.
(vi) Let $t_n = f(x_n)$ from below and $t_n = f(s_n)$ from above.
(vii) $\lim_{t \to y^-} g(t)$ exists from (iv) and equals $g(y)$ from (vi).
Similarly for the limit from above, so $\lim_{t \to y} g(t) = g(y)$ and g is continuous at y by qn 6.89.

26 f only has an inverse on $\mathbb{R}^+ \cup \{0\}$ or on $\mathbb{R} \backslash \mathbb{R}^+$. On $\mathbb{R}^+ \cup \{0\}$, $x \mapsto x^2$ is continuous by qn 6.29, and strictly monotonic by qn 2.14. The range is an interval by the Intermediate Value Theorem and is unbounded above, so f is a bijection.
$f^{-1} : x \mapsto \sqrt{x}$ is continuous and monotonic on $\mathbb{R}^+ \cup \{0\}$, by qn 25.

27 f is continuous by qn 6.28, strictly monotonic by qn 2.20, and unbounded above, so f is a bijection. Then f has a continuous inverse by qn 25.

28 Let $a = \sqrt[n]{x}$, then

$$f(x) = \frac{1-x}{1-x^{m/n}} = \frac{1-a^n}{1-a^m} = \frac{1+a+a^2+\ldots+a^{n-1}}{1+a+a^2+\ldots+a^{m-1}}.$$

As $x \to 1,\ a \to 1$, by the continuity of $x \mapsto \sqrt[n]{x}$.
Now by qns 6.29, 6.54 and 6.86, $\lim_{x\to 1} f(x) = n/m$.

29 The range must be an interval by qn 21.

30 The only possible ranges are a singleton and a closed interval.

31
(i) None whatsoever.
(ii) $a \le x_n \le b$.
(iii) Yes, by qn 4.46, every bounded sequence has a convergent subsequence.
(iv) By qn 3.78, the closed interval property.
(v) Its limit is $f(c)$.
(vi) Take $\varepsilon = 1$.
(vii) Choose $n_i > f(c) + 1$, then, from the definition of x_n, $f(x_{n_i}) > n_i > f(c) + 1$, and this contradicts (vi), so our original hypothesis was wrong.

32 As in qn 31, but starting from a sequence (x_n) in the domain of the function such that $f(x_n) < -n$.

33
(i) Denominator not zero on $\mathbb{Q}$, so the function is continuous by qn 6.51.
(ii) Because $\mathbb{Q}$ is not complete, a bounded sequence in $\mathbb{Q}$ need not contain a subsequence which converges to a point of $\mathbb{Q}$.

34
(i) From qn 21.
(ii) From qn 31.
(iii) Every non-empty set of real numbers which is bounded above has a least upper bound: qns 4.80 and 4.81.
(iv) Question 4.64.
(v) Since $a \le x_n \le b$, (x_n) is bounded, and so must contain a convergent subsequence by qn 4.46, every bounded sequence has a convergent subsequence.
(vi) By qn 3.78, the closed interval property.
(vii) Because f is continuous.
(viii) Since $|f(x_n) - M| < 1/n$, the limit of the subsequence $(f(x_{n_i}))$ is M, so $M = f(c)$.
(ix) If $\inf V = m$, construct a sequence (y_n) such that $m \le f(y_n) < m + 1/n$. Argue as in (iv)–(viii).

35 $\sin x = \pm 1$ only when $x = (n + \frac{1}{2})\pi$ which is irrational. But the domain of the function is dense in $\mathbb{R}$ so it attains values arbitrarily close to ± 1.
Since $\mathbb{Q}$ is not complete, the argument of qn 34(v) fails on $\mathbb{Q}$.

36 For unbounded domains use the functions of qn 8.
For bounded domains consider $x \mapsto 1/x$ on $(0, 1)$ or $(0, 1]$, and for bounded ranges consider $x \mapsto x$ on $(0, 1)$ or $(0, 1]$.

37 $|(x^2 - y^2)/(x - y)| = |x + y| \leq |x| + |y| \leq 20$, so
$|x^2 - y^2| \leq 20|x - y|$.

38 $\delta_1 \leq 1/6$. $\delta_2 \leq 1/20$.
$|1/x - 1/a| < 1 \Leftrightarrow |x - a| < |xa| \Leftrightarrow \delta \leq (a - \delta)a \Leftrightarrow \delta \leq a^2/(1 + a)$. As a gets near to 0, an ever smaller δ is required, so there is no one δ for all a.

39 Taking $y = a$ gives the neighbourhood definition of continuity.

40 For the suggested x and y, $|x - y| < 1/n$, but $|x^2 - y^2| > 2$, so there is no universal δ for $\varepsilon = 2$ or less.

41 For the suggested x and y, $|x - y| < 1/n^2$, but $f(x) = 0$ and $f(y) = 1$, so $|f(x) - f(y)| = 1 > \frac{1}{2}$, so there is no universal δ for $\varepsilon = \frac{1}{2}$.

42 Take $\delta = \varepsilon/L$.

43
(i) None at all.
(ii) $a \leq x_n \leq b$.
(iii) Yes, by qn 4.46, every bounded sequence has a convergent subsequence.
(iv) Yes, use qn 3.54(v), the difference rule, and the fact that $(x_n - y_n)$ is a null sequence.
(v) By qn 3.78, the closed interval property.
(vi) Both tend to $f(c)$.
(vii) For sufficiently large n_i both $|f(x_{n_i}) - f(c)| < \frac{1}{2}\varepsilon$ and $|f(y_{n_i}) - f(c)| < \frac{1}{2}\varepsilon$.

44 f is continuous by qn 26, and therefore uniformly continuous on $[0, 1]$ by qn 43. $|x - y| = |\sqrt{x} - \sqrt{y}| \cdot |\sqrt{x} + \sqrt{y}| > |\sqrt{x} - \sqrt{y}|$ provided either x or $y > 1$. To establish uniform continuity on $\mathbb{R}^+ \cup \{0\}$, for a given ε, choose the lesser of the δ from $[0, 1]$ and $\delta = \varepsilon$.

45 Define $g(a) = \lim\limits_{x \to a^+} f(x)$ and $g(b) = \lim\limits_{x \to b^-} f(x)$ and then g is continuous on $[a, b]$ by qn 6.92. So g is uniformly continuous on $[a, b]$ by qn 43 and therefore uniformly continuous on the subset (a, b) where g coincides with f.

46 (i) $x = 1/\sqrt{2}$, (ii) yes, (iii) yes, (v) if $(a_n) \to 1/\sqrt{2}$ then a continuous g would make $(g(a_n)) \to g(1/\sqrt{2})$. But, if the sequence (a_n) only has rational terms, $g(a_n) = f(a_n)$. However, $(f(a_n)) \to +\infty$.

47
(i) If, given $\varepsilon > 0$, $|f(x) - f(y)| < \varepsilon$ when $|x - y| < \delta$; for sufficiently large n, $|a_{n+m} - a_n| < \delta$, so $|f(a_{n+m}) - f(a_n)| < \varepsilon$.

(ii) From qn 3.54(v), the difference rule, $(a_n - b_n)$ is a null sequence, so, for sufficiently large n, $|a_n - b_n| < \delta$, so $|f(a_n) - f(b_n)| < \varepsilon$. Thus $(f(a_n) - f(b_n))$ is a null sequence.

(iii) By (ii), every rational sequence $(a_n) \to a$ gives a sequence $(f(a_n))$ converging to the same limit. So $g(a)$ is well defined.

(iv) $a \le a_n \le b_n \le b$. $|a - b| < \delta \Rightarrow |a_n - b_n| < \delta \Rightarrow |f(a_n) - f(b_n)| < \frac{1}{3}\varepsilon$. For sufficiently large n, $|g(a) - f(a_n)| < \frac{1}{3}\varepsilon$ and $|f(b_n) - g(b)| < \frac{1}{3}\varepsilon$. So $|a - b| < \delta \Rightarrow |g(a) - g(b)| < \varepsilon$.

48 Take $a > 1$, and work with positive x. Let $p, q, r, s \in \mathbb{Z}^+$.

(i) $p/q < r/s \Leftrightarrow ps < qr \Leftrightarrow a^{ps} < a^{qr} \Leftrightarrow a^{p/q} < a^{r/s}$, using qn 2.20.

(ii) $ps < qr \Leftrightarrow g(ps) < g(qr) \Leftrightarrow g(p/q) < g(r/s)$ changing from a to b. From qn 4.41, $g(1/n) \to L$ (say) as $n \to \infty$, so, since g is strictly increasing, $\lim_{x \to 0+} g(x) = L$. Define $g(0) = L$.

(iii) $\dfrac{a^x - a^y}{x - y} = a^y \left(\dfrac{a^{x-y} - 1}{x - y} \right) < a^B \left(\dfrac{a^B - 1}{B} \right)$, supposing $y < x$.

(v) Define $f(-x) = 1/f(x)$.

8

Derivatives

Tangents

Preliminary reading: Bryant ch. 4, Courant and John ch. 2.
Concurrent reading: Hart, Spivak ch. 9 and ch. 10.
Further reading: Mason, Tall (1982).

Definition of derivative

It is easy enough to say when a straight line is a tangent to a circle or an ellipse. For these curves, a straight line – infinitely extended – meets the curve in 0, 1 or 2 points. Each line with a unique point of intersection is a tangent. However, if we try to use such a test to identify tangents to other curves we are in for a disappointment, and on several counts.

1 At how many points does the line $x = 1$ intersect the parabola $y = x^2$? Draw a sketch. Is this line a tangent to the curve?

2 At how many points does the line $y = -2$ intersect the cubic curve $y = x^3 - 3x$? Draw a sketch. Is this line a tangent to the curve?

From qn 2 we learn that whether a line is a tangent to a curve or not is a *local* question, which must be asked relative to the particular point of intersection, that is, inside a sufficiently small neighbourhood of that point.

3 At how many points does the line $y = mx$ intersect the cubic curve $y = x^3$? For how many values of m might the line be a tangent to the curve?
Check that the line $y = h^2x$ is the chord joining the two points $(0, 0)$ and (h, h^3) on the curve, provided $h \neq 0$.
If $h \to 0$, to what does the slope (or gradient) of the chord tend?

4 Write down the equation of the line with slope m through the point $(a, f(a))$.
Write down the equation of the chord joining the points $(a, f(a))$ and $(a+h, f(a+h))$.
If

$$\lim_{h\to 0} \frac{f(a+h)-f(a)}{h} = m,$$

how would you describe the line $y - f(a) = m(x-a)$?

On the strength of the ideas in qn 4, we will define the *derivative* of a function at a point a of its domain.

DEFINITION OF DERIVATIVE
If a is a cluster point of the domain of the real function f and

$$\lim_{h\to 0} \frac{f(a+h)-f(a)}{h} = m, \text{ for some real number } m,$$

then m is called the *derivative* of f at a, usually denoted by $f'(a)$, and f is said to be *differentiable* at a.

The definition is designed to give a formal definition of the slope of the tangent to $y = f(x)$ at $x = a$, and hence to make it possible to define analytically what is meant by a tangent to a curve. On a distance–time graph the slope of a chord gives the average velocity between two points, while the slope of a tangent gives the velocity at a point.

5 Although the motivation for our study of derivatives has been the geometric notion of tangent, there is still one circumstance when a tangent to the graph of a function may exist without a derivative of the function at the point in question. Examine the definition of derivative carefully in order to identify the circumstance in question.

6 If f is a constant function, what is $f'(a)$?

The attempt to establish a converse to qn 6 exposes some unexpected subtleties, and will be examined in qn 9.17 using the Mean Value Theorem.

7 If $f(x) = mx + c$, what is $f'(a)$?

You will have noticed in your calculations for qns 6 and 7, how critical it is that in finding the limit of a function as $h \to 0$ we pay no regard to the value of that function when $h = 0$. In fact the 'slope of the chord' function is not defined when $h = 0$. An equation like $h/h = 1$ is only valid when $h \neq 0$.

8 Show that the two limits

$$\lim_{h\to 0}\frac{f(a+h)-f(a)}{h} \text{ and } \lim_{x\to a}\frac{f(x)-f(a)}{x-a}$$

are equivalent.

9 If $f(x)=x^2$, what is $f'(a)$?

Sums of functions

10 If two real functions f and g are both differentiable at a, prove using qn 6.93, on the algebra of limits, that the function $f+g$ defined by $f+g: x\mapsto f(x)+g(x)$ is differentiable at a, and that

$$(f+g)'(a)=f'(a)+g'(a).$$

Explain how this result may be generalised to show that the sum of n real functions, each differentiable at a, is also differentiable at a.

11 If f is differentiable at a, and $g(x)=k\cdot f(x)$, prove that

$$g'(a)=k\cdot f'(a).$$

Questions 10 and 11 give linearity in determining derivatives, in the sense that if the functions f and g are differentiable at a, then the function given by $x\mapsto l\cdot f(x)+m\cdot g(x)$ has derivative $l\cdot f'(a)+m\cdot g'(a)$ at a.

The product rule

12 By considering the product

$$f(x)-f(a)=\frac{f(x)-f(a)}{x-a}\cdot(x-a), \text{ when } x\neq a,$$

and using qn 6.93 on the algebra of limits, show that if the real function f is differentiable at a, then f is continuous at a.

13 If two real functions f and g are both differentiable at a, prove using qns 6.93 and 12 that the function $f\cdot g$ defined by $f\cdot g: x\mapsto f(x)\cdot g(x)$ is differentiable at a, and that

$$(f\cdot g)'(a)=f(a)\cdot g'(a)+f'(a)\cdot g(a).$$

14 Prove by induction that, if $f(x)=x^n$, then $f'(a)=n\cdot a^{n-1}$, for any positive integer n.
Obtain the same result by considering the equation

$$x^n-a^n=(x-a)(x^{n-1}+x^{n-2}a+x^{n-3}a^2+\ldots+xa^{n-2}+a^{n-1}).$$

15 By using qns 10, 11 and 14 find $f'(a)$ when

$$f(x)=b_0+b_1x+b_2x^2+\ldots+b_nx^n, \text{ and the } b_i \text{ are constant.}$$

16 If $f(x) = 1/x^n$, prove that $f'(a) = -n/a^{n+1}$, for any positive integer n, provided $a \neq 0$.

The quotient rule

17 If two real functions f and g are both differentiable at a, prove using qns 6.93 and 12 that the function f/g defined by $f/g : x \mapsto f(x)/g(x)$ is differentiable at a, and that

$$\left(\frac{f}{g}\right)'(a) = \frac{g(a) \cdot f'(a) - f(a) \cdot g'(a)}{(g(a))^2},$$

provided $g(a) \neq 0$.

The chain rule

18 If the real function g is differentiable at a, and the real function f is differentiable at $g(a)$, we seek to prove that the composite function $f \circ g$ defined by $f \circ g : x \mapsto f(g(x))$ is differentiable at a, and that $(f \circ g)'(a) = f'(g(a)) \cdot g'(a)$.

(i) First suppose that there is a neighbourhood of a in which $g(x) \neq g(a)$, unless $x = a$, and use the continuity of g at a and the equation

$$\frac{f(g(x)) - f(g(a))}{x - a} = \frac{f(g(x)) - f(g(a))}{g(x) - g(a)} \cdot \frac{g(x) - g(a)}{x - a},$$

for $x \neq a$, to prove the result.

(ii) If in every neighbourhood of a there are points x such that $g(x) = g(a)$, the cancellation on which part (i) is based cannot be carried out. For this case only, we construct an argument establishing the chain rule by showing that each side of the equation $(f \circ g)'(a) = f'(g(a)) \cdot g'(a)$ is 0.
Suppose that in every neighbourhood of a there are points $x \neq a$ for which $g(x) = g(a)$.
Let $A = \{x |\ g(x) = g(a), x \neq a\}$ and let

$$G(x) = \frac{g(x) - g(a)}{x - a}, \quad \text{for } x \neq a.$$

Since g is differentiable at a, for any sequence $(a_n) \to a$ with $a_n \neq a$, in the domain of g, $(G(a_n)) \to g'(a)$.
Construct such a sequence in A and deduce that $g'(a) = 0$, so the right-hand side of the equation is 0. Let

$$F(x) = \frac{f(g(x)) - f(g(a))}{x - a}, \quad \text{for } x \neq a.$$

Let (a_n) be any sequence tending to a, with $a_n \neq a$, in the domain of $f \circ g$. We seek to show

$$(F(a_n)) \to 0 = f'(g(a)) \cdot 0 = f'(g(a)) \cdot g'(a) \text{ as } n \to \infty.$$

Each a_n is either in A or not.
If there is a subsequence of terms of (a_n) in A, use the argument for G, above, to show that for these a_n, $(F(a_n)) \to 0$. If there is a subsequence of terms of (a_n) not in A, use the algebra of part (i) and the product rule for sequences to show that, for these a_n, $(F(a_n)) \to f'(g(a)) \cdot g'(a) = 0$.
Deduce that, in any case, $(F(a_n))$ is a null sequence, and that this establishes the required limit.

Differentiability and continuity

19 Sketch the graph of the real function f given by $f(x) = |x|$. For this function find

$$\lim_{h \to 0^+} \frac{f(h) - f(0)}{h} \text{ and } \lim_{h \to 0^-} \frac{f(h) - f(0)}{h}.$$

Deduce that f is not differentiable at $x = 0$.

20 (*Cauchy*, 1821) Define the function $f : \mathbb{R} \to \mathbb{R}$ by

$$f(x) = \sin(1/x) \text{ when } x \neq 0,$$
$$f(0) = k.$$

By considering the two null sequences $(1/2n\pi)$ and $(1/(2n + \frac{1}{2})\pi)$ prove that the function f is not continuous at $x = 0$ whatever value k may have.

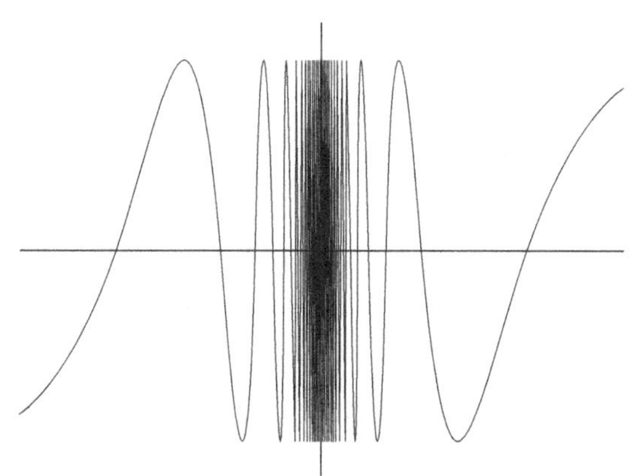

Figure 8.1

21 (*Weierstrass*, 1874) Define the function $f : \mathbb{R} \to \mathbb{R}$ by

$f(x) = x \sin(1/x)$ when $x \neq 0$,
$f(0) = 0$.

Using the fact that $-|x| \leq f(x) \leq |x|$ for all x, and qn 6.36, prove that f is continuous at $x = 0$.
Prove from the definition of derivative that f is not differentiable at $x = 0$.

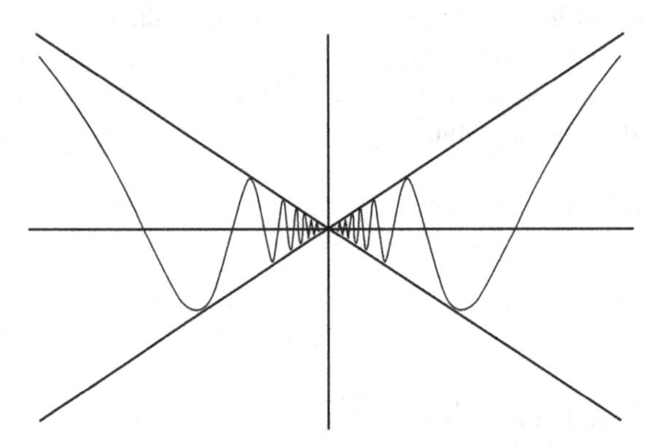

Figure 8.2

22 Define the function $f : \mathbb{R} \to \mathbb{R}$ by

$f(x) = x^2 \sin(1/x)$ when $x \neq 0$,
$f(0) = 0$.

By methods similar to those of qn 21, prove that f is continuous at $x = 0$.
Prove also that f is differentiable at $x = 0$.

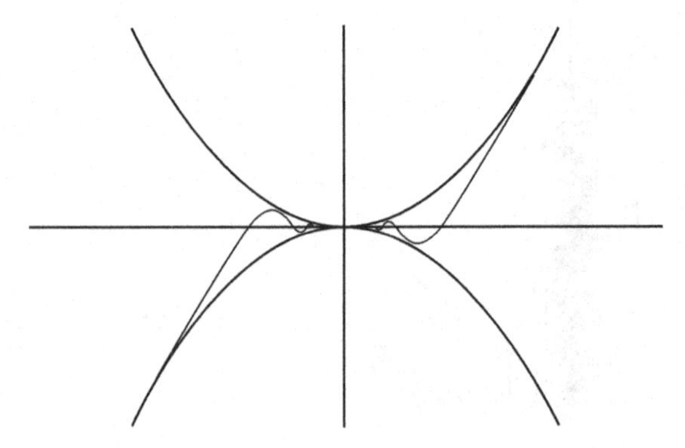

Figure 8.3

In qn 12 we proved that a function that is differentiable at a point is necessarily continuous at that point. We now have two examples (in qns 19 and 21) which illustrate that the converse of this theorem is false. In fact the independence of differentiability and continuity is remarkable, for a function may be continuous at every point and differentiable at no points; such a function, constructed by Tagaki in 1903, with some aspects illustrated by Spivak (page 501) and rediscovered by David Tall (1982), is examined in qn 12.46.

23 If a function f is differentiable at the point a, prove that

$$\lim_{h\to 0} \frac{f(a+h)-f(a-h)}{2h} = f'(a).$$

Give an example to show that this limit may exist when f is not differentiable at a.

When we have a formal definition, however well it seems to match our intuition, it is important to press the meaning of that definition as far as possible. We therefore ask a question which focuses on differentiability at a single point, by considering functions which are, for the most part, not continuous, and therefore not differentiable.

24 The real functions f, g and h are defined on $\mathbb{R}$, such that

Figure 8.4

$$f(x) = \begin{cases} 0 & \text{when } x \text{ is irrational,} \\ 1 & \text{when } x \text{ is rational;} \end{cases}$$

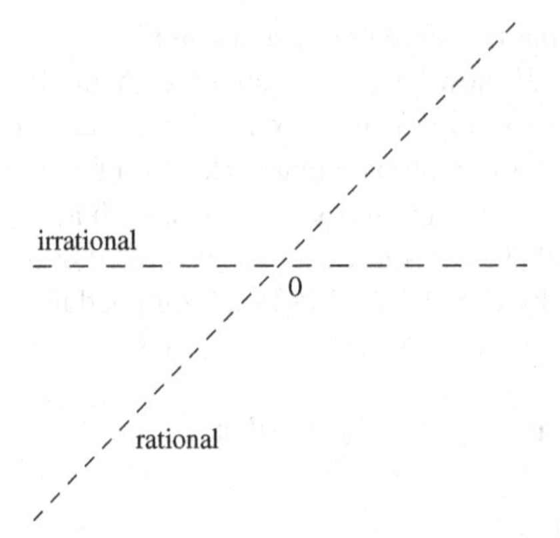

Figure 8.5

$$g(x) = \begin{cases} 0 & \text{when } x \text{ is irrational,} \\ x & \text{when } x \text{ is rational;} \end{cases}$$

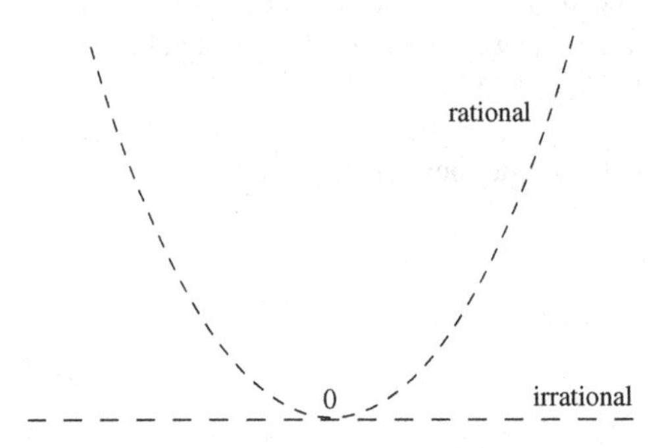

Figure 8.6

$$h(x) = \begin{cases} 0 & \text{when } x \text{ is irrational,} \\ x^2 & \text{when } x \text{ is rational.} \end{cases}$$

Clearly none of these functions is continuous when $x \neq 0$, as in qn 6.20, and therefore none is differentiable when $x \neq 0$.
Determine whether f, g or h is
(i) continuous at $x = 0$; (ii) differentiable at $x = 0$.

Another conflict with intuition comes with the discovery that a function with a positive derivative at a point need not be increasing in any neighbourhood of that point.

25 (i) Use the methods of qn 22 to show, for the function defined by $f(x) = x^2 \sin(1/x^2)$ when $x \neq 0$, and $f(0) = 0$, that $f'(0) = 0$, but that $f'(x)$ takes arbitrarily large values near $x = 0$.

(ii) Draw the graphs of $y = x + 2x^2$ and $y = x - 2x^2$ near the origin. Define the function f by

$$f(x) = x + 2x^2 \cos(1/x) \text{ when } x \neq 0$$
$$f(0) = 0.$$

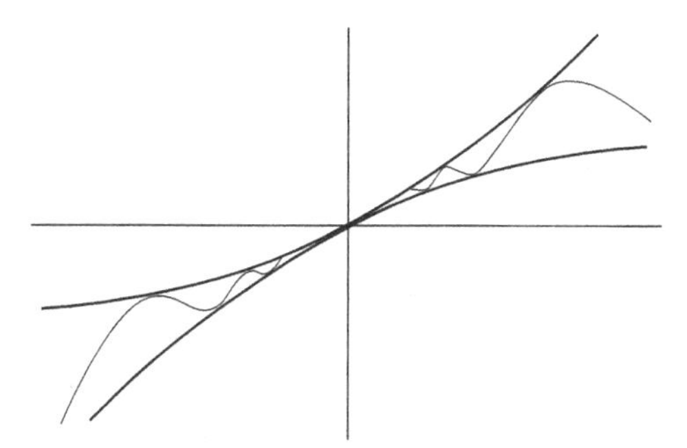

Figure 8.7

Use the method of qn 22 to show that $f'(0) = 1$.
Prove that f is not increasing in any neighbourhood of 0 by showing that $f(1/2n\pi) > f(1/(2n-1)\pi)$ for any $n \in \mathbb{Z}^+$. Show also that $f'(1/(2n - \frac{1}{2})\pi) = -1$.

Derived functions

26 For the real function f given by $f(x) = |x|$, which is defined for all $x \in \mathbb{R}$, find the subset of the domain which consists of those points of $\mathbb{R}$ at which f is differentiable.

If a real function $f: A \to \mathbb{R}$ is differentiable at each point of a subset $B \subseteq A$, with B maximal, we define the *derived function* of f, $f': B \to \mathbb{R}$, by $f': x \mapsto f'(x)$, and we say that f is differentiable on B.

27 Sketch the graph of the derived function of f where f is defined by $f(x) = |x|$.

28 Sketch the graph of the function given by $f(x) = x^2 - 2x + 2$ and of its derived function. How does the point where the graph of the derived function cuts the x-axis relate to the shape of the graph of f?

29 Define the terms *local maximum* and *local minimum* in such a way that they describe respectively the points $(-1, 5)$ and $(1, 1)$ on the graph of the real function defined by $f(x) = x^3 - 3x + 3$. What is $f'(-1)$ and $f'(1)$?

30 If a real function f has a local maximum or a local minimum at a point a of its domain, and f is differentiable at a, prove that $f'(a) = 0$ by considering limits from above and below.

31 Give an example of a function f which shows that it is possible to have $f'(a) = 0$ without the function f having a local maximum or a local minimum at a.

32 Give an example of a function f which shows that it is possible for the function to have a local maximum at $x = a$ without being differentiable at a.

Questions 15 and 17 enable us to find derived functions for polynomials and rational functions. In chapter 11 we will give a formal definition of logarithmic and exponential functions and obtain $\ln'(x) = 1/x$, for positive x, and $\exp'(x) = \exp(x)$, for all real x, and we will give a formal definition of the circular (or trigonometric) functions sine and cosine and obtain $\sin'(x) = \cos(x)$ and $\cos'(x) = -\sin(x)$ for all real x.

According to the Leibnizian description of derived functions, when, for example, $y = x^3$,

$$\frac{dy}{dx} = 3x^2.$$

For the product rule the Leibnizian expression is

$$\frac{d(uv)}{dx} = u \cdot \frac{dv}{dx} + v \cdot \frac{du}{dx}.$$

For the quotient rule the Leibnizian expression is

$$\frac{d\left(\dfrac{u}{v}\right)}{dx} = \frac{v \cdot \dfrac{du}{dx} - u \cdot \dfrac{dv}{dx}}{v^2}.$$

The notation of Leibniz is particularly suggestive when describing the chain rule:

$$\frac{dy}{dx} = \frac{dy}{du} \cdot \frac{du}{dx}.$$

33 Apply the chain rule to find the derived function of f where $f(x) = \sin x^2$ by considering $y = \sin u$ and $u = x^2$.

34 Give algebraic expressions for the derived function f' of qn 22. Is the function f' continuous at $x = 0$?

Second derivatives

35 A real function f is defined by

$$f(x) = \begin{cases} 0 & \text{when } x < 0, \\ x & \text{when } 0 \leq x < 1, \\ \frac{1}{2}x^2 + \frac{1}{2} & \text{when } 1 \leq x. \end{cases}$$

Sketch the graphs of (i) f, (ii) f', (iii) the derived function of f'.

When the real function $f: A \to \mathbb{R}$ has a derived function $f': B \to \mathbb{R}$, and the derived function $f': B \to \mathbb{R}$ has a derived function $(f')': C \to \mathbb{R}$, with $C \subseteq B \subseteq A$, we write $(f')' = f''$ and call f'' the *second derivative* of f.

When $f''(a)$ exists, we say that f is *twice differentiable* at a.

36 How do you know that, when $f''(a)$ exists, f' is continuous at a?

37 A function $f: \mathbb{R} \to \mathbb{R}$ is defined by

$$f(x) = \begin{cases} -x^2 & \text{when } x < 0 \text{ and} \\ x^2 & \text{when } x \geq 0. \end{cases}$$

Determine whether

(i) f is continuous at 0,
(ii) $f'(0)$ exists,
(iii) f' is continuous at 0,
(iv) $f''(0)$ exists.

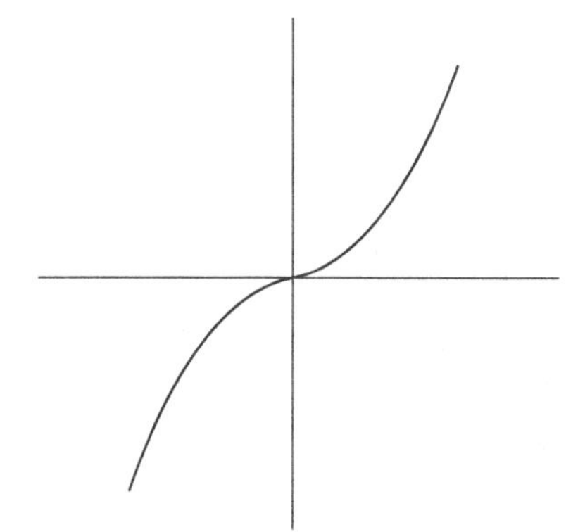

Figure 8.8

38 Define the third derivative of f, and inductively, the nth derivative of f, which is written $f^{(n)}$.

Inverse functions

39 Find the derivative of the function f given by $f(x) = \sqrt{x}$ at some point $a > 0$, from first principles. The function f is continuous at a from qn 6.27.
If $g(x) = x^2$, attempt to relate $f'(a)$ to $g'(f(a))$.

40 If the real function $f: A \to B$ is a bijection, and $g: B \to A$ is its inverse, so that $f(g(b)) = b$ for any $b \in B$, and $g(f(a)) = a$ for any $a \in A$, and if further we suppose that f is differentiable at a, and g is differentiable at $f(a)$, use the chain rule to prove that

$$g'(f(a)) = \frac{1}{f'(a)}.$$

The Leibnizian expression is again particularly suggestive here:

$$\frac{dx}{dy} = 1 \Big/ \left(\frac{dy}{dx}\right).$$

41 If f and g are inverse functions as in qn 40, and $f'(a) = 0$, prove that g cannot be differentiable at $f(a)$.
Indicate by a sketch why this should be expected and relate your answer to qn 5.

42 Let f be a continuous bijection $f: A \to B$ where A and B are open intervals, and let $g: B \to A$ be the inverse of f. We saw in qn 7.25 that g is also a continuous bijection. Now suppose that at some point $a \in A$, f is differentiable and $f'(a) \neq 0$.

(i) Prove that

$$\lim_{x \to a} \frac{g(f(x)) - g(f(a))}{f(x) - f(a)} = \frac{1}{f'(a)}.$$

(ii) Now let $f(a) = b$, and let (b_n) be a sequence in B which tends to b, but which does not contain any term equal to b.
Why must there exist a unique $a_n \in A$, $a_n \neq a$, such that $f(a_n) = b_n$?

(iii) Use the continuity of g to prove that $(a_n) \to a$.

(iv) Use the limit above to prove that

$$\left(\frac{g(b_n) - g(b)}{b_n - b}\right) \to \frac{1}{f'(a)}, \quad \text{as } n \to \infty.$$

(v) Deduce that g is differentiable at $f(a)$.

The only definition or theorem in this chapter which depends on the Completeness Principle is qn 42. Completeness is needed for part (iii). This was necessary because, on rational domains, differentiable bijections do not have to have continuous inverses. See Körner (1991).

43 If the function f is defined by $f(x) = x^n$ for $x \geq 0$ and a positive integer n, identify the inverse function f^{-1}, and find its derived function.
If, for positive x, and positive integers p and q, we define

$$x^{p/q} = (x^p)^{1/q},$$

use the chain rule to evaluate the derived function for $x \mapsto x^{p/q}$. Extend this result to show that for any rational number $r \neq 0$, the derived function of g, where $g(x) = x^r$, for positive x, is given by $g'(x) = rx^{r-1}$.
For what rational numbers r is g differentiable at $x = 0$?

44 Find a maximal interval domain containing zero for which the sine function has a well-defined inverse. Find the derived function of this inverse function (known as arcsine) and state the maximal interval domain for this derived function. You may use any of the properties of the circular functions which are established in chapter 11.

45 Defining $\tan x = (\sin x)/(\cos x)$, provided $\cos x \neq 0$, identify a maximal interval domain containing zero for which the tangent function has a well-defined inverse. Find the derived function of this inverse function (known as arctan) and state the maximal domain for this derived function. You may use the properties of the circular functions established in chapter 11.

Derivatives at end points

We say that a function f is differentiable on a set A when f is differentiable at each point of A. If A is an open interval, the meaning is clear enough, since an open interval contains a neighbourhood of each of its points. But functions such as $x \mapsto \sqrt{x}$ may be defined on $[0, \infty)$ and in such a case it may be important to be able to affirm or deny the differentiability of the function at 0.

46 Look back to the definition of derivative and propose a criterion to determine whether a function $f: [a, b] \to \mathbb{R}$ is differentiable at a. Modify your criterion to determine differentiability (or otherwise) at b.

Summary

Definition qn 4	If a is a cluster point of the domain of a real function f, then, when the limit $\lim_{x \to a} \dfrac{f(x) - f(a)}{x - a}$ exists, f is said to be *differentiable* at a and the value of the limit is denoted by $f'(a)$.
Theorem qn 12	If a function f is differentiable at a, then f is continuous at a.
Theorem	If functions f and g are both differentiable at a, then
The sum rule qn 10	$(f + g)'(a) = f'(a) + g'(a)$,
The product rule qn 13	$(f \cdot g)'(a) = f(a) \cdot g'(a) + f'(a) \cdot g(a)$,
The quotient rule qn 17	and $(f/g)'(a) = (f'(a) \cdot g(a) - f(a) \cdot g'(a))/(g(a))^2$, provided $g(a) \neq 0$.
Theorem qn 15	If $f(x) = b_0 + b_1 x + b_2 x^2 + \ldots + b_n x^n$, then $f'(a) = b_1 + 2b_2 a + 3b_3 a^2 + \ldots + nb_n a^{n-1}$.
The chain rule qn 18	If the function g is differentiable at a and the function f is differentiable at $g(a)$, then $(f \circ g)'(a) = f'(g(a)) \cdot g'(a)$.
Definition qn 29	A function f is said to have a *local maximum* at a if, for all x in some neighbourhood of a, $f(x) \leq f(a)$. A function f is said to have a *local minimum* at a if, for all x in some neighbourhood of a, $f(x) \geq f(a)$.
Theorem qn 30	If the function f has a local maximum or a local minimum at a, and f is differentiable at a, then $f'(a) = 0$.
Definition qns 36, 38	If $f: A \to \mathbb{R}$ is differentiable at each point of $B \subseteq A$, then $f': B \to \mathbb{R}$ is called the *derived function* or *derivative* of f. By induction we define the nth *derivative* of f, denoted by $f^{(n)}$, as the derivative of $f^{(n-1)}$.
Theorem qn 40	If f is a bijection and g is its inverse, and if f is differentiable at a and g is differentiable at $f(a)$, then $g'(f(a)) = 1/f'(a)$.

Historical Note

From one point of view, most of the material of this chapter was known by mathematicians before Newton's creative burst in 1664–66. From another point of view, none of the results of this chapter were proved in the sense in which we take them today until the time of Weierstrass, 1860–70.

The essential backcloth to the calculus is Descartes' method of converting geometric problems into algebraic ones (coordinate geometry) and his notation for exponents (1637), which we use today. Fermat (about 1640) knew how to find the slope of the tangent to the graph of a polynomial using the limit of the slope of a chord, and found maxima and minima by examining the slope of the chord near the point. There were in fact a host of special methods for dealing with special curves.

Both Newton (when he was away from Cambridge for fear of the plague) in 1665 and 1666, and Leibniz in Paris in 1675, working independently, systematically established the product, quotient and chain rules and discovered the inverse relationship between slopes of tangents and areas under curves. Having discovered the Binomial Theorem for rational index, Newton was strongly motivated by what he could do by differentiating and integrating power series term by term. Newton considered his variables as real entities which were changing with time, and he expressed the derivative of the 'fluent' x, as the 'fluxion' $\dot{x}$. Leibniz had originally been motivated by the use of differences in summing series (as in qns 5.3 and 5.4). In this notation, dx originally meant the difference between x and its next value, and dy/dx was originally a ratio of infinitesimals. Leibniz' great discovery was of the results he could generate using the characteristic triangle with sides dx, dy and ds. Leibniz' dy/dx was expressed by Newton in the form $\dot{y}/\dot{x}$.

The first calculus book was published in 1696. This was *Analyse des Infiniments Petits* by the Marquis de l'Hôpital. It was repeatedly reprinted during the eighteenth century. The book consists of lectures on the differential calculus which Johann Bernoulli gave to de l'Hôpital in 1691, on Leibniz' work. A little of Newton's calculus appeared in his *Principia* (1687), much more was in the appendix to his *Opticks* (1704), but his original exposition *De Analysi* (written in 1669 and circulated to friends) was not published until 1711. Both Newton and Leibniz originally thought of limits in terms of infinitesimals, but later in life tried to avoid them. Newton spoke of the ratio of evanescent quantities becoming an ultimate ratio, as the quantities vanish. Leibniz eventually came to speak of infinitesimals as convenient fictions.

In Bishop Berkeley's tract of 1734, the apparently self-contradictory nature of limits was highlighted. In affirming that, as x tends to a, $(x^2 - a^2)/(x - a)$ tends to $2a$, x is not equal to a at the beginning of the computation, and this allows the division to be performed, and then x is put equal to a at the end of the computation to obtain the answer. This paradox was not resolved during the eighteenth century, but it did not discourage mathematicians in their work. Indeed, an immediate chronological successor to Bishop Berkeley was L. Euler, the most ingenious manipulator of infinitesimals there has ever been. Berkeley's criticisms were not forgotten and in 1764 d'Alembert offered the useful notion of the value of the function differing from the limit 'by as little as one wishes', a phrase which was taken up by Cauchy.

The origin of both Newton's and Leibniz' calculus was geometric and it was only late in the eighteenth century that J. L. Lagrange (1797) tried to make algebraic definitions of its fundamental notions. Having, as he thought, proved that every function has a Taylor series he showed that $f(a+h) = f(a) + hf'(a) + hr(h)$ where $r(h)$ tends to 0 as h tends to 0, seemingly without reference to infinitesimals, and used this to prove (the need for proof being itself a remarkable insight) that a differentiable function with positive derivatives was increasing. It was a small step from this to Cauchy's methods. It is to Lagrange that we owe the notion of the derived function $f'(x)$ as against the ratio of infinitesimals dy/dx, and also the word *derivative*.

In 1823, Cauchy offered an $\varepsilon - \delta$ neighbourhood description of derivatives, and this largely resolved the problem of the definition of limits and therefore made it possible to construct rigorous proofs using limits. Cauchy used infinitesimals, but he defined them as variables which were tending to zero, not as indivisibles as they had been considered in the seventeenth century. However, continuous functions were the context of Cauchy's work on differentiation, so that the proposition that a differentiable function was necessarily continuous was for Cauchy a matter of definition, not a theorem. Indeed, some contemporaries of Cauchy believed that they had successfully proved that continuous functions must be differentiable! Also Cauchy's functions were continuous, or differentiable, on intervals, so that he did not conceive of a function that was defined everywhere, but continuous at only one point. Defining continuity and differentiability in terms of the behaviour of functions at a single point followed the work of Riemann (1854) and Weierstrass (1860). Likewise until the time of Weierstrass, with the awareness of the completeness of the real numbers and of the possibility of continuous but non-differentiable functions, it was not realised that the result of our qn 42 was distinct from that of qn 40 on the invertibility of derivatives.

The advent of rigour did not come to the calculus all at once. Distinguished mathematicians in the twentieth century have failed to provide a rigorous treatment of the chain rule. (See Hardy, 9th edition, ch. 6, §114.)

Answers

1 One. No.

2 Two. Yes.

3 Three if $m > 0$. One if $m \le 0$. Only $m = 0$. Slope $\to 0$.

4 $y - f(a) = m(x - a)$.
$y - f(a) = \dfrac{f(a+h) - f(a)}{h}(x - a)$. With the given m this is a tangent.

5 There may still be a tangent when the slope of the chord $\to \infty$.

6 $f'(a) = 0$.

7 $f'(a) = m$.

8 If, given $\varepsilon > 0$, $0 < |h| < \delta \Rightarrow \left|\dfrac{f(a+h) - f(a)}{h} - l\right| < \varepsilon$,

then $0 < |x - a| < \delta \Rightarrow \left|\dfrac{f(x) - f(a)}{x - a} - l\right| < \varepsilon$; and conversely.

9 $2a$.

10 Use 6.93 and generalise by induction as in qn 6.25.

11 $\dfrac{g(x) - g(a)}{x - a} = k \cdot \dfrac{f(x) - f(a)}{x - a}$ and use qn 6.93, the algebra of limits.

12 By qn 6.93, the algebra of limits, $\lim_{x \to a} f(x) - f(a) = f'(a) \cdot 0$.
So $\lim_{x \to a} f(x) = f(a)$ and f is continuous at a by qn 6.89.

13 $\dfrac{f(x) \cdot g(x) - f(a) \cdot g(a)}{x - a} = f(x) \cdot \dfrac{g(x) - g(a)}{x - a} + g(a) \cdot \dfrac{f(x) - f(a)}{x - a}$.
This tends to $f(a) \cdot g'(a) + g(a) \cdot f'(a)$ by qn 6.93, the algebra of limits, and the continuity of f.

14 Trivial for $n = 1$. If $f(x) = x^n \Rightarrow f'(a) = na^{n-1}$, then for $g(x) = x^n x$,
$g'(a) = a^n \cdot 1 + na^{n-1} \cdot a = (n+1)a^n$ by qn 13.

$\lim_{x \to a} x^{n-1} + x^{n-2}a + \ldots + a^{n-1} = a^{n-1} + a^{n-1} + \ldots + a^{n-1} = na^{n-1}$

by qn 6.93, the algebra of limits.

15 $f'(a) = b_1 + 2b_2 a + 3b_3 a^2 + \ldots + nb_n a^{n-1}$.

16 $\dfrac{f(x) - f(a)}{x - a} = \dfrac{1/x^n - 1/a^n}{x - a} = \dfrac{a^n - x^n}{a^n x^n (x - a)} \to \dfrac{-na^{n-1}}{a^n a^n}$, using qn 14.

17
$$\frac{f(x)/g(x) - f(a)/g(a)}{x - a} = \frac{f(x) \cdot g(a) - f(a) \cdot g(x)}{g(x) \cdot g(a) \cdot (x - a)}$$
$$= \frac{g(a)(f(x) - f(a)) - f(a)(g(x) - g(a))}{g(x) \cdot g(a) \cdot (x - a)}$$

which gives the required result by qn 6.93, the algebra of limits, and the continuity of $1/g$ at a.

18 (ii) $\lim_{x \to a} \frac{g(x) - g(a)}{x - a} = g'(a)$.

If, in every neighbourhood of a, there are points $x \neq a$ such that $g(x) = g(a)$ then $g'(a) = 0$: because $G(a_n) = 0$ for all $a_n \in A$

19 First limit $= 1$, second limit $= -1$, so two-sided limit does not exist.

20 If $a_n = 1/(2n\pi)$ then $f(a_n) = 0$. If $b_n = 1/((2n + \frac{1}{2})\pi)$ then $f(b_n) = 1$.
So $(f(a_n))$ and $(f(b_n))$ have different limits and therefore f is not continuous at 0 whatever the value of $f(0)$. Compare with qn 6.19.

21 $x \mapsto |x|$ is continuous, by qn 6.32.
$x \mapsto -|x|$ is continuous, by qns 6.32 and 6.26.
So f is continuous, by qn 6.36. But

$$\frac{f(x) - f(0)}{x - 0} = \sin\frac{1}{x}, \text{ for } x \neq 0,$$

and this has no limit as $x \to 0$.

22 $x \mapsto x^2$ and $x \mapsto -x^2$ are both continuous, by qn 6.29.

$-x^2 \leq f(x) \leq x^2$ so f is continuous at $x = 0$, by qn 6.36.

$\frac{f(x) - f(0)}{x - 0} = x \sin\frac{1}{x}$ when $x \neq 0$ and from qn 21 this $\to 0$ as $x \to 0$.

23 If $\lim_{h \to 0} \frac{f(a+h) - f(a)}{h} = f'(a)$,

then $\lim_{h \to 0} \frac{f(a-h) - f(a)}{-h} = f'(a)$. Add.

The result follows from qn 6.93, the algebra of limits.
The limit exists when $a = 0$ for $f(x) = |x|$, but this function is not differentiable at 0 by qn 19.

24 None of these functions is continuous when $x \neq 0$, using arguments like that of qn 6.20.
Because f is not continuous at 0 by qn 6.20, f is not differentiable at 0 by qn 12.
The function g is continuous at 0 by qn 6.35. But g is not differentiable at $x = 0$ because

$$\frac{g(x) - g(0)}{x - 0} = f(x) \text{ when } x \neq 0,$$

and f has no limit as $x \to 0$.

Since $0 \le h(x) \le x^2$ for all x, h is continuous at 0 by qn 6.36.

$$\frac{h(x)-h(0)}{x-0} = g(x) \text{ when } x \neq 0, \text{ and } \lim_{x\to 0} g(x) = g(0)$$

so h is differentiable at 0.

25 (i) $(f(x)-f(0))/(x-0) = x\sin(1/x^2)$ which tends to 0 as x tends to 0, as in qn 21. When $x \neq 0$, $f'(x) = 2x\sin(1/x^2) - 2/x\cos(1/x^2)$. $f'(1/\sqrt{(2n+1)\pi}) = 2\sqrt{((2n+1)\pi)}$.

(ii) $\dfrac{f(x)-f(0)}{x-0} = 1 + 2x\cos\dfrac{1}{x} \to 1$ as $x \to 0$, as in qn 21.

$$f\left(\frac{1}{2n\pi}\right) > f\left(\frac{1}{(2n-1)\pi}\right) \Leftrightarrow \frac{1}{2n} + \frac{2}{4n^2\pi} > \frac{1}{2n-1} - \frac{2}{(2n-1)^2\pi}$$

$$\Leftrightarrow \frac{2}{\pi}\left(2 + \frac{1}{4n^2-2n}\right) > 1.$$

When $x \neq 0$, $f'(x) = 1 + 4x\cos(1/x) + 2\sin(1/x)$.
When f' is positive throughout an interval, this counter-intuitive possibility cannot occur, as we shall see in the next chapter.

26 $\mathbb{R}\backslash\{0\}$.

27 When $x > 0$, $f'(x) = 1$. When $x < 0$, $f'(x) = -1$.

28 $f'(1) = 0$. $f(x) \ge f(1)$ for all x. So $f(1)$ is the minimum value of the function.

29 If for all x in some neighbourhood of a, $f(x) \le f(a)$, then f has a *local maximum* at a. If for all x in some neighbourhood of a, $f(x) \ge f(a)$, then f has a *local minimum* at a. $f'(-1) = f'(1) = 0$.

30 If f has a local maximum at a then in some neighbourhood of a

$$\frac{f(x)-f(a)}{x-a} \ge 0 \text{ when } x < a, \text{ and } \frac{f(x)-f(a)}{x-a} \le 0 \text{ when } x > a.$$

Therefore $\displaystyle\lim_{x\to a^-} \frac{f(x)-f(a)}{x-a} \ge 0$ and $\displaystyle\lim_{x\to a^+} \frac{f(x)-f(a)}{x-a} \le 0$.

Since the two-sided limit exists it equals 0.

31 $f(x) = x^3$ at $x = 0$.

32 For example, $x \mapsto -|x|$ at $x = 0$.

33 $f'(x) = (\cos x^2)(2x)$.

34 $f'(x) = 2x\sin(1/x) - \cos(1/x)$, when $x \neq 0$,
$f'(0) = 0$.
f' is not continuous at 0, since $x \mapsto \cos(1/x)$ is like the function in qn 20.
So f' is not differentiable at 0.

35 f has domain $\mathbb{R}$. f' has domain $\mathbb{R}\backslash\{0\}$. f'' has domain $\mathbb{R}\backslash\{0, 1\}$.

36 Apply qn 12 to f'.

37 Consider limits from below and limits from above at 0.
(i) Yes. (ii) Yes. (iii) Yes. (iv) No.

38 f''' is the derived function of f''.
$f^{(n)}$ is the derived function of $f^{(n-1)}$.

39 $$\frac{f(x)-f(a)}{x-a} = \frac{1}{\sqrt{x}+\sqrt{a}} \to \frac{1}{2\sqrt{a}} \text{ as } x \to a.$$

So $f'(a) = 1/(2\sqrt{a})$ and $g'(f(a)) = 2\sqrt{a}$.

41 $g'(f(a)) \cdot f'(a) = 1$. But if $f'(a) = 0$, there is no possible value for $g'(f(a))$.

Consider the graphs of the functions f and g in qn 39 at 0.

42 (i) $$\frac{g(f(x))-g(f(a))}{f(x)-f(a)} = \frac{x-a}{f(x)-f(a)} = 1\Big/\left(\frac{f(x)-f(a)}{x-a}\right).$$

Use the fact that $f'(a)$ exists and is non-zero, with qn 6.93, the algebra of limits.

(ii) Because f is a bijection.

(iii) $(b_n) \to b$ by definition. This may be rewritten $(f(a_n)) \to f(a)$. By the continuity of g at $b = f(a)$, $(g(f(a_n))) \to g(f(a))$. So $(a_n) \to a$.

(iv) From the limit in (i),

$$\left(\frac{g(f(a_n))-g(f(a))}{f(a_n)-f(a)}\right) \to \frac{1}{f'(a)},$$

so

$$\left(\frac{g(b_n)-g(b)}{b_n-b}\right) \to \frac{1}{f'(a)}.$$

(v) From the definition of (b_n), limits from above and below b exist and are equal, so the two-sided limit

$$\lim_{y\to b} \frac{g(y)-g(b)}{y-b} = \frac{1}{f'(a)}.$$

Thus $g'(f(a))$ exists and equals $1/f'(a)$.

43 If $f(x) = x^n$, then $f^{-1}(x) = \sqrt[n]{x}$. From qn 42, f^{-1} is differentiable except at 0.

$f^{-1\prime}(a) = f^{-1\prime}(f(\sqrt[n]{a})) = 1/f'(\sqrt[n]{a}) = 1/(n(\sqrt[n]{a})^{n-1}) = (1/n)a^{(1/n)-1}$.
Now let $P(x) = x^p$ and $Q(x) = x^{1/q}$, then $Q(P(x)) = x^{p/q}$.

$(Q \circ P)'(x) = Q'(P(x)) \cdot P'(x) = (1/q)(x^p)^{(1/q)-1} p x^{p-1} = (p/q)x^{(p/q)-1}$.

Use qn 17 to extend to negative rational indices.

g is differentiable at $x = 0$ provided $r \geq 1$.

44 $\left[-\frac{1}{2}\pi, \frac{1}{2}\pi\right] \to [-1, 1]$.

$\sin'(\arcsin x) \cdot \arcsin' x = 1$, so

$\arcsin' x = 1/\cos(\arcsin x) = 1/\sqrt{(1 - x^2)}$.

45 $\left(-\frac{1}{2}\pi, \frac{1}{2}\pi\right)$. $\tan' x = (\cos^2 x + \sin^2 x)/(\cos^2 x) = 1 + \tan^2 x$.

$\tan'(\arctan x) \cdot \arctan' x = 1$, so $\arctan' x = 1/(1 + x^2)$.

46 $f\colon [a, b] \to \mathbb{R}$ is differentiable at a if $\lim_{x \to a^+} \dfrac{f(x) - f(a)}{x - a}$ exists and is finite.
Use the limit from below for b.

9

Differentiation and completeness

Mean Value Theorems, Taylor's Theorem

Concurrent reading: Burkill ch. 4 and §5.8, Courant and John ch. 5.
Further reading: Mason, Spivak chs 11, 20.

Both Rolle's Theorem and the Mean Value Theorem are geometrically transparent. Each claims, with slightly more generality in the case of the Mean Value Theorem, that for the graph of a differentiable function, there is always a tangent parallel to a chord. It is something of a surprise to find that such intuitive results can lead to such powerful conclusions: namely, de l'Hôpital's rule and the existence of power series convergent to a wide family of functions.

Rolle's Theorem

1 Sketch the graphs of some differentiable functions

$$f\colon [a, b] \to \mathbb{R},$$

for which $f(a) = f(b)$.
Can you find a point c, with $a < c < b$, such that $f'(c) = 0$, for each function f which you have sketched?
What words could you use to describe the point c in relation to the function f?

2 Must *any* differentiable function

$$f\colon [a, b] \to \mathbb{R}$$

have a maximum and a minimum value? Why?
Must *any* such function, for which $f(a) = f(b)$, have a maximum *and* a minimum value in the *open* interval (a, b)?

3 If $f\colon [a, b] \to \mathbb{R}$ is differentiable and, for some c such that $a < c < b$, $f(x) \le f(c)$ for all $x \in [a, b]$, must $f'(c) = 0$? Why?

4 If $f\colon [a, b] \to \mathbb{R}$ is differentiable,
$f(a) = f(b)$,
$M = \sup\{f(x) |\ a \le x \le b\}$,
and $m = \inf\{f(x) |\ a \le x \le b\}$,
use qn 2 to show that there is a c with $a < c < b$ for which $f(c) = M$, or a c for which $f(c) = m$, or possibly distinct cs satisfying each of these conditions.
Deduce from qn 3 that $f'(c) = 0$.

5 Sketch the graph of a continuous function $f\colon [a, b] \to \mathbb{R}$ which is differentiable on the open interval (a, b), but not differentiable at the points a or b.

6 Identify the two places in the proof of the result of qn 4 at which the differentiability of the function f has been invoked.

7 *Rolle's Theorem*
If the function $f\colon [a, b] \to \mathbb{R}$ satisfies the conditions

(i) f is continuous on $[a, b]$,
(ii) f is differentiable on (a, b),
(iii) $f(a) = f(b)$,

show that there is a point c in the open interval (a, b) such that $f'(c) = 0$.

8 Sketch the graphs of functions for which one or more of the conditions (i), (ii) and (iii) in qn 7 fail, illustrating how the conditions are necessary for Rolle's Theorem, and also how, even in their absence, there may still exist a point $c \in (a, b)$ for which $f'(c) = 0$.

9 If $a_0 + a_1/2 + a_2/3 + \ldots + a_n/(n + 1) = 0$, prove that the polynomial function

$$x \mapsto a_0 + a_1x + a_2x^2 + \ldots + a_nx^n$$

has a zero in the interval $[0, 1]$.

10 If the function $f\colon \mathbb{R} \to \mathbb{R}$ is twice differentiable and $f(a) = f(b) = f(c) = 0$, with $a < b < c$, prove that for some $d \in (a, c)$, $f''(d) = 0$. Generalise.

11 (i) The function $f\colon \mathbb{Q} \to \mathbb{Q}$ is defined by $f(x) = 1/(x^2 - 2)^2$. Show that the function f satisfies the conditions (i), (ii) and (iii) of qn 7 on $[0, 2] \cap \mathbb{Q}$. Why does the conclusion of Rolle's Theorem fail in this case?

(ii) The function $g : \mathbb{Q} \to \mathbb{Q}$ is defined by $g(x) = x^3 - 21x + 20$. Show that the function g satisfies the conditions (i), (ii) and (iii) of qn 7 on $[1, 4] \cap \mathbb{Q}$. Why does the conclusion of Rolle's Theorem fail in this case?

An intermediate value theorem for derivatives

Although we have seen that differentiable functions do not necessarily have continuous derivatives (in qn 8.34), derivatives have an intermediate value property which we might only expect from a continuous function. So the intermediate value property cannot be used to define continuity.

12 (*Darboux's Theorem*, 1875) If $[a, b]$ is contained in the domain of a real differentiable function f and $f'(a) < k < f'(b)$, determine whether the function g defined by

$$g(x) = f(x) - kx$$

is

(i) a real function which is continuous on $[a, b]$,
(ii) a function which is bounded and attains its bounds on $[a, b]$,
(iii) a function which is differentiable on $[a, b]$,
(iv) a function which is not constant on $[a, b]$.

Determine the derivative of g at a point where it attains a bound, and deduce that, for some c in the open interval (a, b), $f'(c) = k$. Check the compatibility of this result with the differentiable function in qn 8.22 having discontinuous derivative, using graph-drawing facilities on a computer.

The Mean Value Theorem

The next question looks at a slanted version of Rolle's Theorem, and identifies conditions under which an arc has a tangent parallel to a chord joining two points of the arc.

13 A function $f : [a, b] \to \mathbb{R}$ is continuous on the closed interval $[a, b]$ and differentiable on the open interval (a, b).

(i) Construct a function D which measures the distance (in a direction parallel to the y-axis) between the graph of the function f and the chord joining $(a, f(a))$ to $(b, f(b))$. Does D satisfy the conditions of Rolle's Theorem? What is the result of applying Rolle's Theorem to D, when expressed in terms of the function f?

(ii) Let the slope of the chord from $(a, f(a))$ to $(b, f(b))$ be

$$\frac{f(b) - f(a)}{b - a} = K.$$

Define the function $F: [a, b] \to \mathbb{R}$ by

$$F(x) = f(b) - f(x) - K(b - x).$$

Check that F satisfies the conditions of Rolle's Theorem on $[a, b]$. Apply Rolle's Theorem to the function F, to obtain a value for K in terms of the function f'.

You should have obtained the same result from qn 13(i) and 13(ii).
This result is called the *Mean Value Theorem*. One application is that, on a continuous and smooth journey, the average speed of the journey must actually be reached at least once during the journey. While this result looks obvious enough, it has powerful and significant applications.

14 Show that the conclusion of the Mean Value Theorem applied to a function f on an interval $[a, a + h]$ may be expressed by saying that, under the appropriate conditions, there is a θ, with $0 < \theta < 1$, such that $f(a + h) - f(a) = hf'(a + \theta h)$.

15 A function $f: [a, b] \to \mathbb{R}$ is continuous on the closed interval $[a, b]$ and differentiable on the open interval (a, b).
Prove that if $f'(x) > 0$ for all values of x in (a, b), then the function f is strictly monotonic increasing.
Show also that if $f'(x) < 0$ for all values of x, then the function f is strictly monotonic decreasing.

(16) Functions f_1, f_2, f_3 and f_4 are defined on $[0, 2\pi]$ by

$$f_1(x) = x - \sin x, \quad f_2(x) = \cos x - 1 + \frac{x^2}{2},$$

$$f_3(x) = \sin x - x + \frac{x^3}{6}, \quad f_4(x) = 1 - \frac{x^2}{2} + \frac{x^4}{24} - \cos x.$$

Prove that each of these four functions is strictly increasing on the given domain.
Deduce that

$$x > \sin x > x - \frac{x^3}{6},$$

and

$$1 - \frac{x^2}{2} + \frac{x^4}{24} > \cos x > 1 - \frac{x^2}{2}, \text{ on } (0, 2\pi).$$

Use graph-drawing facilities on a computer to compare the graphs of sine and cosine with those of the polynomials given here.

17 (*Cauchy*, 1823) A function $f: [a, b] \to \mathbb{R}$ is continuous on the closed interval $[a, b]$ and differentiable on the open interval (a, b).
Prove that if $f'(x) = 0$ for all values of x, then the function f is constant.
Construct a real function f which is not constant, but for which $f'(x) = 0$ for every point x of its domain.

18 A function $f: [a, b] \to [a, b]$ is continuous on the closed interval $[a, b]$ and differentiable on the open interval (a, b).
If $|f'(x)| \leq L$ for all values of x, prove that

$$|f(x) - f(y)| \leq L \cdot |x - y| \text{ for all } x \text{ and } y \text{ in the domain,}$$

so that f satisfies a Lipschitz condition, and is uniformly continuous as in qn 7.42.
If a sequence (a_n) is defined in $[a, b]$ by $a_{n+1} = f(a_n)$, prove that

$$|a_{n+m} - a_{n+1}| \leq L^n |a_m - a_1| \leq L^n |b - a|.$$

If, further, $L < 1$, deduce that (a_n) is a Cauchy sequence, and hence convergent.
If (b_n) is a sequence in $[a, b]$ defined by $b_{n+1} = f(b_n)$ and $(a_n) \to A$, and $(b_n) \to B$, explain why A and B lie in $[a, b]$, why $f(A) = A$ and $f(B) = B$, and prove that $A = B$.

The condition $|f'(x)| \leq L < 1$ makes the function f, of qn 18, an example of a *contraction mapping* of the domain, which, as we have shown, has a unique fixed point, satisfying $f(x) = x$. This result is particularly useful for calculating better approximations to the roots of given equations and provides a reason for the convergence of sequences defined by an iterative formula. The condition $f'(x) < 1$ (without the L and without the modulus) is sufficient to guarantee the uniqueness of the fixed point. At least one fixed point may be deduced from qn 7.20, but two or more fixed points would give an x such that $f'(x) = 1$ by the Mean Value Theorem. Though if $\mathbb{R}$ is the domain of f, there need be no fixed point as with $f(x) = \frac{1}{2}(x + \sqrt{(x^2 + 4)})$.

19 Prove that the equation $x = \cos x$ has exactly one root in the interval $[0, \frac{1}{2}\pi]$.

20 Prove that the function defined by $f(x) = \sqrt{(x + 2)}$ is a contraction mapping on the interval $[0, \infty)$. Deduce that the sequence defined in qn 3.79 is convergent when the first term is positive.

21 For a continuous function $f: [a, b] \to [a, b]$ the existence of a root for $f(x) = x$ is guaranteed, as we found in qn 7.20. Illustrate how, in the absence of the condition $|f'(x)| < 1$,

either (a) multiple roots for $f(x) = x$ may exist,
or (b) the sequence (a_n), defined in qn 18, may not converge.

22 A function $f: [a, b] \to \mathbb{R}$ is continuous on the closed interval $[a, b]$ and differentiable on the open interval (a, b).
If, for some c, with $a < c < b$, $f'(c) = 0$, and $f''(c)$ exists and > 0, prove that there is a neighbourhood of c within which $f'(x) > 0$ when $x > c$, and $f'(x) < 0$ when $x < c$. Deduce that f has a local minimum at c.

The result of qn 22 may seem deceptively straightforward. However, the two conditions, $f'(c) = 0$ and $f''(c) > 0$, do not, by themselves, imply the continuity of derivatives other than at $x = c$.

(23) Let real functions $f_n : \mathbb{R} \to \mathbb{R}$ be defined by

$$f_n(x) = 1 - x + \frac{x^2}{2!} - \frac{x^3}{3!} + \ldots + (-1)^n \frac{x^n}{n!}.$$

Check that $f'_{n+1} = -f_n$.
If n is even and $f_n(x) > 0$ for all x, prove that f_{n+1} is monotonic decreasing and has one and only one root. The existence of at least one root comes from the Intermediate Value Theorem as in qn 7.19.
Deduce that there is exactly one value of x at which $f'_{n+2}(x) = 0$, that f_{n+2} is minimal at this point, and that $f_{n+2}(x) > 0$ at this point so that f_{n+2} is positive everywhere.
Now use induction to prove that f_n is positive when n is even and has a unique root when n is odd.

24 (i) If the domain of a real function f contains at least the closed interval $[x, y]$ and $x \neq y$, show that the point $(\theta x + (1-\theta)y, \theta f(x) + (1-\theta)f(y))$ lies on the line segment joining the two points $(x, f(x))$ and $(y, f(y))$, provided $0 < \theta < 1$.
Draw a diagram.
How would you describe the inequality $f(\theta x + (1-\theta)y) < \theta f(x) + (1-\theta)f(y)$ according to your diagram?
When such an inequality holds throughout the domain of definition of a function and for all θ, with $0 < \theta < 1$, the function is said to be *concave upwards* (or *convex downwards*).

(ii) Suppose that $x < y$, and let $s = \theta x + (1-\theta)y$.
Show that the inequality in (i) is equivalent to

$$\frac{f(s) - f(x)}{s - x} < \frac{f(y) - f(s)}{y - s}.$$

(iii) If the function f is differentiable on an interval including x and y, use the Mean Value Theorem to show that the inequality in (ii) would follow if it was known that f' was strictly increasing.

(iv) If $f''(x) > 0$ on an interval, use qn 15 to deduce that f is concave upwards on that interval.

Cauchy's Mean Value Theorem

25 (*Cauchy*, 1823) Functions $f\colon [a,b] \to \mathbb{R}$ and $g : [a,b] \to \mathbb{R}$ are continuous on the closed interval $[a,b]$, and differentiable on the open interval (a,b), and $g'(x) \neq 0$ for any $x \in (a,b)$. Let

$$\frac{f(b) - f(a)}{g(b) - g(a)} = K.$$

Prove that the conditions of Rolle's Theorem hold for the function $F\colon [a,b] \to \mathbb{R}$ defined by

$$F(x) = f(b) - f(x) - K(g(b) - g(x)).$$

Apply Rolle's Theorem to F to show that, for some c such that $a < c < b$,

$$K = \frac{f'(c)}{g'(c)}.$$

de l'Hôpital's rule

26 Functions $f\colon [a,b] \to \mathbb{R}$ and $g : [a,b] \to \mathbb{R}$ are continuous on the closed interval $[a,b]$ and differentiable on the open interval (a,b). We further assume that $f(a) = g(a) = 0$ and $g'(x) \neq 0$ for any x in (a,b). What does Cauchy's Mean Value Theorem give in this case?
Now suppose that

$$\lim_{x \to a^+} \frac{f'(x)}{g'(x)} = l.$$

Prove that, given $h > 0$,

$$\left| \frac{f(a+h)}{g(a+h)} - l \right| = \left| \frac{f'(a+\theta h)}{g'(a+\theta h)} - l \right|$$

for some θ, with $0 < \theta < 1$.
Deduce that if $0 < k < \delta$ implies

$$\left| \frac{f'(a+k)}{g'(a+k)} - l \right| < \varepsilon,$$

then $0 < h < \delta$ implies

$$\left| \frac{f(a+h)}{g(a+h)} - l \right| < \varepsilon,$$

so that

$$\lim_{x \to a^+} \frac{f(x)}{g(x)} = l.$$

It is easy to see that a result analogous to that of qn 26 can be obtained using limits from the left if appropriate conditions apply. When the functions f and g are differentiable at a, either result may be used.

27 Use de l'Hôpital's rule, qn 26, to find

(i) $\lim_{x \to 0} \dfrac{\sqrt{(x+2)} - \sqrt{2}}{\sqrt{(x+1)} - 1}$,

(ii) $\lim_{x \to 0} \dfrac{e^x - 1}{x}$

(iii) $\lim_{x \to 0} \dfrac{\sin x}{x}$, hence $\lim_{x \to 0} \dfrac{1 - \cos x}{x^2}$,

and hence $\lim_{x \to 0} \dfrac{x - \sin x}{x^3}$.

28 If, in qn 26, we had had $\lim_{x \to a^+} \dfrac{f'(x)}{g'(x)} = +\infty$, instead of $\lim_{x \to a^+} \dfrac{f'(x)}{g'(x)} = l$, would we have been able to deduce that $\lim_{x \to a^+} \dfrac{f(x)}{g(x)} = +\infty$?

29 A function $f : [a, b] \to \mathbb{R}$ is continuous on the closed interval $[a, b]$ and differentiable on the open interval (a, b).
By applying de l'Hôpital's rule to

$$\frac{f(c+h) + f(c-h) - 2f(c)}{h^2},$$

prove that if f'' exists at $c \in (a, b)$ then

$$f''(c) = \lim_{h \to 0} \frac{f(c+h) + f(c-h) - 2f(c)}{h^2}.$$

By considering the function f for which $f(x) = x^2$ when $x \geq 0$, and $f(x) = -x^2$ when $x < 0$, show that this limit may exist when f is not twice differentiable.

It may happen that $\lim_{x \to a^+} f'(x)/g'(x)$ does not exist, when $\lim_{x \to a^+} f(x)/g(x)$ does.

30 Let $f(x) = x^2 \sin(1/x)$, when $x \neq 0$, $f(0) = 0$ and $g(x) = x$.
Show that $\lim_{x \to 0^+} f(x)/g(x) = 0$.
Show also that $\lim_{x \to 0^+} f'(x)/g'(x)$ does not exist.
Check that the conditions for Cauchy's Mean Value Theorem hold here for any interval $[0, x]$, with $x \neq 0$. Why may the argument in the proof of de l'Hôpital's rule *not* be used to establish the existence of $\lim f'(x)/g'(x)$ when $\lim f(x)/g(x)$ is known?

Summary: Rolle's Theorem and Mean Value Theorem

Rolle's Theorem

qn 7 If $f : [a, b] \to \mathbb{R}$
(i) is continuous on $[a, b]$,
(ii) is differentiable on (a, b), and
(iii) $f(a) = f(b)$,
then $f'(c) = 0$ for some c, with $a < c < b$.

An intermediate value theorem for derivatives

qn 12 If $f : [a, b] \to \mathbb{R}$ is differentiable on $[a, b]$ and $f'(a) < k < f'(b)$,
then $f'(c) = k$ for some c, with $a < c < b$.

Mean Value Theorem

qn 7 If $f : [a, b] \to \mathbb{R}$ is
(i) continuous on $[a, b]$, and
(ii) differentiable on (a, b), then

$$\frac{f(b) - f(a)}{b - a} = f'(c) \text{ for some } c,$$

with $a < c < b$.

Theorem qn 15 If $f : [a, b] \to \mathbb{R}$ satisfies the conditions of the Mean Value Theorem and $f'(x) > 0$ for all x, with $a < x < b$, then f is strictly monotonic increasing.

Theorem qn 17 If $f : [a, b] \to \mathbb{R}$ satisfies the conditions of the Mean Value Theorem and $f'(x) = 0$ for all x, with $a < x < b$, then f is constant.

Theorem qn 18 If $f : [a, b] \to [a, b]$ satisfies the conditions of the Mean Value Theorem and $|f'(x)| < 1$ for all x, with $a < x < b$, then $f(x) = x$ has exactly one solution. If, further, $|f'(x)| \leq L < 1$, then this solution is the limit of any sequence (a_n) defined by $a_{n+1} = f(a_n)$.

Theorem qn 22 If $f: [a, b] \to \mathbb{R}$ satisfies the conditions of the Mean Value Theorem and if, for some c, with $a < c < b$, $f'(c) = 0$ and $f''(c) > 0$, then f has a local minimum at c.

Cauchy's Mean Value Theorem

qn 25 If $f: [a, b] \to \mathbb{R}$ and $g : [a, b] \to \mathbb{R}$ are functions such that
(i) both are continuous on $[a, b]$,
(ii) both are differentiable on (a, b), and
(iii) $g'(x) \neq 0$ for any x, with $a < x < b$, then
$$\frac{f(b) - f(a)}{g(b) - g(a)} = \frac{f'(c)}{g'(c)}$$
for some c, with $a < c < b$.

de l'Hôpital's rule

qn 26 If $f: [a, b] \to \mathbb{R}$ and $g : [a, b] \to \mathbb{R}$ are functions such that
(i) both are continuous on $[a, b]$,
(ii) both are differentiable on (a, b),
(iii) $f(a) = g(a) = 0$, and
(iv) $\displaystyle\lim_{x \to a^+} \frac{f'(x)}{g'(x)} = l$,
then $\displaystyle\lim_{x \to a^+} \frac{f(x)}{g(x)} = l$.

The Second and Third Mean Value Theorems

31 *Second Mean Value Theorem*
A function $f: [a, b] \to \mathbb{R}$ has a continuous derivative on the closed interval $[a, b]$ and is twice differentiable on the open interval (a, b). Let
$$\frac{f(b) - f(a) - (b - a)f'(a)}{(b - a)^2} = K.$$
Apply Rolle's Theorem to the function $F: [a, b] \to \mathbb{R}$ given by
$$F(x) = f(b) - f(x) - (b - x)f'(x) - K(b - x)^2$$
to show that, for some c with $a < c < b$, $K = \frac{1}{2}f''(c)$. Deduce that
$$f(b) = f(a) + (b - a)f'(a) + \frac{(b - a)^2}{2}f''(c).$$

32 Rewrite the Second Mean Value Theorem replacing b by $a + h$. Then, by a judicious choice of a and h, show that for any twice differentiable function $f: \mathbb{R} \to \mathbb{R}$
$$f(x) = f(0) + xf'(0) + \frac{x^2}{2}f''(c) \text{ for some } c \in (0, x).$$

33 *Third Mean Value Theorem*
If a function $f : [a, b] \to \mathbb{R}$ has a continuous second derivative on the closed interval $[a, b]$ and is three times differentiable on the open interval (a, b), then there exists a c, with $a < c < b$, such that

$$f(b) = f(a) + (b-a)f'(a) + \frac{(b-a)^2}{2!}f''(a) + \frac{(b-a)^3}{3!}f'''(c).$$

Indicate how to prove the Third Mean Value Theorem by defining a constant K and a function F analogous to those in qn 31, and applying Rolle's theorem.

Reformulate the theorem, replacing b by $a + h$.

34 If $f(x) = a_0 + a_1x + a_2x^2 + \ldots + a_nx^n$, prove that

$$a_0 = f(0), \quad a_1 = f'(0), \quad a_2 = \frac{1}{2}f''(0), \quad \ldots, \quad a_n = \frac{f^{(n)}(0)}{n!}.$$

We now extend the Second and Third Mean Value Theorems to an nth Mean Value Theorem to compare functions, which may be differentiated any number of times, with polynomials.

Taylor's Theorem

Taylor's Theorem or nth Mean Value Theorem

35 The function $f : [a, b] \to \mathbb{R}$ has a continuous $(n-1)$th derivative on the closed interval $[a, b]$ and is differentiable n times on the open interval (a, b). Let

$$\frac{f(b) - f(a) - (b-a)f'(a) - \ldots - \dfrac{(b-a)^{n-1}}{(n-1)!}f^{(n-1)}(a)}{(b-a)^n} = K.$$

Apply Rolle's Theorem to the function $F : [a, b] \to \mathbb{R}$ given by

$$F(x) = f(b) - f(x) - (b-x)f'(x) - \frac{(b-x)^2}{2!}f''(x) - \\ \cdots - \frac{(b-x)^{n-1}}{(n-1)!}f^{(n-1)}(x) - K(b-x)^n,$$

to show that, for some c with $a < c < b$, $K = f^{(n)}(c)/n!$. So

$$f(b) = f(a) + (b-a)f'(a) + \frac{(b-a)^2}{2!}f''(a) + \\ \cdots + \frac{(b-a)^{n-1}}{(n-1)!}f^{(n-1)}(a) + \frac{(b-a)^n}{n!}f^{(n)}(c).$$

36 Rewrite the conclusion of Taylor's Theorem substituting $a + h$ for b.

Maclaurin's Theorem

37 By putting $a = 0$ and $h = x$ in qn 36 show that, for any function $f\colon [0, x] \to \mathbb{R}$ which is differentiable n times,

$$f(x) = f(0) + xf'(0) + \frac{x^2}{2}f''(0) + \ldots + \frac{x^{n-1}}{(n-1)!}f^{(n-1)}(0) + \frac{x^n}{n!}f^{(n)}(\theta x),$$

for some $\theta \in (0, 1)$.

38 Use Maclaurin's Theorem to say what you can about $R_n(x)$ where

$$\exp(x) = 1 + x + \frac{x^2}{2!} + \ldots + \frac{x^{n-1}}{(n-1)!} + R_n(x),$$

assuming that $\exp(0) = 1$ and $\exp'(x) = \exp(x)$.
Prove that, for any given x, $(R_n(x)) \to 0$ as $n \to \infty$.
Deduce that

$$\exp(x) = \sum_{n=0}^{\infty} \frac{x^n}{n!}.$$

Use computer graphics to observe the connection between the graphs of $y = 1$, $y = 1 + x$, $y = 1 + x + \frac{1}{2}x^2, \ldots$, and $y = \exp(x)$.

39 Use Maclaurin's Theorem to say what you can about $R_{2n}(x)$ where

$$\sin x = x - \frac{x^3}{3!} + \ldots + (-1)^{n+1}\frac{x^{2n-1}}{(2n-1)!} + R_{2n}(x).$$

Prove that, for any given x, $(R_{2n}(x)) \to 0$ as $n \to \infty$.
Deduce that

$$\sin x = \sum_{n=1}^{\infty}(-1)^{n+1}\frac{x^{2n-1}}{(2n-1)!}.$$

Observe the connection between the graphs you drew for qn 16 and this result.

40 Use Maclaurin's Theorem to say what you can about $R_{2n+1}(x)$, where

$$\cos x = 1 - \frac{x^2}{2!} + \ldots + (-1)^n\frac{x^{2n}}{(2n)!} + R_{2n+1}(x).$$

Prove that, for any given x, $(R_{2n+1}(x)) \to 0$ as $n \to \infty$.
Deduce that

$$\cos x = \sum_{n=0}^{\infty}(-1)^n\frac{x^{2n}}{(2n)!}.$$

Observe the connection between the graphs you drew for qn 16 and this result.

41 If $f(x) = \ln(1+x)$, use qn 1.8(i) to show that

$$f^{(n)}(x) = (-1)^{n+1}\frac{(n-1)!}{(1+x)^n}.$$

Use Maclaurin's Theorem to say what you can about $R_n(x)$ where

$$\ln(1+x) = x - \frac{x^2}{2} + \frac{x^3}{3} + \ldots + (-1)^n\frac{x^{n-1}}{n-1} + R_n(x).$$

Prove that $(R_n(x)) \to 0$ as $n \to \infty$

(i) when $0 \le x < 1$;
(ii) when $x = 1$;
(iii) when $-\frac{1}{2} \le x < 0$.

Deduce that

$$\ln(1+x) = \sum_{n=1}^{\infty}(-1)^{n+1}\frac{x^n}{n}, \text{ when } -\frac{1}{2} \le x \le 1.$$

In fact, the series expansion is valid for $-1 < x \le 1$ as we will show in qns 46 and 11.38.

42 Use d'Alembert's ratio test (qn 5.69 or 5.95) to prove that

$$x - \frac{1}{2}x^2 + \frac{1}{3}x^3 - \ldots$$

is not convergent when $x > 1$. Deduce that, in this case, the sequence $(R_n(x))$ in qn 41 is not a null sequence.

In qn 36,

$$f(a) + hf'(a) + \frac{h^2}{2!}f''(a) + \ldots + \frac{h^{n-1}}{(n-1)!}f^{(n-1)}(a)$$

is called the *polynomial expansion* of $f(a+h)$ and

$$R_n(h) = \frac{h^n}{n!}f^{(n)}(a+\theta h),$$

is called the *Lagrange form of the remainder*. In qn 37,

$$f(0) + xf'(0) + \frac{x^2}{2!}f''(0) + \ldots + \frac{x^{n-1}}{(n-1)!}f^{(n-1)}(0)$$

is called the *polynomial expansion* of $f(x)$ and

$$R_n(x) = \frac{x^n}{n!}f^{(n)}(\theta x),$$

is called the *Lagrange form of the remainder*.

(i) If we regard the polynomial expansions as the partial sums of a power series (known as the Taylor series in the case of qn 36 and as the Maclaurin series in the case of qn 37) the power series may be convergent or not. In any case Taylor's Theorem is still a valid theorem and the remainder term measures how good an approximation to the function is given by the polynomial. There is an example of a non-convergent Taylor series in qn 42.

(ii) If the Taylor series is convergent and the sequence of remainders tends to 0 as n tends to infinity, then the Taylor series converges to the function as we have seen in qns 38–41. In fact, when the remainders tend to 0, the Taylor series is necessarily convergent.

(iii) Somewhat surprisingly, a Taylor series may be convergent *without* the remainders tending to zero. In this case, the Taylor series cannot converge to the function. There is an example of this in qn 44.

43 Write down the Maclaurin series for the function $f(x) = \sqrt{(1+x)}$. Does it coincide with the binomial expansion of qn 5.98 for $a = \frac{1}{2}$? If the Maclaurin series is $f(x) = 1 + a_1x + a_2x^2 + a_3x^3 + \ldots$, check that Σa_n is an alternating series and that $a_{n+1}/a_n = (\frac{1}{2} - n)/(n+1)$, so that the absolute values of the terms are monotonic decreasing. Now check that $|a_n| = \frac{1}{2} \cdot \frac{3}{4} \cdot \frac{5}{6} \cdot \frac{7}{8} \cdot \ldots \cdot \frac{2n-3}{2n-2} \cdot \frac{1}{2n}$, so that $|a_n| \leq 1/2n$ and $(|a_n|)$ is a null sequence. Use the alternating series test to prove that the Maclaurin series for $f(1)$ is convergent. How many terms of this series are needed before $\sqrt{2}$ has been found correct to one place of decimals, that is, until the error is less than 0.05?

44 (*Cauchy*, 1823) The function $f : \mathbb{R} \to \mathbb{R}$ defined by

$$f(x) = \begin{cases} e^{-1/x^2} & \text{when} \quad x \neq 0, \\ 0 & \text{when} \quad x = 0, \end{cases}$$

has the curious property that $f^{(n)}(0) = 0$ for every value of n. [A proof is given in Scott and Tims, p. 335, and another in Bressoud, pp. 90–91.] What is the Maclaurin series for this function? Is there a non-zero value of x for which the remainders can form a null sequence? Construct an infinity of different functions each of which has the same Maclaurin series.

A Taylor series is a valid description of the function from which it originates if and only if its sequence of remainders is null. It is so important to be able to prove that the sequence of remainders of a Taylor series is null that a number of different forms of the remainder have been devised since the Lagrange form of the remainder, which we have been using, is not always sufficient to decide the matter, as we found in qn 41.

(**45**) With the notation and conditions of qn 35 let

$$\frac{f(b)-f(a)-(b-a)f'(a)-\ldots-\dfrac{(b-a)^{n-1}}{(n-1)!}f^{(n-1)}(a)}{b-a}=K.$$

Apply Rolle's Theorem to the function $F\colon[a,b]\to\mathbb{R}$ given by

$$F(x)=f(b)-f(x)-(b-x)f'(x)-\frac{(b-x)^2}{2!}f''(x)-$$
$$\ldots-\frac{(b-x)^{n-1}}{(n-1)!}f^{(n-1)}(x)-K(b-x)$$

to show that, for some c with $a<c<b$,

$$K=\frac{(b-c)^{n-1}}{(n-1)!}f^{(n)}(c).$$

So

$$f(b)=f(a)+(b-a)f'(a)+\frac{(b-a)^2}{2!}f''(a)+\ldots$$
$$+\frac{(b-a)^{n-1}}{(n-1)!}f^{(n-1)}(a)+\frac{(b-c)^{n-1}(b-a)}{(n-1)!}f^{(n)}(c).$$

This last term is called *Cauchy's form of the remainder.*

(**46**) Use Cauchy's form of the remainder to prove that the Taylor series for $\ln(1+x)$ converges to the function when $-1<x<1$.

Summary: Taylor's Theorem

Taylor's Theorem with Lagrange's form of the remainder, or nth Mean Value Theorem

qn 35 If $f\colon[a,b]\to\mathbb{R}$
(i) has a continuous $(n-1)$th derivative on $[a,b]$,
(ii) is differentiable n times on (a,b),
then

$$f(b)=f(a)+(b-a)f'(a)+\frac{(b-a)^2}{2!}f''(a)+$$
$$\ldots+\frac{(b-a)^{n-1}}{(n-1)!}f^{(n-1)}(a)+\frac{(b-a)^n}{n!}f^{(n)}(c)$$

for some c, with $a<c<b$.

Taylor's Theorem with Cauchy's form of the remainder

qn 45 If $f: [a, b] \to \mathbb{R}$

(i) has a continuous $(n-1)$th derivative on $[a, b]$,
(ii) is differentiable n times on (a, b),

then

$$f(b) = f(a) + (b-a)f'(a) + \frac{(b-a)^2}{2!}f''(a) + \ldots + \frac{(b-a)^{n-1}}{(n-1)!}f^{(n-1)}(a) + \frac{(b-c)^{n-1}(b-a)}{(n-1)!}f^{(n)}(c).$$

for some c, with $a < c < b$.

Maclaurin's Theorem

qn 37 If $f: [0, x] \to \mathbb{R}$ is differentiable n times, then

$$f(x) = f(0) + xf'(0) + \frac{x^2}{2!}f''(0) + \ldots + \frac{x^{n-1}}{(n-1)!}f^{(n-1)}(0) + \frac{x^n}{n!}f^{(n)}(\theta x),$$

for some θ, with $0 < \theta < 1$.

Definition The power series

$$f(a+h) = f(a) + hf'(a) + \frac{h^2}{2!}f''(a) + \ldots + \frac{h^n}{n!}f^{(n)}(a) + \ldots$$

is called the Taylor series of f at a.

Definition The power series

$$f(x) = f(0) + xf'(0) + \frac{x^2}{2!}f''(0) + \ldots + \frac{x^n}{n!}f^{(n)}(0) + \ldots$$

is called the Maclaurin series for f.

Theorem The Taylor series expansion of a function at a

qn 38 converges to $f(a+h)$ if and only if the difference between the nth partial sum of the series and $f(a+h)$ is a null sequence as $n \to \infty$; or, in other words, if the remainders form a null sequence.

Historical Note

Early in the seventeenth century Cavalieri affirmed that there was always a tangent parallel to a chord, and this must count as an embryonic form of the Mean Value Theorem.

In 1691, Michel Rolle proved that between two adjacent roots of a polynomial $f(x)$ there was a root of $f'(x)$. His definition of $f'(x)$ was algebraic (so that, if $f(x) = x^n$, then $f'(x) = nx^{n-1}$) and his proof was like an application of the Intermediate Value Theorem in that he claimed that f' had a different sign at adjacent roots of f and therefore must be zero somewhere between the roots.

Lagrange was the first mathematician to bring inequalities into theorems and proofs about the foundations of calculus. In 1797 he argued (wrongly as it turned out) that every function had a power series expansion. He obtained $f(a+h) = f(a) + hf'(a) + hr(h)$, where $r(h)$ tends to 0 as h tends to 0, from his power series for f and deduced firstly that functions with positive derivatives were increasing. He then deduced that $(f(a+h) - f(a))/h$ lay between the greatest and least values of $f'(x)$ on the interval $[a, a+h]$. Presuming the Intermediate Value Theorem for derivatives gave him the Mean Value Theorem. He applied the same argument to obtain the Second and Third Mean Value Theorems. In 1823, Cauchy worked from an $\varepsilon - \delta$ description of derivative on an interval. He divided his interval $[a, b]$ into lengths shorter than δ and obtained a chain of inequalities of the form

$$f'(x_i) - \varepsilon \leq \frac{f(x_i) - f(x_{i-1})}{x_i - x_{i-1}} \leq f'(x_i) + \varepsilon,$$

implicitly assuming that the derivatives were uniformly continuous.

The sum of the numerators is $f(b) - f(a)$ and the sum of the denominators is $b - a$, and as their quotient forms a sort of mean of the quotients, it lies between the greatest and least of the derivatives. Now Cauchy claimed the Intermediate Value Theorem (assuming continuous derivatives) and obtained the Mean Value Theorem. The proof of the Mean Value Theorem directly from Rolle's Theorem, as in qn 13, appears in Serret's *Calculus* (1868), but in a pre-Weierstrassian context. Rolle's theorem and the Mean Value Theorem, as we know them today, depend on the results of Weierstrass which distinguish precisely between properties of continuous functions and properties of differentiable functions and therefore expose exactly what is assured according to the number of times a function may be differentiable. In a letter which Schwarz wrote to Cantor in 1870 (both Schwarz and Cantor had been students of Weierstrass) Schwarz proved that $f'(x) = 0 \Rightarrow f(x)$ is constant, and said that this was the foundation of the differential and integral calculus (see the end of appendix 1). In 1878, U. Dini published the first proof of Rolle's Theorem with Weierstrassian rigour.

De l'Hôpital's rule first appeared in his *Analyse des infiniments petits* (1696) where de l'Hôpital singularly failed to give the credit for this result where it was due, namely to his teacher, Johann Bernoulli. In 1823, working with functions with continuous derivatives, Cauchy proved his extension of the Mean Value Theorem and deduced de l'Hôpital's rule from it.

The expansion of binomial, trigonometric and exponential expressions in the form of power series goes back to Newton and James Gregory (1669). Brook Taylor constructed the series which bears his name in 1712. His method was based on the Gregory–Newton construction of a polynomial from finite differences which appeared in Newton's *Principia* (1687), to which Taylor applied a limiting process. Maclaurin (1742) obtained his series by repeated differentiation of a presumed infinite series expansion, putting $x = 0$ at each stage recognising it to be a special case of Taylor's. Lagrange (1797) having 'proved' that every function has a Taylor series expansion established the First, Second and Third Mean Value Theorems as we have described above. In each case he obtained bounds on the remainder (really on K, in qns 13, 31 and 33) after one, two and three terms of the Taylor series (respectively) and claimed, correctly, that his method extended to n terms giving the Lagrange form of the remainder as we know it today.

In 1823 Cauchy claimed that all the derivatives of $\exp(-1/x^2)$ are zero at $x = 0$ so that this function does not have a Taylor series at this point. This undermined Lagrange's programme for a theory of analytic functions based entirely on Taylor series. Cauchy derived his form of the Taylor series remainder from the remainder in integral form, which he also established in 1823. See qn 10.59.

In Weierstrass' lectures in 1861 he claimed the Fundamental Theorem of Analysis to be that, when $f^{(i)}(x_0) = 0$ for $i = 1, 2, 3, \ldots, n-1$,

$$f(x) - f(x_0) = \frac{(x - x_0)^n}{n!} \cdot f^{(n)}(x_0 + \theta(x - x_0)),$$

for some θ, $0 < \theta < 1$, a form of Taylor's Theorem with Lagrange remainder which had been obtained by Cauchy in 1829.

Answers

1 Yes. Either a local maximum or a local minimum for the function.

2 A differentiable function is continuous and a continuous function on a closed interval is bounded and attains its bounds, from qns 7.31 and 7.34.
If $f(a) = f(b)$ is not a maximum then the maximum occurs on the open interval. If $f(a) = f(b)$ is not a minimum then the minimum occurs on the open interval. If $f(a) = f(b)$ is both a maximum and a minimum, then the function is constant. Yes.

3 Yes, the function has a local maximum at c, by qn 8.30.

4 If $f(a) = f(b) = M > m$, there is a c with $f(c) = m$.
If $f(a) = f(b) = m < M$, then there is a c with $f(c) = M$.
If $f(a) = f(b) = M = m$, then $f(x) = M = m$ for all x, and so $f'(x) = 0$ from qn 8.6.

5 See Figure 9.1.

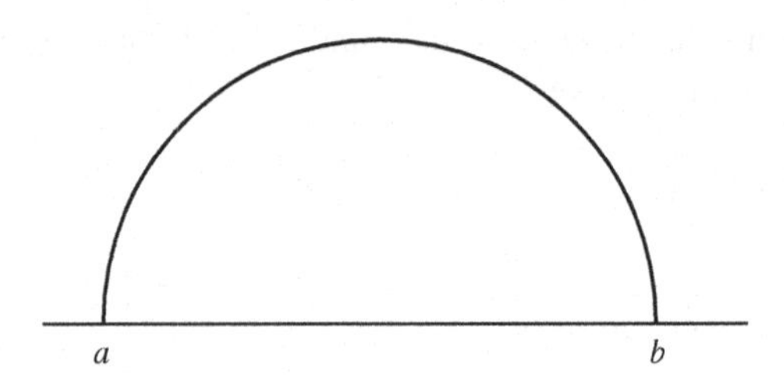

Figure 9.1

6 Differentiable $\Rightarrow$ Continuous.

At a local maximum or minimum $f'(x) = 0$.

7 (i) $\Rightarrow$ f is bounded and attains its bounds.
(ii) $\Rightarrow$ when $f(a)$ is not a maximum, there is a maximum $f(c)$ with $a < c < b$; when $f(a)$ is not a minimum there is a minimum $f(c)$ with $a < c < b$; and when $f(a)$ is a maximum and a minimum $f(c)$ is a maximum and a minimum.
(iii) $\Rightarrow$ $f'(c) = 0$.

8 See Figure 9.2.

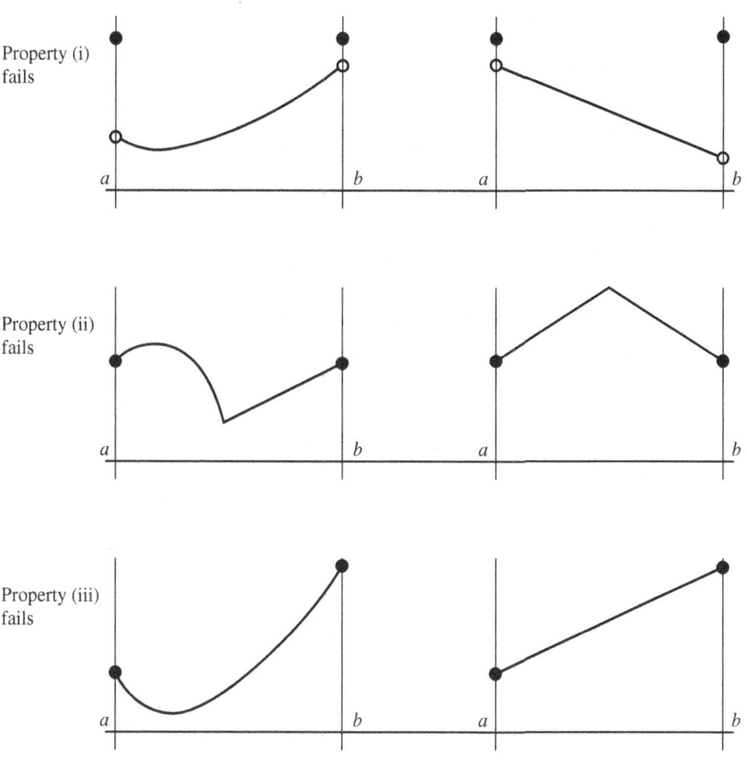

Figure 9.2

9 Apply Rolle's Theorem to the function

$$x \mapsto a_0 x + a_1 x^2/2 + a_2 x^3/3 + \ldots + a_n x^{n+1}/(n+1)$$

on the interval $[0, 1]$.

10 Apply Rolle's Theorem to $[a, b]$ and $[b, c]$, to find $f'(x) = 0$ with $a < x < b$ and $f'(y) = 0$ with $b < y < c$. Then apply Rolle's Theorem to the function f' on the interval $[x, y]$.
If an n-times differentiable function has $n + 1$ zeros, then between the extreme roots there is a point c at which $f^{(n)}(c) = 0$.

11 (i) Continuous and differentiable by qns 8.15 and 8.17 since the denominator is never 0. $f(0) = f(2) = \frac{1}{4}$. Rolle's Theorem fails. Completeness is needed for the Maximum–Minimum Theorem.
(ii) Polynomials continuous and differentiable. $g(1) = g(4) = 0$. g is bounded below but does not attain its lower bound. Domain not an interval, i.e. not complete.

12 (i) By qn 6.23.
(ii) By qns 7.31, 7.32 and 7.34.
(iii) By qns 8.10 and 8.11.
(iv) By qn 8.6 since two derivatives are unequal.
Use (ii) and qn 8.30 to show that for some c, $g'(c) = 0$.
The function f in qn 8.22 is differentiable at every point. For this function, $f'(x) = 2x \cdot \sin(1/x) - \cos(1/x)$ when $x \neq 0$, and $f'(0) = 0$. The derived function f' is discontinuous at 0. Examine the graph of f' on a graphics calculator to see how the intermediate value property for derivatives holds on any interval $[0, a]$. Near to zero, the function oscillates infinitely many times between positive and negative values.

13 (i) If the point (x, y) lies on the chord joining $(a, f(a))$ to $(b, f(b))$ then

$$\frac{y - f(a)}{x - a} = \frac{f(b) - f(a)}{b - a},$$

so

$$y = f(a) + \frac{f(b) - f(a)}{b - a}(x - a)$$

and

$$D(x) = f(x) - f(a) - \frac{f(b) - f(a)}{b - a}(x - a).$$

D satisfies the conditions for Rolle's Theorem on $[a, b]$.

$$D'(c) = 0 \Leftrightarrow f'(c) = \frac{(b) - f(a)}{b - a}.$$

(ii) $F(a) = 0$ from the definition of K. $F(b) = 0$, trivially. Differentiability follows from qn 8.10. $F'(c) = 0 \Leftrightarrow f'(c) = K$. For some c in (a, b),

$$f'(c) = K = \frac{f(b) - f(a)}{b - a}.$$

14 If $b = a + h$, any c in (a, b) has the form $a + \theta h$.

15 Let $a \leq x < y \leq b$; then, for some c in the interval (x, y),

$$\frac{f(y) - f(x)}{y - x} = f'(c) > 0.$$

So $x < y \Leftrightarrow f(x) < f(y)$.

16 $f_1'(x) = 1 - \cos x > 0$ on $(0, 2\pi)$, so f_1 is strictly increasing by qn 15. Since $f_1(0) = 0$, $f_1(x) > 0$ on $(0, 2\pi)$.

$f_2'(x) = f_1(x) > 0$ on $(0, 2\pi)$, so f_2 is strictly increasing.
Since $f_2(0) = 0$, $f_2(x) > 0$ on $(0, 2\pi)$. $f_3'(x) = f_2(x)$ and $f_4'(x) = f_3(x)$.
The argument may be repeated.

17 Let $a < x \leq b$; then

$$\frac{f(x) - f(a)}{x - a} = f'(c) = 0 \text{ for some } c \text{ in } (a, x).$$

So $f(x) = f(a)$.
For non-constant f, the domain must be disconnected.

18 $$\frac{f(y) - f(x)}{y - x} = f'(c).$$

$|f'(c)| \leq L$ gives the result.

$$|a_{m+1} - a_2| = |f(a_m) - f(a_1)| \leq L \cdot |a_m - a_1| \leq L \cdot |b - a|.$$
$$|a_{m+2} - a_3| = |f(a_{m+1}) - f(a_2)| \leq L \cdot |a_{m+1} - a_2| \leq L^2 \cdot |a_m - a_1|$$
$$\leq L^2 \cdot |b - a|.$$

Further steps by induction. Now $0 < L < 1$, so (L^n) is a null sequence.
So for sufficiently large n, $|a_{n+m} - a_{n+1}| < \varepsilon$, and (a_n) is a Cauchy sequence and so, by qn 4.57, is convergent.
A and B lie in $[a, b]$ by qn 3.78, since the interval is closed.
$(a_n) \to A$ implies $(a_{n+1}) \to A$, so $(f(a_n)) \to A$. But f is continuous at A, so $f(A) = A$ and similarly $f(B) = B$.

$$|b_{n+1} - a_{n+1}| = |f(b_n) - f(a_n)| \leq L \cdot |b_n - a_n| \leq \ldots \leq L^n \cdot |b_1 - a_1|.$$

So $(b_n - a_n)$ is a null sequence and by qn 3.54(v) (a_n) and (b_n) have the same limit.

19 Let $f(x) = \cos x$. Then if $0 \leq x \leq \frac{1}{2}\pi$, $0 \leq \cos x \leq 1$.
$f'(x) = -\sin x$. So f is a contraction mapping on $[0, 1]$ with $L = \sin 1$, and, by qn 18, $f(x) = x$ has a unique solution. Obviously no solution on $[1, \frac{1}{2}\pi]$.

20 For $x \geq 0$, $f'(x) = \frac{1}{2\sqrt{x+2}}$. So, $0 < f'(x) \leq \frac{1}{2\sqrt{2}}$, and f is increasing with $|f'(x)| \leq \frac{1}{2\sqrt{2}} < 1$.
For $c > 2$, let $c = 2 + \delta$ (say), then $c^2 = 4 + 4\delta + \delta^2 > c + 2 > 4$, so $c > \sqrt{(c+2)} > 2$. Thus $f\colon [0, c] \to [0, c]$ and on this domain, f is a contraction mapping. So the sequence defined by $a_{n+1} = f(a_n)$ is convergent provided $a_1 \geq 0$, by qn 18.

21 See Figure 9.3.

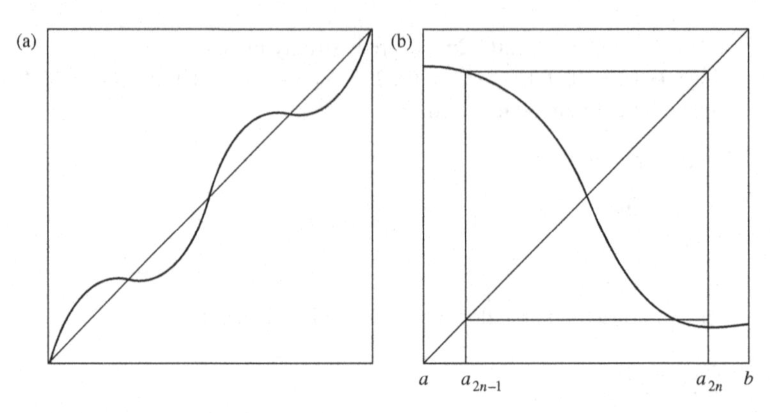

Figure 9.3

22 Let $f''(c) = l > 0$. Then, for some δ,

$$0 < |x - c| < \delta \Rightarrow \left| \frac{f'(x) - f'(c)}{x - c} - l \right| < \tfrac{1}{2}l \Leftrightarrow \tfrac{1}{2}l < \frac{f'(x)}{x - c} < \tfrac{3}{2}l,$$

since $f'(c) = 0$.
Thus $c - \delta < x < c \Rightarrow f'(x) < 0$, and $c < x < c + \delta \Rightarrow f'(x) > 0$.
Now, by the Mean Value Theorem, for any x satisfying $c - \delta < x < c$,

$$\frac{f(x) - f(c)}{x - c} = f'(d)$$

for some $d \in (x, c)$, so $f'(d) < 0$ and $f(x) > f(c)$.
Likewise for $c < x < c + \delta$, $f(x) > f(c)$, so f has local minimum at c.

23 f_{n+1} is strictly monotonic decreasing by qn 15, and so has at most one root. $n + 1$ odd $\Rightarrow f_{n+1}$ has at least one root from qn 7.19 since x^{n+1} changes sign. $f'_{n+2}(x) = 0 \Leftrightarrow f_{n+1}(x) = 0$ and this occurs once, at $x = a \neq 0$. For any value of x, $f''_{n+2}(x) = -f'_{n+1}(x) = f_n(x) > 0$.
So f_{n+2} is minimal at $x = a$ by qn 22.

$$f_{n+2}(a) = f_{n+1}(a) + \frac{a^{n+2}}{(n+2)!} = \frac{a^{n+2}}{(n+2)!} > 0,$$

since n is even and $a \neq 0$.
At its only minimum value f_{n+2} is positive, so the function is positive everywhere. Thus f_n positive implies f_{n+2} positive.

$$f_2(x) = 1 - x + \tfrac{1}{2}x^2 = \tfrac{1}{2}[1 + (1 - x)^2] \geq \tfrac{1}{2}.$$

So by induction f_n is positive when n is even. We have also shown that, if this is the case, then f_{n+1} has a unique root.

24 (i) Point $= (y, f(y)) + \theta(x - y, f(x) - f(y))$. Inequality indicates that the graph of the function lies below the line segment joining the two points. (iii)

If f is differentiable, then for some c and d, the inequality is equivalent to $f'(c) < f'(d)$, where $x < c < s < d < y$, applying the Mean Value Theorem twice. (iv) By qn 15, $f''(x) > 0 \Rightarrow f'(x)$ is strictly monotonic increasing. This implies (iii) which implies (ii) which implies the function is concave.

25 Note that $g(b) \neq g(a)$ because, if not, Rolle's Theorem applied to g on $[a, b]$ would contradict $g'(x) \neq 0$.

26 For some c, $a < c < b$,

$$\frac{f(b)}{g(b)} = \frac{f'(c)}{g'(c)}.$$

First equation follows directly from this.
For the conditions we have established $0 < k < h$. So if $h < \delta$, then certainly $k < \delta$.
What this establishes is a convenient rule for finding difficult limits. If $f(a) = g(a) = 0$, then our ordinary procedures for finding $\lim_{x \to a} f(x)/g(x)$ may not be used, but we can try to find $\lim_{x \to a} f'(x)/g'(x)$, and, if we are successful, that is the answer we are looking for.

27 (i) $1/\sqrt{2}$. (ii) 1, 1/2, 1/6. The first result in (iii) is of great importance and is variously justified in different treatments. Some make it plausible from geometric considerations and then use it to find the derivative of sine. In such a case, de l'Hôpital's rule may not be used, for that would make a circular argument. Some treatments define sine by a power series, and then de l'Hôpital's rule is needed for a sound argument. We will define sine analytically (but with geometric motivation) in chapter 11, and that provides a proper basis for the argument here.

28 The basis of the question is the application of Cauchy's Mean Value Theorem in case $f(a) = g(a) = 0$. In this case, for $b > a$, we have

$$\frac{f(b)}{g(b)} = \frac{f'(c)}{g'(c)} \text{ for some } c, \text{ with } a < c < b.$$

If, given B, there exists a δ such that

$$a < x < a + \delta \Rightarrow \frac{f'(x)}{g'(x)} > B,$$

then

$$a < y < a + \delta \Rightarrow \frac{f(y)}{g(y)} = \frac{f'(x)}{g'(x)} > B$$

for some x, with $a < x < y < a + \delta$. The result is proved.

29 After applying de l'Hôpital's rule, use qn 8.23.
$f'(x) = 2x$ for $x \geq 0$, and $f'(x) = -2x$ for $x < 0$. So $f'(x) = 2|x|$. This f' is not differentiable at 0. But the limit exists and equals 0.

30 For the first result see qn 8.21.
$f'(x)/g'(x) = 2x \sin(1/x) - \cos(1/x)$, of which the first component tends to 0 as in qn 8.21, but the second component oscillates near 0 like the function of qn 8.20.
Cauchy's Mean Value Theorem says that, if $a < b$, then there exists a c with $a < c < b$, such that .... But if $b \to a$ there is nothing to indicate that the consequent cs for each b exhaust the possibilities.

31 $F(a) = 0$ from the definition of K. $F(b) = 0$, trivially. F is differentiable from qns 8.10 and 8.13.

32 $f(a+h) = f(a) + hf'(a) + \frac{1}{2}h^2 f''(a+\theta h)$, for some θ, $0 < \theta < 1$. Put $a = 0$ and $h = x$.

33 Let

$$\frac{f(b) - f(a) - (b-a)f'(a) - \frac{1}{2}(b-a)^2 f''(a)}{(b-a)^3} = K,$$

and let $F(x) = f(b) - f(x) - (b-x)f'(x) - \frac{1}{2}(b-x)^2 f''(x) - K(b-x)^3$.
The application of Rolle's Theorem to F gives $K = f'''(c)/6$ for some c, with $a < c < b$.

$$f(a+h) = f(a) + hf'(a) + (h^2/2!)f''(a) + (h^3/3!)f'''(a+\theta h).$$

36 For some θ, $0 < \theta < 1$,

$$f(a+h) = f(a) + hf'(a) + \frac{h^2}{2!}f''(a) + \\ \dots + \frac{h^{n-1}}{(n-1)!}f^{(n-1)}(a) + \frac{h^n}{n!}f^{(n)}(a+\theta h).$$

38 $R_n(x) = \frac{x^n}{n!}\exp(\theta x)$, for some θ, $0 < \theta < 1$.
The function exp is monotonic (see chapter 11) so if $k = \max(1, \exp(x))$, $|R_n(x)| \leq |x^n/n!| \cdot k$. Now $(|x^n/n!|)$ is a null sequence by qn 3.74(ii). So $(R_n(x)) \to 0$ as $n \to \infty$ by qn 3.34 (squeeze rule). The nth partial sum of the power series $= \exp(x) - R_n(x) \to \exp(x)$ as $n \to \infty$.

39 If $f(x) = \sin x$, $f'(x) = \cos x = \sin(x + \frac{1}{2}\pi)$, so $f^{(n)}(x) = \sin(x + \frac{1}{2}n\pi)$.
$R_{2n}(x) = (x^{2n}/2n!) \cdot \sin(\theta x + n\pi)$. So $|R_{2n}(x)| \leq |x^{2n}/2n!|$. This shows that $(R_{2n}(x)) \to 0$ as $n \to \infty$. As in qn 38, this enables us to prove that the series converges to the function for each value of x.

40 If $f(x) = \cos x$, $f'(x) = -\sin x = \cos(x + \frac{1}{2}\pi)$, so $f^{(n)} = \cos(x + \frac{1}{2}n\pi)$.
The argument proceeds as in qn 38.

41 $R_n(x) = \dfrac{x^n \cdot (-1)^{n+1}(n-1)!}{n!(1+\theta x)^n} = (-1)^{n+1}\dfrac{x^n}{n(1+\theta x)^n}$.

(i) For $0 \le x < 1$, $|R_n(x)| \le |x^n/n|$ and so $(R_n(x))$ is a null sequence, and the series converges to the function, as in qn 38.

(ii) $|R_n(1)| = 1/n(1+\theta)^n < 1/n$, and again $(R_n(x))$ is a null sequence.

(iii) For $-\frac{1}{2} \le x < 0$, $|x| \le \frac{1}{2} < |1+\theta x|$, so $|x/(1+\theta x)| < 1$ and therefore $|R_n(x)| < 1/n$, so that $(R_n(x))$ is a null sequence.

42 From qn 5.95 (or 5.69) the radius of convergence is 1. Now consider a particular $x > 1$.
The nth partial sum of series $= \ln(1+x) - R_n(x)$. If $(R_n(x))$ were a null sequence, the sequence of nth partial sums would converge.

43 The two series coincide. Radius of convergence $= 1$.
For $n \ge 2$, | remainder after the nth term | $= a_{n+1}/(1+\theta)^{n-3/2} < a_{n+1}$, $a_4 = 0.06\ldots$, $a_5 = 0.039\ldots$. So four terms needed.

44 The Maclaurin series is $0 + 0 + 0 + \ldots$.
The remainder always equals the whole value of the function, so remainders are constant and not null.
For each choice of the constant k, the function $x \mapsto \sin x + k \cdot f(x)$ has the same Maclaurin series as $\sin x$.

45 As in qn 35, $F(a) = F(b) = 0$, and F continuous on $[a, b]$ and differentiable on (a, b),

$$F'(c) = 0 \Rightarrow -\frac{(b-c)^{n-1}}{(n-1)!} f^{(n)}(c) + K = 0.$$

46 Using the work of qn 41, Cauchy's form of the remainder

$$R_n(x) = \frac{(x-\theta x)^{n-1}x}{(n-1)!}(-1)^{n+1}\frac{(n-1!)}{(1+\theta x)^n}$$
$$= (-1)^{n+1}x^n\left(\frac{1-\theta}{1+\theta x}\right)^{n-1} \cdot \frac{1}{1+\theta x}.$$

Now $-1 < x \Rightarrow -\theta < \theta x \Rightarrow 0 < 1-\theta < 1+\theta x \Rightarrow (1-\theta)/(1+\theta x) < 1$.
So $-1 < x \Rightarrow |R_n(x)| < |x^n/(1+\theta x)| < |x^n/(1-|x|)|$. But, if $|x| < 1$, this last expression gives a null sequence, so $(R_n(x))$ is a null sequence.

10

Integration

The Fundamental Theorem of Calculus

Preliminary reading: Gardiner ch. III.3, Toeplitz ch. 2.
Concurrent reading: Bryant ch. 5, Courant and John ch. 2.
Further reading: Mason, Spivak chs 13 and 14.

In the first nine questions of this chapter we will apply the methods of seventeenth-century mathematicians to find areas. Having used limits so much, we will not avoid them now, but it is interesting to see how much of the work in qns 1 to 9 could be completed without limits, using Archimedes' method as in qn 3.25.

Areas with curved boundaries

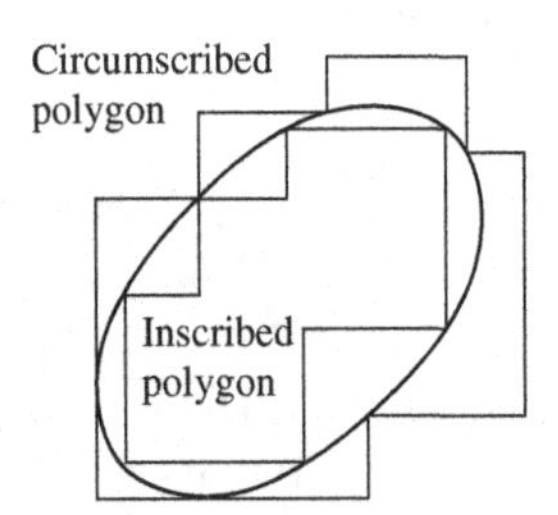

Figure 10.1

The idea of integration comes from the need to measure areas. The measurement of area is the measurement of the quantity of units of surface needed for an exact covering. From the use of a unit square comes the area of a rectangle as length times breadth. Since a triangle can be dissected and reassembled to form a rectangle, the area of a triangle can be calculated, and since any area bounded by a polygon can be dissected into triangles, the area of any polygon can also be calculated. However, when the boundary is curved, the measurement of area is more difficult. The areas of inscribed and circumscribed polygons provide lower and upper bounds on the area (see Figure 10.1), and if the construction of inscribed and circumscribed polygons can be done progressively in such a way that the areas of inscribed polygons and the areas of circumscribed polygons tend to the same limit, then that common limit will be the area to be determined, provided the boundary is continuous.

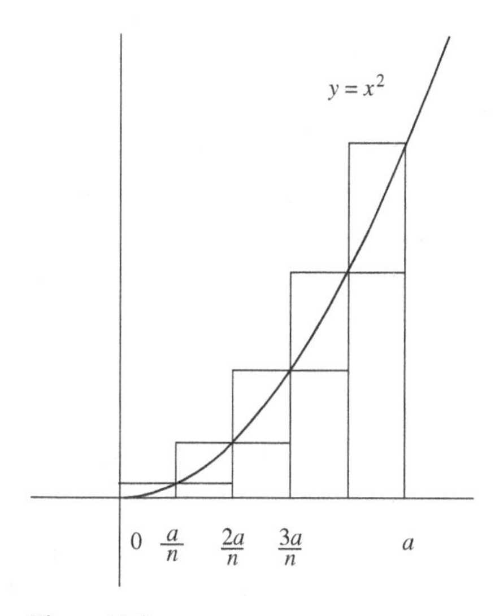

Figure 10.2

1 (*Fermat*, 1636, with later enhancement with limits) We investigate the area, A, bounded by the parabola $y = x^2$, the x-axis and the line $x = a$, where a is positive. In Fig. 10.2 the area is covered by parallel strips between the lines $x = ia/n$, where $i = 0, 1, 2, \ldots, n$.
Between these strips, the greatest rectangles within A have areas 0, $(a/n)(a/n)^2$, $(a/n)(2a/n)^2, \ldots, (a/n)((n-1)a/n)^2$. These are rectangles inscribed in A. If we add these, we get

$$I_n = (a/n)^3(1^2 + 2^2 + 3^2 + \ldots + (n-1)^2).$$

Between the same strips, the least rectangles which contain A have areas $(a/n)(a/n)^2$, $(a/n)(2a/n)^2, \ldots, (a/n)((n-1)a/n)^2$, $(a/n)a^2$. These are rectangles circumscribed about A. If we add these we get

$$C_n = (a/n)^3(1^2 + 2^2 + 3^2 + \ldots + n^2).$$

So $I_n < A < C_n$.

Now use qn 1.1 to show

$$\left(\frac{a}{n}\right)^3 \frac{(n-1)n(2n-1)}{6} < A < \left(\frac{a}{n}\right)^3 \frac{n(n+1)(2n+1)}{6}.$$

Simplify to get

$$\frac{1}{3}a^3\left(1-\frac{1}{n}\right)\left(1-\frac{1}{2n}\right) < A < \frac{1}{3}a^3\left(1+\frac{1}{n}\right)\left(1+\frac{1}{2n}\right).$$

Now consider the limits of I_n and C_n as $n \to \infty$, to determine the only possible value of A.

When, in qns 2–9, inscribed and circumscribed rectangles are referred to, maximal inscribed and minimal circumscribed rectangles will be meant.

2 (*Fermat*, 1636) Let A be the area bounded by the curve $y = x^3$, the x-axis and the line $x = a$, where we take a to be positive. Using the method of qn 1, and appealing to qn 1.3(iii), prove that, for all values of the positive integer n,

$$\frac{1}{4}a^4(1-1/n)^2 < A < \frac{1}{4}a^4(1+1/n)^2.$$

Deduce that $A = \frac{1}{4}a^4$ is the only value which satisfies all of these inequalities.

While Fermat claimed that the method of qns 1 and 2 could be extended to higher integral powers of x in 1636, a few years later he found that he could calculate the areas under $y = x^k$ for a greater range of values of k if he used the recently admitted sum of an infinite geometric progression, qn 5.9.

3 (*Fermat*, 1658) Let A be the area between the curve $y = 1/x^2$, the x-axis and the line $x = a$. Let $r > 1$. Divide the area A into parallel strips between the lines $x = ar^n$, where $n = 0, 1, 2, \ldots$, and show that the areas of inscribed rectangles are

$(r-1)/ar^2, (r^2-r)/ar^4, (r^3-r^2)/ar^6, \ldots$

which sum to $I = 1/ar$.

Show further that the areas of circumscribed rectangles are

$(r-1)/a, (r^2-r)/ar^2, (r^3-r^2)/ar^4, \ldots$

which sum to $C = r/a$.

Thus $I < A < C$ gives $1/ar < A < r/a$, and since this holds for all $r > 1$, $A = 1/a$ from an argument like that of qn 2.65(a). It follows that the area between $x = a$ and $x = b$ is $1/a - 1/b$.

The method used here can be extended to find the area between the curve $y = 1/x^k$, the x-axis and the line $x = a$, for a positive integer $k > 2$.

4 Let A be the area between the curve $y = x^k$, the x-axis and the line $x = a$, where k is a positive integer and a is positive. Let $0 < r < 1$. Divide the area A into parallel strips between the lines $x = ar^n$, where $n = 0, 1, 2, \ldots$, and show that the areas of inscribed rectangles are

$$(a - ar)(ar)^k, (ar - ar^2)(ar^2)^k, (ar^2 - ar^3)(ar^3)^k, \ldots$$

which sum to $I = \dfrac{a^{k+1}(1-r)r^k}{1-r^{k+1}}$.

Show further that the areas of circumscribed rectangles are

$$(a - ar)a^k, (ar - ar^2)(ar)^k, (ar^2 - ar^3)(ar^2)^k, \ldots$$

which sum to $C = \dfrac{a^{k+1}(1-r)}{1-r^{k+1}}$.

Thus $I < A < C$.

Simplify the expressions for I and C using qn 1.3(vi), and consider the effect of letting $r \to 1$ to show that $A = a^{k+1}/(k+1)$.

5 (Optional) Modify question 4 by taking k to be a positive rational, p/q. At some stage in the solution, you will have to substitute s^q for r.

Monotonic functions

We revert to the method of qns 1 and 2 and generalise the argument.

6 (*Newton*, 1687)

(i) We start with a lemma to help us with qns 6(ii), 7, 8 and 9. Suppose $I_n \le A \le C_n$ for all positive integers n, and that $(C_n - I_n)$ is a null sequence. Prove that $(I_n) \to A$, and $(C_n) \to A$. Use qns 3.34 and 3.45.

(ii) Let $f: [a, b] \to \mathbb{R}$ be a function which is positive, monotonic increasing and continuous. Let A be the area bounded by $y = f(x)$, the x-axis, $x = a$ and $x = b$. Let $h = (b - a)/n$. Divide the area A into n parallel strips between the lines $x = a + ih$, where $i = 0, 1, 2, \ldots, n$.
The inscribed rectangle between $x = a + (i - 1)h$ and $x = a + ih$ has area $f(a + (i - 1)h)\cdot h$. Denote the sum of the areas of the inscribed rectangles by I_n. The circumscribed rectangle between $x = a + (i - 1)h$ and $x = a + ih$ has area $f(a + ih)\cdot h$. Denote the sum of the areas of the circumscribed rectangles by C_n. Draw a figure to illustrate this and mark a rectangular area equal to $C_n - I_n$. Since $I_n \le A \le C_n$ and $C_n - I_n = (f(b) - f(a))(b - a)/n$, deduce that $(I_n) \to A$ and $(C_n) \to A$ as $n \to \infty$.

7 In qn 6(ii), had f been monotonic decreasing, how might the argument have been affected?

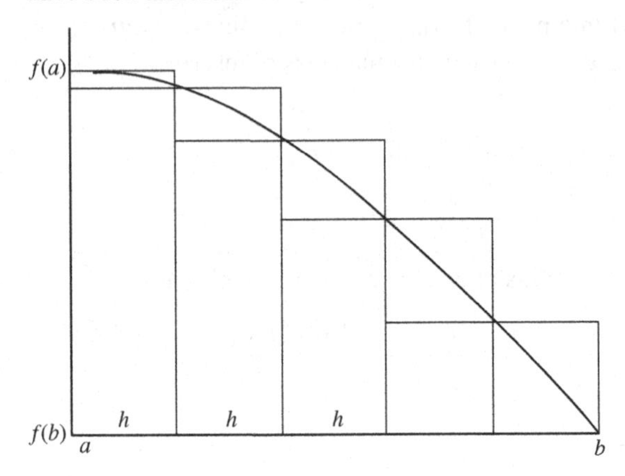

Figure 10.3

8 (*Gregory of St Vincent*, 1647, after *Toeplitz*, 1963)
Let $A[a, b]$ denote the area bounded by the curve $y = 1/x$, the x-axis and the lines $x = a$ and $x = b$, where a and b are positive. Let $h = (b - a)/n$. Divide the area $A[a, b]$ into n parallel strips between the lines $x = a + ih$, where $i = 0, 1, 2, \ldots, n$.
Denote the sum of the areas of the circumscribed rectangles within these strips by C_n. Taking t to be positive, divide the area $A[ta, tb]$ into n parallel strips between the lines $x = ta + ith$, where $i = 0, 1, 2, \ldots, n$.
Denote the sum of the areas of the circumscribed rectangles within these strips by D_n. Prove that $C_n = D_n$.
Now from qn 7, $(C_n) \to A[a, b]$ and $(D_n) \to A[ta, tb]$, so $A[a, b] = A[ta, tb]$. Use the fact that $A[1, ab] = A[1, a] + A[a, ab]$ and the equality we have just obtained to prove that $A[1, ab] = A[1, a] + A[1, b]$, the logarithmic property.

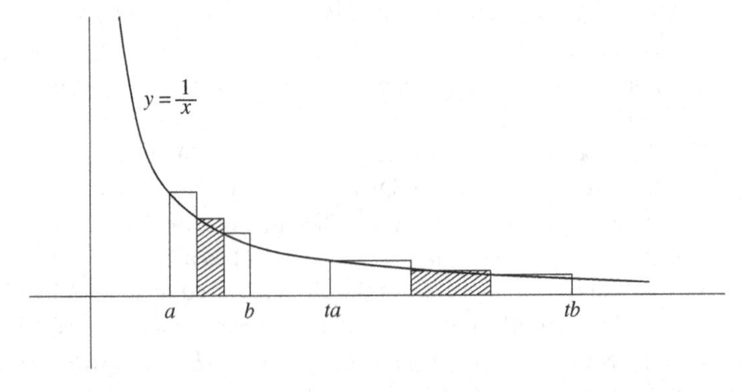

Figure 10.4

9 With the notation of qn 8, divide the area $A[1, a]$ into n parallel strips between the lines $x = \sqrt[n]{a^i}$, where $i = 0, 1, 2, 3, \ldots, n$. Let $a > 1$. Prove that the area of each inscribed rectangle is $1 - 1/\sqrt[n]{a}$ and (that the area of each circumscribed rectangle is $\sqrt[n]{a} - 1$.

So $\dfrac{n(\sqrt[n]{a} - 1)}{\sqrt[n]{a}} < A[1, a] < n(\sqrt[n]{a} - 1)$.

From qn 2.50 we know that $(n(\sqrt[n]{a} - 1))$ is a decreasing sequence, so $n(\sqrt[n]{a} - 1) \le a - 1$ for all n. The difference

$$n(\sqrt[n]{a} - 1) - n(\sqrt[n]{a} - 1)/\sqrt[n]{a} \le (a - 1)(1 - 1/\sqrt[n]{a})$$

which from qns 3.57(d) and 3.65 is a null sequence. So the limit of $(n(\sqrt[n]{a} - 1))$ is $A[1, a]$ from qn 6(i).

Notice that the values of x in this question are all powers of $\sqrt[n]{a}$, and therefore increase in geometric progression, while the corresponding areas increase in arithmetic progression.

The definite integral

It is now appropriate to recognise that the argument in qn 6(ii) did not require the function f to have positive values except for the purpose of using the word 'area'. The inequalities of qn 6(ii) are independent of any notion of area.

In qn 6(ii) we can define

$$I_n = \sum_{i=1}^{i=n} h \cdot f(a + (i - 1)h) \text{ and } C_n = \sum_{i=1}^{i=n} h \cdot f(a + ih),$$

without reference to area, and we obtain convergent sequences as before with equal limits. The common limit is called the *integral* of the function f on $[a, b]$, and is denoted by $\int_a^b f$. The symbol $\int$ is a large squashed S (for sum).

10 Give examples of monotonic functions f and g which are both continuous on $[a, b]$, for which the equation

$$\int_a^b (f + g) = \int_a^b f + \int_a^b g$$

would fail if the integral were always to denote the positive value of the area between the graph and the x-axis.

A further distinction between the notion of integral and the measurement of area under a graph seems natural when we recognise that the argument of qn 6, as modified before qn 10, does not depend on the continuity of the function f. An area must have a boundary, and in order to give *boundaries* to the areas under discussion in qn 6, the function f was taken to be continuous. The graph of a discontinuous function may have gaps in it and so does not bound an area. But the limiting arguments of qn 6 as modified before qn 10 hold for a monotonic function f even if f is discontinuous at many points of the interval $[a, b]$. So such a function can have a well-defined integral.

11 (*Dirichlet*, 1829) If even discontinuous functions can have integrals, are there any functions which cannot? Find the area of the smallest circumscribed rectangle and the largest inscribed rectangle for the area bounded by $x = 0$, the x-axis, $x = 1$ and the function defined by

$$f(x) = \begin{cases} 1 & \text{when } x \text{ is rational,} \\ 0 & \text{when } x \text{ is irrational;} \end{cases}$$

on the interval $[0, 1]$.

As we search for a definition of $\int_a^b f$, we will limit our attention to functions which are bounded, so that they may be covered, above and below, by rectangles, and we seek a definition with the following two properties which we retain from our experience of finding areas with curved boundaries using inscribed and circumscribed polygons:

(i) if $m \leq f(x) \leq M$ on $[a, b]$, then $m(b-a) \leq \int_a^b f \leq M(b-a)$;
(ii) $\int_a^c f + \int_c^b f = \int_a^b f$, when these integrals exist.

Step functions

In qns 12–15, we construct a family of functions, called *step functions*, with convenient, and obvious, integrals, which we define in qn 15. These functions are not generally either positive or continuous, but they will be able to play the rôle of inscribed and circumscribed polygons for us in an analytical setting. The use of properties (i) and (ii), following qn 11, in qns 12–15 is to establish the reasonableness of the definition of the integral of a step function when it comes in qn 15.

12 If $f(x) = A$ on $[a, b]$, show that there is a unique definition of $\int_a^b f$ which is compatible with (i) above.

13 If $f(x) = A$ on $[a, c]$ and $f(x) = B$ on $(c, b]$, with $A \leq B$, use the equation

$$\int_a^b f = \int_a^c f + \int_c^{c+h} f + \int_{c+h}^b f$$

to show that

$$(c-a)A + hA + (b-c-h)B \le \int_a^b f$$
$$\le (c-a)A + hB + (b-c-h)B$$

so

$$(c-a)A + (b-c)B - h(B-A) \le \int_a^b f$$
$$\le (c-a)A + (b-c)B$$

and, if this is to hold for all h, however small, $\int_a^b f$ must be defined as

$(c-a)A + (b-c)B$.

Is the same definition necessary if $A \ge B$?

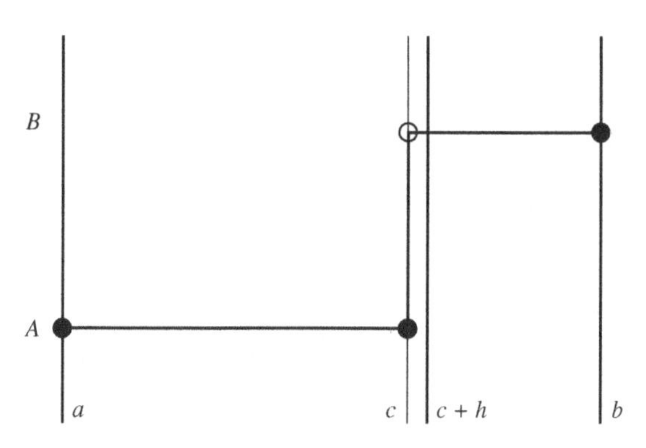

Figure 10.5

14 Show that the value of the integral found in qn 13 must also be the value of the integral for the function, f, defined by

$f(x) = A$ on $[a, c)$,
$f(c) = C$,
$f(x) = B$ on $(c, b]$,

by using the equation

$$\int_a^b f = \int_a^{c-h} f + \int_{c-h}^{c+h} f + \int_{c+h}^b f.$$

Start working with $A < B < C$.

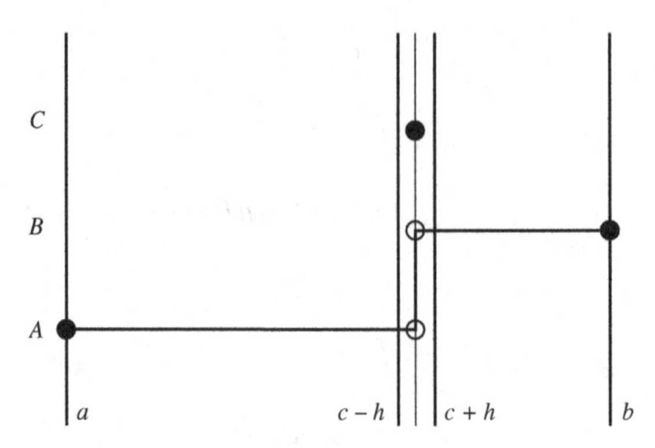

Figure 10.6

15 Let $a = x_0 < x_1 < x_2 < \ldots < x_n = b$, and a real function f be defined on $[a, b]$ such that

$$f(x) = A_i, \text{ for } x_{i-1} < x < x_i, i = 1, 2, \ldots, n; \text{ and}$$
$$f(x_i) = B_i, i = 0, 1, 2, \ldots, n.$$

Such a function is called a *step function.*
Use qn 14 to prove that if this step function has an integral compatible with properties (i) and (ii) after qn 11, then its value is

$$\sum_{i=1}^{i=n}(x_i - x_{i-1})A_i.$$

We now adopt this result as the *definition* of the integral of a step function.

Lower integral and upper integral

16 Consider a function $f: [a, b] \to [m, M]$. Let $s: [a, b] \to \mathbb{R}$ be a step function such that $s(x) \leq f(x)$ for all values of x. Such a step function is called a *lower step function* for f.
We call $\int_a^b s$ a *lower sum* for f.
Why must $m(b - a)$ be a lower sum for f?
Why must every lower sum be $\leq M(b - a)$?
Since, for the given bounded function f, the set of all lower sums is non-empty and bounded above it has a supremum, which is called the *lower integral* for f, and is denoted by $\underline{\int_a^b} f$.

17 Every bounded function on a closed interval has a lower integral. Find the lower integral for the function in qn 11.

18 Consider a function $f: [a, b] \to [m, M]$. Let $S: [a, b] \to \mathbb{R}$ be a step function such that $S(x) \geq f(x)$ for all values of x. Such a step function is called an *upper step function* for f.
We call $\int_a^b S$ an *upper sum* for f.
Why must $M(b - a)$ be an upper sum for f?
Why must every upper sum be greater than or equal to $m(b - a)$? Since, for the given bounded function f, the set of all upper sums is non-empty and bounded below it has an infinimum, which is called the *upper integral* for f, and is denoted by $\overline{\int_a^b} f$.

19 Find the upper integral for the function in qn 11.

It is in proving the existence of upper and lower integrals for bounded functions that completeness is needed in building a theory of integration.

20 (*Darboux*, 1875) For a given bounded function $f: [a, b] \to \mathbb{R}$, let S be an upper step function which is constant on the intervals (x_{i-1}, x_i) where $a = x_0 < x_1 < x_2 < \ldots < x_n = b$.
Let s be a lower step function which is constant on the intervals (y_{j-1}, y_j) where $a = y_0 < y_1 < y_2 < \ldots < y_m = b$.
Let $\{z_0, z_1, z_2, \ldots, z_l\} = \{x_0, \ldots, x_n\} \cup \{y_0, \ldots, y_m\}$ where $a = z_0 < z_1 < z_2 < \ldots < z_l = b$.
Why are both S and s constant on each of the intervals (z_{k-1}, z_k)?
Prove that $s(x) \leq S(x)$ on such an interval.
Does $\int_a^b S$ have the same value whether it is calculated on the intervals (x_{i-1}, x_i) or on the intervals (z_{k-1}, z_k)? What is the corresponding claim for $\int_a^b s$? Deduce that the lower sum from the lower step function $s \leq$ the upper sum from the upper step function S. The idea of uniting the two dissections is due to Cauchy, 1823. You have proved that every lower sum $\leq$ every upper sum.

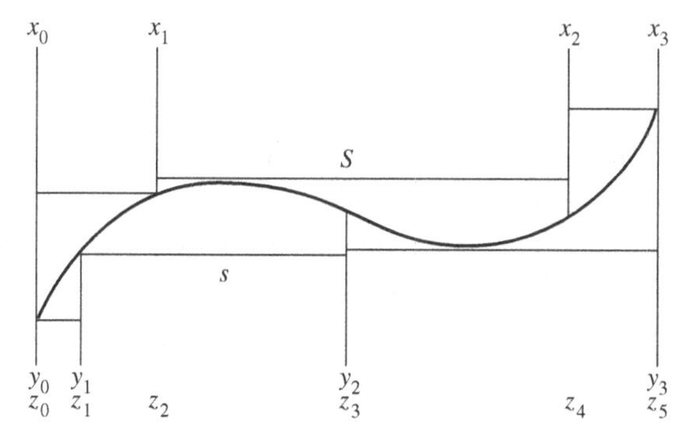

Figure 10.7

21 Show that for any bounded function $f : [a, b] \to \mathbb{R}$, the lower integral must be less than or equal to the upper integral, that is $\underline{\int_a^b} f \le \overline{\int_a^b} f$, by showing that if this were false, we could use the basic property of supremum and infimum (see qns 4.64 and 4.72) to contradict the result of qn 20.

Putting together the definitions of upper and lower integral with the results of qns 20 and 21, we obtain

$$\int_a^b s \le \underline{\int_a^b} f \le \overline{\int_a^b} f \le \int_a^b S,$$

for all lower step functions s and all upper step functions S.

22 Verify that any number lying between the upper integral and the lower integral would satisfy the conditions (i) and (ii) which we stated after qn 11 for the integral $\int_a^b f$.

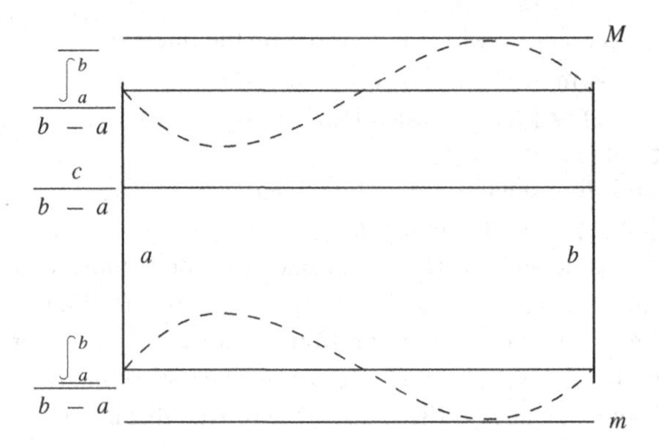

Figure 10.8

The Riemann integral

Question 22 implies that no unique integral satisfying the conditions (i) and (ii), as stated after qn 11, can exist unless $\underline{\int_a^b} f = \overline{\int_a^b} f$, and then $\int_a^b f$ must equal the common value. When this is the case the bounded function f is defined to be *integrable* (or more precisely *Riemann integrable*) on $[a, b]$, and the common value of $\underline{\int_a^b} f$ and $\overline{\int_a^b} f$ is called the *Riemann integral* of the function.

When $\underline{\int_a^b} f < \overline{\int_a^b} f$, we say that the function f is *not integrable* on $[a, b]$. A function which is not bounded cannot have an integral in this sense.

23 (*du Bois Reymond*, 1875, *Darboux*, 1875) Use the basic property of an infimum (see qn 4.72), and the analogous property of a supremum, to show that if the function f is integrable on $[a, b]$ (that is, $\underline{\int_a^b} f = \int_a^b f = \overline{\int_a^b} f$) then it is possible to find an upper sum and a lower sum for f which are arbitrarily close to each other; that is, given $\varepsilon > 0$, it is possible to find an upper step function S and a lower step function s such that

$$\int_a^b S - \int_a^b s < \varepsilon.$$

24 (*du Bois Reymond*, 1875, *Darboux*, 1875) (Converse of qn 23) If for a function $f\colon [a, b] \to \mathbb{R}$ it is possible to find upper sums and lower sums whose differences are arbitrarily small, use the chain of inequalities before qn 22 to prove that the upper and lower integrals are equal, so that the function is integrable.

Question 24 provides the standard test for integrability. We proved in qns 12–15 that a step function was necessarily integrable. In qns 6 and 7 we have an argument which can be generalised to show that a bounded monotonic function is integrable even if it is not continuous.

25 Prove that the function $f\colon [0, 1] \to [0, 1]$, given by

$$f(1/n) = 1 \quad \text{when } n = 1, 2, 3, \ldots,$$
$$f(x) = 0 \quad \text{otherwise},$$

is integrable.

Summary: Definition of the Riemann integral

Step functions

Definition qn 15 — A function $s\colon [a, b] \to \mathbb{R}$ is called a *step function* if, for some subdivision $a = x_0 < x_1 < x_2 < \ldots < x_n = b$, $s(x) = A_i$ when $x_{i-1} < x < x_i$, $i = 1, 2, \ldots, n$; $s(x_i) = B_i$ when $i = 0, 1, \ldots, n$.
The integral of this step function is

$$\int_a^b s = \sum_{i=1}^{i=n} A_i(x_i - x_{i-1}).$$

Definition qns 16, 18 — If $s\colon [a, b] \to \mathbb{R}$ is a step function and $f\colon [a, b] \to \mathbb{R}$ is a function, s is called a *lower step function* for f when $s(x) \le f(x)$ for all values of x, and s is called an *upper step function* for f when

$f(x) \le s(x)$ for all values of x.
When s is a lower step function, its integral is called a *lower sum* for f. When s is an upper step function, its integral is called an *upper sum* for f.

Theorem (qns 16, 18) Any bounded function $f: [a, b] \to [m, M]$ has an upper sum $M(b - a)$ and a lower sum $m(b - a)$.

Theorem (qn 20) Each lower sum of a bounded function $f: [a, b] \to \mathbb{R}$ is less than or equal to each upper sum.

The Riemann integral

Definition (qns 16, 18) The supremum of the lower sums of a bounded function $f: [a, b] \to \mathbb{R}$ is called the *lower integral* of f and denoted by $\underline{\int}_a^b f$.
The infimum of the upper sums of f is called the *upper integral* of f and denoted by $\overline{\int}_a^b f$.

Theorem (qn 21) For a given bounded function $f: [a, b] \to \mathbb{R}$,
each lower sum $\le$ the lower integral
$\le$ the upper integral
$\le$ each upper sum.

Definition (qn 22) A bounded function $f: [a, b] \to \mathbb{R}$ is said to be *Riemann integrable* when its upper integral is equal to its lower integral. Their common value is called the *Riemann integral* of the function and denoted by $\int_a^b f$.

Theorem (qns 23, 24) A function $f: [a, b] \to \mathbb{R}$ is Riemann integrable if and only if, given $\varepsilon > 0$, there exists an upper sum Σ and a lower sum σ, such that $\Sigma - \sigma < \varepsilon$.

Theorem (qns 6, 7, 10+) Every function which is monotonic on a closed interval is integrable on that interval.

Theorems on integrability

While obtaining the basic algebraic properties of integrals (in qns 26–31) it will be useful to have notation available that will not have to be repeatedly defined. We suppose that the integrable function $f: [a, b] \to \mathbb{R}$ has integral F, and, following qn 23, given $\varepsilon > 0$, f has a lower step function s with lower sum σ and an upper step function S with an upper sum Σ, such that $\Sigma - \sigma < \varepsilon$.

26 If, for $k > 0$, the function $k \cdot f$ is defined by

$k \cdot f : x \mapsto k \cdot f(x)$,

show that $k \cdot s$ is a lower step function and $k \cdot S$ is an upper step function for $k \cdot f$ and deduce from qn 24 that $k \cdot f$ is integrable, and has integral $k \cdot F$.

27 If the function $-f$ is defined by

$$-f : x \mapsto -f(x),$$

show that $-S$ is a lower step function and $-s$ is an upper step function for $-f$, and deduce from qn 24 that $-f$ is integrable and has integral $-F$.

28 From qns 26 and 27 extend the result of qn 26 to show that the result in that question holds for any real value of k.

We suppose further that the integrable function $g : [a, b] \to \mathbb{R}$ has integral G and, following qn 23, given $\varepsilon > 0$, g has a lower step function s' with lower sum σ' and an upper step function S' with upper sum Σ', such that $\Sigma' - \sigma' < \varepsilon$.

29 If the function $f + g : [a, b] \to \mathbb{R}$ is defined by

$$f + g : x \mapsto f(x) + g(x),$$

show that $s + s'$ is a lower step function for $f + g$ and that $S + S'$ is an upper step function for $f + g$, and that the difference between the upper and lower sums for these step functions is $(\Sigma - \sigma) + (\Sigma' - \sigma')$. Deduce that $f + g$ is integrable and has integral $F + G$.

Putting together the results of qns 28 and 29, we can show that, if $f : [a, b] \to \mathbb{R}$ is integrable with integral F and $g : [a, b] \to \mathbb{R}$ is integrable with integral G, then the function $k \cdot f + l \cdot g$ is integrable with integral $k \cdot F + l \cdot G$. This enables the obtaining of an integral a linear function of the family of integrable functions on the domain $[a, b]$.

30 Use qns 28 and 29 with qn 4 to determine the value of $\int_0^a f$ when:

(i) $f(x) = x^2 + x^3$;
(ii) $f(x) = 2x + 3x^2 + 4x^3$;
(iii) $f(x) = c_0 + c_1 x + c_2 x^2 + \ldots + c_n x^n$.

31 If the function f^+ is defined by

$$f^+(x) = \begin{cases} f(x) & \text{when } f(x) \geq 0, \\ 0 & \text{when } f(x) < 0, \end{cases}$$

prove that s^+ is a lower step function for f^+ and that S^+ is an upper step function for f^+. Then by considering s^+ and S^+ on the intervals on which both are constant, as in qn 20, show that the difference between the

upper and lower sums which they give is less than or equal to $\Sigma - \sigma$.
Deduce that, if f is integrable, then so is f^+.

32 If the function f^- is defined by

$$f^-(x) = \begin{cases} 0 & \text{when } f(x) > 0, \\ f(x) & \text{when } f(x) \le 0, \end{cases}$$

show that $f^-(x) = -(-f)^+(x)$ for all values of x.
Deduce from qns 27 and 31 that, if f is integrable, then so is f^-.

33 If the function $|f|$ is defined by

$$|f|(x) = \begin{cases} f(x) & \text{when } f(x) \ge 0, \\ -f(x) & \text{when } f(x) < 0, \end{cases}$$

check that $|f| = f^+ - f^-$, and deduce that, if f is integrable, then so is $|f|$.

34 Give an example to show that $|f|$ may be integrable even when f is not.

35 Use the inequality

$$\left| \int_a^b f^+ + \int_a^b f^- \right| \le \left| \int_a^b f^+ \right| + \left| \int_a^b f^- \right|,$$

which follows from the triangle inequality of qn 2.61, to prove that $|\int_a^b f| \le \int_a^b |f|$ when f is integrable on $[a, b]$.

36 (This is a hard question on which nothing further in this book depends, but it opens up the question as to what functions may be Riemann integrable in a most dramatic way.)
(*Thomae*, 1875) In qn 6.72 we examined the continuity of the ruler function f defined on [0, 1] by

$f(x) = 0$ when x is irrational or zero,
$f(p/q) = 1/q$ when p and q are non-zero integers with no common factor.

We found that this function was continuous at each irrational point and discontinuous at each rational point, except 0.
We will consider the integrability of this function on the interval [0, 1] by seeking an upper sum and a lower sum which differ by less than $\varepsilon > 0$.

(i) Identify a lower step function for which the lower sum = 0.

(ii) Choose a positive integer m such that $1/m < \frac{1}{2}\varepsilon$. Show that there cannot be more than $\frac{1}{2}m(m-1)$ points $x \in [0, 1]$ such that $f(x) > 1/m$.

(iii) Suppose that there are exactly N such points, which we denote in order of magnitude by $c_1, c_2, c_3, \ldots, c_N$.
Find a value of δ such that an upper step function defined on the subdivision

$$0 \leq c_1 - \delta < c_1 + \delta < c_2 - \delta < c_2 + \delta < \ldots < c_N + \delta \leq 1$$

by

$$\begin{aligned} S(x) &= 1 && \text{when } c_i - \delta < x < c_i + \delta, \\ S(x) &= 1/m && \text{otherwise,} \end{aligned}$$

gives an upper sum $< \varepsilon$, where points are suppressed in the subdivision where there is overlap.

(iv) Deduce that the function *is* Riemann integrable on [0, 1].

Integration and continuity

37 (i) Why is the function f defined by $f(x) = 1/x$ when $x \neq 0$ and $f(0) = 0$ not integrable on [0, 1]?

(ii) Give an example to show that a function which is not continuous at one point of a closed interval may none the less be integrable on that interval.

38 A real function f is continuous on $[a, b]$. How do you know that f is bounded and that therefore f has an upper and a lower integral?

39 (*Cauchy*, 1823) The real function f is continuous on $[a, b]$ and $a = x_0 < x_1 < x_2 < \ldots < x_n = b$.

(i) Must the function f be bounded and attain its bounds on $[x_{i-1}, x_i]$?

(ii) Let $m_i = \inf\{f(x) |\ x_{i-1} \leq x \leq x_i\}$,
and $M_i = \sup\{f(x) |\ x_{i-1} \leq x \leq x_i\}$.
Explain why $\Sigma_{i=1}^{i=n} m_i(x_i - x_{i-1})$ is a lower sum for f and why $\Sigma_{i=1}^{i=n} M_i(x_i - x_{i-1})$ is an upper sum for f.

(iii) Now suppose that the division of the interval $[a, b]$ has been into n equal lengths, so that $x_i - x_{i-1} = (b-a)/n$ for all values of $i = 1, 2, \ldots, n$. Check that in this case the difference between the upper sum and the lower sum is equal to

$$\frac{b-a}{n} \sum_{i=1}^{i=n} (M_i - m_i).$$

If it were possible to prove that this may be made less than any given $\varepsilon > 0$, we would have succeeded in proving that f was integrable by qn 24.
Now use the fact that a function which is continuous on a closed interval is uniformly continuous on that interval to show that, for any given $\varepsilon > 0$, there exists a $\delta > 0$ such that

$$|x - y| < \delta \Rightarrow |f(x) - f(y)| < \varepsilon/(b - a).$$

Show how to choose n so that the difference of upper and lower sums previously calculated is less than ε.

This establishes that a continuous function on a closed interval is integrable.

40 Does the converse of the result at the end of qn 39 hold? If a function is integrable must it be continuous?

41 $\int_a^b f = 0$ and $f(x) \geq 0$ for all $x \in [a, b]$. If f is continuous, prove that $f(x) = 0$ for all $x \in [a, b]$. What if f were not necessarily continuous?

42 If f and g are integrable on $[a, b]$ and $\int_a^b (k \cdot f + l \cdot g)^2 = 0$, under what conditions could you deduce that $k \cdot f(x) = -l \cdot g(x)$ for all $x \in [a, b]$?

43 (i) If f is a continuous function on $[a, b]$, prove that

$$\frac{b-a}{n} \sum_{i=1}^{i=n} f\left(a + \frac{i}{n}(b-a)\right)$$

lies between the upper sum and the lower sum calculated in qn 39.
(ii) Deduce that its limit as $n \to \infty$ is $\int_a^b f$.
(iii) Check that the limit obtained coincides with the measurement of area in qns 1 and 2.

The notation for integrals, proposed by Leibniz, which is in standard use today, namely $\int f(x)dx$, is deliberately suggestive of the limit of the sum $\Sigma f(x_i)\delta x_i$, where δx_i denotes the difference $x_i - x_{i-1}$, and the limit is taken as the greatest of the $\delta x_i \to 0$.

44 Determine whether the functions f and g defined below are integrable on the interval $[0, 1]$.

$f(x) = x \sin(1/x)$ when $x \neq 0$, and $f(0) = 0$.
$g(x) = \sin(1/x)$ when $x \neq 0$, and $g(0) = 0$.

For g you can consider $[0, 1] = [0, \frac{1}{n}] \cup [\frac{1}{n}, 1]$.

Mean Value Theorem for integrals

45 (*Cauchy*, 1823) If f is a continuous function on $[a, b]$, and $m = \inf\{f(x)|a \leq x \leq b\}$ and $M = \sup\{f(x)|a \leq x \leq b\}$, explain why

$$m(b-a) \leq \int_a^b f \leq M(b-a).$$

Deduce that, for some $c \in [a, b]$,

$$\int_a^b f = f(c) \cdot (b-a).$$

This is called the *Mean Value Theorem for integrals.*
If $a = x_0 < x_1 < x_2 < \ldots < x_n = b$, and $x_i - x_{i-1} = (b-a)/n$, show that

$$\left(\left(\sum_{i=1}^{i=n} f(x_i)\right)/n\right) \to f(c) \text{ as } n \to \infty.$$

Integration on subintervals

46 If a function f is integrable on $[a, b]$ and $a \leq c < d \leq b$, prove that f is integrable on $[c, d]$.

47 If a function f is integrable on $[a, b]$ and $a < c < b$, prove that $\int_a^b f = \int_a^c f + \int_c^b f$. We started using this as an intuitively desirable property for integrals, but now that we have a formal definition of what an integral is we need to prove that it is a formal consequence of the definition.

48 Make a definition of $\int_a^a f$ for any function f, which is compatible with the result of qn 47 in case $c = a$ or $c = b$.
Define $\int_a^b f$ in terms of $\int_b^a f$ in a way which will extend the validity of the result in qn 47 whatever the order of the numbers a, b and c on the number line.

Summary: Properties of the Riemann integral

Theorem qns 1, 2, 4, 5	If $f(x) = x^r$ for r a positive rational, then $\int_0^a f = a^{r+1}/(r+1)$.
Theorem qns 8, 9	If $f(x) = 1/x$, then $\int_b^{ba} f = \int_1^a f = \lim_{n\to\infty} n(\sqrt[n]{a} - 1)$.

Theorem If $f\colon [a,b] \to \mathbb{R}$ has integral F, then $k \cdot f$ has
qns 26, 27, 28 integral $k \cdot F$ for any real number k.
Theorem If $f\colon [a,b] \to \mathbb{R}$ is integrable, then $|f|$ is also
qns 33, 35 integrable, and $|\int_a^b f| \le \int_a^b |f|$.
Theorem If $f\colon [a,b] \to \mathbb{R}$ has integral F and $g\colon [a,b] \to \mathbb{R}$
qn 29 has integral G, then $f+g$ has integral $F+G$.
Theorem If $f\colon [a,b] \to \mathbb{R}$ is continuous, then f is integrable.
qn 39
Mean Value Theorem for integrals
qn 45 If $f\colon [a,b] \to \mathbb{R}$ is continuous,
then $\int_a^b f = f(c)\cdot(b-a)$ for some $c \in [a,b]$.
Theorem If $f\colon [a,b] \to \mathbb{R}$ is integrable and $a \le c < d \le b$,
qn 46 then $f\colon [c,d] \to \mathbb{R}$ is integrable.
Theorem If $f\colon [a,b] \to \mathbb{R}$ is integrable and $a < c < b$, then
qn 47
$$\int_a^b f = \int_a^c f + \int_c^b f.$$
Definition
$$\int_a^a f = 0 \text{ and } \int_b^a f = -\int_a^b f.$$
qn 48

Indefinite integrals

49 Give a reason why the integer function defined by $f(x) = \lfloor x \rfloor$ is integrable on any closed interval.
Give a formula for $F(x) = \int_0^x f$

(i) when $0 \le x \le 1$,
(ii) when $1 \le x \le 2$,
(iii) when $2 \le x \le 3$.

Is the function F continuous at $x = 1$ and $x = 2$?

50 (*Darboux*, 1875) The function f is given to be integrable on the interval $[a,b]$ and a function F is defined on the interval $[a,b]$ by

$$F(x) = \int_a^x f.$$

Prove that F is continuous at every point of $[a,b]$.
Identify a real number L such that $|F(x) - F(y)| \le L \cdot |x - y|$ for all x and y in $[a,b]$.
The function F is called an *indefinite integral* for f.

51 (*Darboux*, 1875) Let f be a function which is integrable on $[a,b]$, and suppose there exists a function F such that $F'(x) = f(x)$ for all $x \in [a,b]$. Suppose also that $a = x_0 < x_1 < x_2 < \ldots < x_n = b$ is a subdivision of $[a,b]$.

(i) How do you know that there exists a $c_i \in (x_i, x_{i-1})$ such that

$$\frac{F(x_i) - F(x_{i-1})}{x_i - x_{i-1}} = F'(c_i)?$$

(ii) By adding equations of the type

$$F(x_i) - F(x_{i-1}) = f(c_i)(x_i - x_{i-1}),$$

show that

$$F(b) - F(a) = \sum_{i=1}^{i=n} f(c_i)(x_i - x_{i-1}).$$

(iii) Deduce that $F(b) - F(a) = \int_a^b f$.

It is customary to write $F(b) - F(a) = [F(x)]_a^b$.

There was no requirement in qn 51 that f be continuous.

52 Apply qn 51 on $[0, x]$ to

$$F(x) = x^2 \sin(1/x) \text{ for } x \neq 0,$$
$$F(0) = 0.$$

In qn 51, two conditions were postulated, namely

(i) f is integrable on $[a, b]$, and
(ii) $F'(x) = f(x)$.

This second condition appears to imply that f has an anti-derivative as will be defined after qn 54, and therefore might be thought to imply that f is integrable, but this is not so. Conditions (i) and (ii) are independent, as qn 53 will show. There are even bounded derivatives which are not integrable as Volterra found in 1881.

53 Suppose $F(x) = x^2 \sin(1/x^2)$ when $x \neq 0$ and $F(0) = 0$. Check that F is differentiable on $[0, 1]$. Use computer graphics to examine the graph of F'. Show that F' is unbounded on $[0, 1]$ and so not integrable.

The Fundamental Theorem of Calculus

54 (*Cauchy*, 1823) Let f be a continuous function on $[a, b]$ and let $F(x) = \int_a^x f$.

(i) Why, for some θ, $0 < \theta < 1$, must

$$F(x + h) - F(x) = h \cdot f(x + \theta h)?$$

(ii) Deduce that

$$\frac{F(x+h)-F(x)}{h}-f(x)=f(x+\theta h)-f(x).$$

(iii) Use the continuity of f to show that $F'(x)=f(x)$.
This result is the *Fundamental Theorem of Calculus*.

When f is continuous and $F'(x)=f(x)$, it is customary, following Leibniz, to write $F(x)=\int f(x)dx$ (echoing $\Sigma f(x_i)(x_i-x_{i-1})$); F is called an *anti-derivative* for f. F may be differentiable even when f is discontinuous, as in qn 36. Questions 51 and 54 show why it is that when integrals are to be found, reversing the process of differentiation, when it can be done, will be effective.

55 (*Cauchy*, 1823) If both F and G are anti-derivatives for f on a given interval, use the Mean Value Theorem to deduce that $F(x)-G(x)$ is independent of x. Use this to verify that $F(b)-F(a)=G(b)-G(a)$.

As a consequence of qn 55, it follows that, if F is an anti-derivative of f, then every other anti-derivative of f has the form $x\mapsto F(x)+c$, for some constant real number c.

(**56**) A function f is integrable on a neighbourhood of a and $\lim_{x\to a^+} f(x)=L$. Show that

$$\lim_{h\to 0^+}\frac{F(a+h)-F(a)}{h}=L,$$

where F is defined as in qn 54. Claim the corresponding result for left-hand limits. Deduce that if f is continuous at a, then F is differentiable at a, and $F'(a)=f(a)$, but that if f has a jump discontinuity at a (see qn 6.79), then F is not differentiable at a.

Integration by parts

57 If both the functions f and g have continuous derivatives, use the fact that $f\cdot g$ is an anti-derivative of $f\cdot g'+f'\cdot g$ to prove that

$$\int_a^b f'\cdot g=[f\cdot g]_a^b-\int_a^b f\cdot g'.$$

58 If the function f has a continuous second derivative on $[a,b]$, prove that

$$f(b)=f(a)+(b-a)f'(a)+\int_a^b (b-x)f''(x)dx.$$

59 (*Cauchy*, 1823, *Taylor's Theorem with integral form of the remainder*) If the function f is infinitely differentiable on $\mathbb{R}$, prove by induction, using integration by parts, that for any real numbers a and b

$$f(b) = f(a) + (b-a)f'(a) + \frac{(b-a)^2 f''(a)}{2!} + \ldots + \frac{(b-a)^n f^{(n)}(a)}{n!} + \frac{1}{n!}\int_a^b (b-x)^n f^{(n+1)}(x)dx.$$

Apply the Mean Value Theorem for integrals (qn 45) to deduce the Cauchy form of the remainder for a Taylor series (qn 9.45).
If $f(x) = \sin x$ and $a = 0$, show that the last term of this expansion tends to 0, as $n \to \infty$.

Integration by substitution

60 If f is a continuous function on $[g(a), g(b)]$ with $F' = f$, and the function g has a continuous derivative on $[a, b]$, identify an anti-derivative for the function $(f \circ g) \cdot g'$.
Justify each of the equations

$$\int_{g(a)}^{g(b)} f = [F]_{g(a)}^{g(b)} \quad \text{and} \quad \int_a^b f(g(x)) \cdot g'(x)dx = [F(g(x))]_a^b.$$

What further condition must the function g satisfy if we are to be sure that all four expressions are equal?

Leibniz' notation is helpfully suggestive here as it takes the form

$$\int f(g)dg = \int f(g(x))\frac{dg}{dx}dx.$$

Improper integrals

61 If a and b are positive numbers, find the value of $\int_a^b f$ where $f(x) = 1/x^2$.
Determine $\lim\limits_{b\to\infty} \int_a^b f$.

When a function f is integrable on the interval $[a, b]$ with integral I, and $I \to L$ as $b \to \infty$, we write $\int_a^\infty f = L$. When the limit exists, $\int_a^\infty f$ is called an *improper integral*.

In qns 5.56–5.61 there are theorems which show that the existence of an improper integral may be equivalent to the convergence of an infinite series.

62 Give a definition for $\int_{-\infty}^{b} f$ analogous to that above. Illustrate your definition with an example. When the limit exists, $\int_{-\infty}^{b} f$ is called an *improper integral*.

63 If a and b are positive numbers, find the value of $\int_a^b f$ where $f(x) = 1/\sqrt{x}$.
Determine $\lim_{a \to 0^+} \int_a^b f$.

When a function f is integrable on the interval $[a, b]$ with integral I, and $I \to L$ as $a \to c^+$, we write $\int_c^b f = L$, even when f is not integrable on $[c, b]$. When this limit exists, $\int_c^b f$ is called an *improper integral*.

64 Find $\displaystyle\int_{-1}^{0} \frac{dx}{\sqrt{(1+x)}}$ as an improper integral.
If $-1 < a < 0$, show that $\displaystyle 0 < \int_a^0 \frac{dx}{\sqrt{(1-x^2)}} < \int_a^0 \frac{dx}{\sqrt{(1+x)}}$.
Deduce that $\displaystyle\int_{-1}^{0} \frac{dx}{\sqrt{(1-x^2)}}$ exists as an improper integral.

65 Give a definition for $\int_a^c f$ when f is integrable on $[a, b]$ but not on $[a, c]$, analogous to that above.
Illustrate your definition with an example.

66 Find $\displaystyle\int_{0}^{1} \frac{dx}{\sqrt{(1-x)}}$ as an improper integral.
If $0 < b < 1$, show that $\displaystyle 0 < \int_0^b \frac{dx}{\sqrt{(1-x^2)}} < \int_0^b \frac{dx}{\sqrt{(1-x)}}$.
Deduce that $\displaystyle\int_{0}^{1} \frac{dx}{\sqrt{(1-x^2)}}$ exists as an improper integral.

67 A function f is defined for $x \geq 0$ by

$$f(x) = \frac{(-1)^n}{n+1} \text{ when } n \leq x < n+1;\ n = 0, 1, 2, \ldots.$$

Do either (or both) of $\displaystyle\int_0^{\infty} f$ or $\displaystyle\int_0^{\infty} |f|$ exist? Compare this result with qn 33.

68 Find the limit of $\displaystyle\frac{1}{\sqrt{n}}\left(\frac{1}{1} + \frac{1}{\sqrt{2}} + \frac{1}{\sqrt{3}} + \ldots + \frac{1}{\sqrt{n}}\right)$ as $n \to \infty$.
Use qn 5.59.

Summary: The Fundamental Theorem of Calculus

Theorem qn 50	If $f\colon [a,b] \to \mathbb{R}$ is integrable, then the function F, defined by $F(x) = \int_a^x f$, is continuous on $[a,b]$.
Theorem qn 51	If $f\colon [a,b] \to \mathbb{R}$ is integrable and $F'(x) = f(x)$, then $\int_a^b f = F(b) - F(a)$.

The Fundamental Theorem of Calculus

qn 54	If $f\colon [a,b] \to \mathbb{R}$ is continuous and $F(x) = \int_a^x f$, then $F'(x) = f(x)$.

Improper integrals – definitions

qn 61	If $I(b) = \int_a^b f$ and $I(b) \to L$ as $b \to \infty$, then we write $\int_a^\infty f = L$.
qn 65	If $I(b) \to L$ as $b \to c^-$, then we write $\int_a^c f = L$.
qn 62	If $I(a) = \int_a^b f$ and $I(a) \to L$ as $a \to -\infty$, then we write $\int_{-\infty}^b f = L$.
qn 63	If $I(a) \to L$ as $a \to c^+$, then we write $\int_c^b f = L$.

Historical Note

In Archimedes' book on the *Quadrature of the Parabola* (c. 250 BC) he finds the area bounded by a parabolic segment by inscribing a succession of triangles and summing their areas with a geometric series. In his later work *On Spirals* he included a summation of segments of areas which led to the same algebraic calculations and argument as in our qn 1. The idea for the method of Newton, in qn 6, goes back to Archimedes' determination of the volume of a spheroid, by constructing increasingly thin inscribed and circumscribed cylinders. Cavalieri, a pupil of Galileo, published further studies of area in 1635 and 1647, in which by ingenious use of certain cases of the Binomial Theorem he obtained the area under $y = x^3$ and $y = x^4$, and suggested a formula for the area under $y = x^n$ for positive integral n. Proving this result and extending it to all rational $n \neq -1$ (as in qns 1–5) is commonly credited to Fermat who claimed the result in general form, about 1643, though Torricelli also obtained the result about this time. The connection between the area under a rectangular hyperbola and logarithms was identified by Gregory of St Vincent (1647) and A. A. de Sarasa (1649).

In a book written by Isaac Barrow, Newton's predecessor as Lucasian professor at Cambridge, and published in 1669, while discussing continuous and monotonic distance–time and velocity–time graphs, Barrow described the Fundamental Theorem of Calculus. The description is geometrical (and therefore, from a modern viewpoint, hard to recognise) and the proof uses infinitesimals. Newton helped Barrow to write the book, and there was a copy of it in Leibniz' personal library. From his papers, we know that Newton came to understand the Fundamental Theorem in the year 1665–6. He considered two variables x and y (which he called fluents) changing with time. Their rates of change he denoted by $\dot{x}$ and $\dot{y}$ (which he called fluxions) and he obtained $\dot{A}/\dot{x} = y$, where A is the area under the graph of y against x. Newton's discovery of the inverse relationship between integration and differentiation, or as he would have put it, between the method of quadratures and the method of fluxions, transformed the study of areas and generated a range of powerful applications. Newton's readiness to differentiate and integrate power series term by term (without tests of convergence) brought a further host of significant results relating to the binomial series and trigonometric functions. Leibniz obtained Newton's fundamental theorem independently in 1675. In his *Principia* (1687), Newton included a proof that any monotonic function was integrable. His illustrations show that he was considering only differentiable monotonic functions, but all the functions he was concerned with were piecewise monotonic, so his proof dealt satisfactorily with his field of concern. Quite trivially, his proof may be extended to show that discontinuous monotonic functions are integrable, but this was not done until the integrability of such functions was considered in the second half of the nineteenth century.

The wide application of this result and the fact that, in the century following Newton and Leibniz, functions were considered to be infinitely differentiable meant that, during the eighteenth century, the study of integration consisted of the search for anti-derivatives. It was Cauchy (1823) who shifted his attention from the indefinite to the definite integral for the purpose of defining the integral of a continuous function (a more general notion of function than had been considered in the eighteenth century) as the limit of $\Sigma f(x_i)(x_i - x_{i-1})$ as the lengths of the subintervals $|x_i - x_{i-1}|$ tend to 0. Although Cauchy presumed uniform continuity for his functions, the definition enabled him to prove the Fundamental Theorem of Calculus as we know it. In the same lectures, Cauchy obtained the Taylor series with integral form of the remainder. In 1822, Fourier's work on the conduction of heat raised the possibility of the integrability of discontinuous functions obtained from trigonometric series. Cauchy acknowledged that his definition of the integral could be applied to a function with a finite number of discontinuities and gave formal definitions for improper integrals. Dirichlet, who had discussions with Fourier and perhaps Cauchy, in Paris, published his famous example (qn 11) of a function which is *not* equal to a Fourier series and is *not* integrable in Cauchy's sense in 1829 and asserted in the same paper that a function was integrable in Cauchy's sense provided its discontinuities were nowhere dense.

Riemann tackled the subject of what functions could be integrable under the direction of Dirichlet, and in 1854 prepared a lecture in Göttingen (which was not published until 1867, after his death) in which the possibility of the integration of trigonometric series was considered to a depth that was to be definitive for the rest of the century. Riemann's definition of integral was the limit of $\Sigma f(t_i)(x_i - x_{i-1})$, where $x_{i-1} \le t_i \le x_i$, as the lengths of the subintervals $|x_i - x_{i-1}|$ tend to 0, which is equivalent to Cauchy's definition but without the restriction to continuous functions, and he constructed a function which was integrable in this sense but which was discontinuous on a set of points that was everywhere dense. We have given an example of such a function in qn 36 (using a construction due to Thomae, 1875). Riemann's example was put together rather differently: let $g(x) = 0$ when $x - \lfloor x \rfloor = \frac{1}{2}$ and let $g(x) = x - \lfloor x + \frac{1}{2} \rfloor$ otherwise; then Riemann's function is

$$f(x) = g(x) + \frac{g(2x)}{2^2} + \frac{g(3x)}{3^2} + \ldots + \frac{g(nx)}{n^2} + \ldots,$$

which is discontinuous at every point $x = p/2q$, where p and q are odd numbers without a common divisor. The function f is none the less integrable because the series is uniformly convergent everywhere.

The treatment of the Riemann integral that we have given, with step functions, upper and lower sums, and upper and lower integrals, is due to Darboux (1875) and, contemporaneously, to du Bois Reymond (1875). Darboux showed that the integral of Riemann's function was continuous and not differentiable at the points where the original function was discontinuous.

Answers

1 Since $(1/n) \to 0$, $(I_n) \to \frac{1}{3}a^3$ and $(C_n) \to \frac{1}{3}a^3$ so $A = \frac{1}{3}a^3$.

4 $A = a^{k+1}/(k+1)$.

5 With the notation of qn 4, $I = \dfrac{a^{k+1}(1-r)r^k}{1-r^{k+1}}$, where $k = p/q$.

Let $r = s^q$ and we have $I = \dfrac{a^{k+1}(1-s^q)r^k}{1-s^{p+q}}$. If $r \to 1$. then $s \to 1$, and so by qn 7.28 or de l'Hôpital's rule, qn 9.26, $I \to \dfrac{a^{k+1}q}{p+q} = \dfrac{a^{k+1}}{k+1}$. Similarly for C.

6 (i) $0 \le A - I_n \le C_n - I_n$, so by the squeeze rule, qn 3.34, $(A - I_n)$ is a null sequence, so $(I_n) \to A$ by definition.
Also by the difference rule, qn 3.45. $((C_n - I_n) - (A - I_n))$ is a null sequence, so $(C_n - A)$ is a null sequence and $(C_n) \to A$ by definition.
(ii) The rectangular area between $x = a$ and $x = a + h$, and between $y = f(a)$ and $y = f(b)$, is equal to $C_n - I_n$.

8 The circumscribed rectangle of $A[a, b]$ between $x = a + ih$ and $x = a + (i+1)h$ has area $h/(a+ih)$. The circumscribed rectangle of $A[ta, tb]$ between $x = ta + ith$ and $x = ta + (i+1)th$ has area $th/(ta + ith) = h/(a + ih)$. Since corresponding circumscribed rectangles have equal areas, $C_n = D_n$.
The area from $x = 1$ to $x = ab$ is equal to the area from $x = 1$ to $x = a$ together with the area from $x = a$ to $x = ab$. But this last area is equal to the area from $x = 1$ to $x = b$.

9 The inscribed rectangle between $x = \sqrt[n]{a^i}$ and $x = \sqrt[n]{a^{i+1}}$ has area

$$(\sqrt[n]{a^{i+1}} - \sqrt[n]{a^t})/\sqrt[n]{a^{i+1}} = 1 - 1/\sqrt[n]{a},$$

which is independent of i.
The circumscribed rectangle between $x = \sqrt[n]{a^i}$ and $x = \sqrt[n]{a^{i+1}}$ has area

$$(\sqrt[n]{a^{i+1}} - \sqrt[n]{a^t})/\sqrt[n]{a^i} = \sqrt[n]{a} - 1,$$

which is also independent of i.

10 Let $f(x) = x^2$ and $g(x) = -x^2$: both functions are monotonic for positive x. On a positive interval $\int_a^b f + g = 0$, while $\int_a^b f = \int_a^b g \neq 0$ if all integrals were positive.

11 Inscribed area $= 0$. Circumscribed area $= 1$. Not equal, so no integral.

12 $A \le f(x) \le A$, so $A(b-a) \le \int_a^b f \le A(b-a)$ and $\int_a^b f = A(b-a)$.

13 The same definition is necessary when $A \ge B$.

14 Suppose that $A < B < C$ (the argument can be easily adapted for other orderings of these numbers).

$$A(c-h-a) + 2hA + B(b-c-h) \leqslant \int_a^b f$$
$$\leqslant A(c-h-a) + 2hC + B(b-c-h)$$

which gives

$$A(c-a) + B(b-c) - h(B-A) \leqslant \int_a^b f$$
$$\leqslant A(c-a) + B(b-c) + h(2C-B-A).$$

If this must hold for all h, however small, $\int_a^b f = A(c-a) + B(b-c)$.

15 The argument of qn 14 shows that a distinct value at an isolated point does not affect the value of the integral. Questions 12 and 13 establish the rest of the result.

16 Since $m \leq f(x)$, $s(x) = m$ defines a lower step function, so $m(b-a)$ is a lower sum.
Now $s(x) \leq f(x) \leq M$ for all x, so if $s(x) = A_i$ when $x_{i-1} < x < x_i$, $A_i \leq M$.
So $A_i(x_i - x_{i-1}) \leq M(x_i - x_{i-1})$, and $\int_a^b s \leq M(b-a)$.

17 Every lower step function ≤ 0, so every lower sum ≤ 0. But 0 is a lower sum, so the lower integral is 0.

18 Since $f(x) \leq M$, $S(x) = M$ defines an upper step function, so $M(b-a)$ is an upper sum.
$m \leq f(x) \leq S(x)$ for all x so, using the argument in qn 16, $m(b-a) \leq \int_a^b S$.

19 Every upper step function ≥ 1, so every upper sum ≥ 1. But 1 is an upper sum, so the upper integral is 1.

20 Since $z_k = x_i$ for some i or y_j for some j, the open interval (z_{k-1}, z_k) is contained both within an interval of the type (x_{i-1}, x_i), so that S is constant on (z_{k-1}, z_k), and within an interval of the type (y_{j-1}, y_j), so that s is constant on (z_{k-1}, z_k).
If $z_{k-1} < x < z_k$, then $s(x) \leq f(x) \leq S(x)$, so $s(x) \leq S(x)$ on each of the open intervals (z_{k-1}, z_k). So, calculated on the z-intervals, $\int_a^b s \leq \int_a^b S$.
If $(x_{i-i}, x_i) = (z_{k-p}, z_k)$, then

$$(x_i - x_{i-1})A = (z_k - z_{k-p})A$$
$$= (z_k - z_{k-1})A + (z_{k-1} - z_{k-2})A + \ldots + (z_{k-p+1} - z_{k-p})A,$$

and so $\int_a^b S$ has the same value whichever intervals it is defined on. Likewise for $\int_a^b s$.
So $\int_a^b s \leq \int_a^b S$ for any upper and lower step functions S and s for the function f.

21 Suppose $\overline{\int_a^b} f < \underline{\int_a^b} f$ and that $\underline{\int_a^b} f - \overline{\int_a^b} f = \varepsilon > 0$. Then there is an upper step function S and a lower step function s such that

$$\overline{\int_a^b} f \le \int_a^b S < \overline{\int_a^b} f + \tfrac{1}{2}\varepsilon = \underline{\int_a^b} f - \tfrac{1}{2}\varepsilon < \int_a^b s \le \underline{\int_a^b} f.$$

This gives $\int_a^b S < \int_a^b s$, which contradicts qn 20.

22 A constant function is a step function and therefore the inequality $\int_a^b s \le \underline{\int_a^b} f \le \overline{\int_a^b} f \le \int_a^b S$ guarantees the first condition after qn 11 for any supposed 'integral' lying between the upper and the lower integral. The second condition also holds because it holds for step functions.

23 Since $\int_a^b f = \sup \int_a^b s$, there is a lower step function s such that $\int_a^b f - \frac{1}{2}\varepsilon < \int_a^b s \le \int_a^b f$ and, since $\int_a^b f = \inf \int_a^b S$, there is an upper step function S such that $\int_a^b f \le \int_a^b S < \int_a^b f + \frac{1}{2}\varepsilon$.

24 If the difference between the upper and lower integrals were ε, then to find an upper step function S and a lower step function s such that $\int_a^b S - \int_a^b s < \varepsilon$ would contradict the chain of inequalities before qn 22.

25 If we consider upper and lower step functions on the intervals $(0, 1/n)$, $(1/n, 1/(n-1))$, $\ldots$, $(\frac{1}{4}, \frac{1}{3})$, $(\frac{1}{3}, \frac{1}{2})$, $(\frac{1}{2}, 1)$, then we have an upper sum $= 1/n$ and a lower sum $= 0$. Since these are arbitrarily close, the function is integrable, and has integral 0.

26 $s(x) \le f(x) \le S(x) \Rightarrow k \cdot s(x) \le k \cdot f(x) \le k \cdot S(x) \Rightarrow k\sigma$ is a lower sum and $k\Sigma$ is an upper sum for $k \cdot f$. Now $k \cdot \Sigma - k \cdot \sigma < k \cdot \varepsilon$, so the difference between an upper sum and a lower sum may be made arbitrarily small and so $k \cdot f$ is integrable. Clearly the integral is $k \cdot F$.

27 $s(x) \le f(x) \le S(x) \Rightarrow -S(x) \le -f(x) \le -s(x) \Rightarrow -S$ is a lower step function and $-s$ is an upper step function for $-f$. So $-\Sigma$ is a lower sum and $-\sigma$ is an upper sum for $-f$. So $(-\sigma) - (-\Sigma) = \Sigma - \sigma < \varepsilon$. Clearly the integral is $-F$.

28 From qn 26, $k \cdot f$ is integrable so, from qn 27, $-k \cdot f$ is integrable, with integral $-k \cdot F$.

29 $s(x) \le f(x) \le S(x)$ and $s'(x) \le g(x) \le S'(x)$
$\Rightarrow s(x) + s'(x) \le f(x) + g(x) \le S(x) + S'(x)$.
Now $s + s'$ is a step function on the union of the subintervals for s and s', in the sense of qn 20, and is therefore a lower step function for $f + g$. Moreover, $\int_a^b (s + s') = \int_a^b s + \int_a^b s'$ using the definition of the integral of a step function on the union of the subintervals. So $\sigma + \sigma'$ is a lower sum for $f + g$, and likewise $\Sigma + \Sigma'$ is an upper sum.

$(\Sigma + \Sigma') - (\sigma + \sigma') = (\Sigma - \sigma) + (\Sigma' - \sigma') < 2\varepsilon$, which may be made arbitrarily small, so $f + g$ is integrable.
$\sigma \le F \le \Sigma$ and $\sigma' \le G \le \Sigma' \Rightarrow \sigma + \sigma' \le F + G \le \Sigma + \Sigma'$,
so the integral is $F + G$.

30 $\int_0^a x^k = a^{k+1}/(k+1)$ from qn 4. Some caution is called for in the appeal to qn 4, since the dissection of $[0, a]$ in qn 4 is infinite, and so the approximating functions are not step functions. However, by choosing finite dissections with an arbitrarily small last interval with end point 0, we can obtain upper and lower sums as close as we like to those used in qn 4. Questions 1 and 2 use step functions.

(i) $\frac{1}{3}a^3 + \frac{1}{4}a^4$ from qn 29.
(ii) $a^2 + a^3 + a^4$ from qns 28 and 29.
(iii) $c_0 a + c_1 a^2/2 + c_2 a^3/3 + \ldots + c_n a^{n+1}/(n+1)$.

31 $s^+(x) > 0 \Rightarrow 0 < s^+(x) = s(x) \le f(x) = f^+(x)$.
$s^+(x) = 0 \Rightarrow s^+(x) = 0 \le f^+(x)$.
Likewise $f(x) > 0 \Rightarrow 0 < f^+(x) = f(x) \le S(x) = S^+(x)$.
$f(x) \le 0 \Rightarrow f^+(x) = 0 \le S^+(x)$.
So $s^+(x) \le f^+(x) \le S^+(x)$.
Evidently since s and S are step functions, so are s^+ and S^+. So we have established that s^+ is a lower step function and S^+ is an upper step function for f^+.
Now clearly, if $S^+(x) = 0$, then $s^+(x) = 0$, so
$S(x) - s(x) \ge S^+(x) - s^+(x) \ge 0$.
Thus the difference between the upper sums and lower sums from S^+ and $s^+ \le \Sigma - \sigma < \varepsilon$. So the difference between upper and lower sums may be made arbitrarily small and so f^+ is integrable.

32 $f^-(x) = 0 \Leftrightarrow f(x) \ge 0 \Leftrightarrow -f(x) \le 0 \Leftrightarrow (-f)^+(x) = 0$.
$f^-(x) < 0 \Leftrightarrow f(x) < 0 \Leftrightarrow -f(x) > 0 \Rightarrow (-f)^+(x) = -f(x)$
$\Rightarrow -(-f)^+(x) = f^-(x)$.
f integrable $\Rightarrow -f$ integrable by qn 27.
$-f$ integrable $\Rightarrow (-f)^+$ integrable by qn 31 $\Rightarrow f^-$ integrable by qn 27.

33 Examine $f(x) > 0, = 0, < 0$, to show $|f| = f^+ - f^-$.
Integrability of $|f|$ follows from qns 31, 32, 27 and 29.

34 On the interval $[0, 1]$ consider $f(x) = 1$ when x is irrational and $f(x) = -1$ when x is rational.

35 From the definitions, $f = f^+ + f^-$, so the left-hand side of the inequality equals $|\int_a^b f|$.
$|f| = f^+ - f^- = |f^+| + |f^-|$, so
$\int_a^b |f| = \int_a^b |f^+| + \int_a^b |f^-|$

36 (i) $s(x) = 0$.

(ii) $f(x) > 1/m \Leftrightarrow x = p/q$, where $q < m$. So $m - 1$ possible values of q. Now $1 \le p \le q$, so at most $(m-1) + (m-2) + \ldots + 2 + 1 = \frac{1}{2}m(m-1)$ possible values of x.

(iii) Take $\delta < \frac{1}{4}\varepsilon/N$, then upper sum $\le 2\delta N + (1/m)(1 - 2\delta N)$ $< \frac{1}{2}\varepsilon + \frac{1}{2}\varepsilon = \varepsilon$.
To avoid overlap, take $\delta < \min(\frac{1}{2}(c_i - c_{i-1}), \frac{1}{4}\varepsilon/N)$, then the formula for the upper sum is exact.

(iv) Lower sum $= 0$, upper sum $< \varepsilon$. Function integrable, integral $= 0$.

37 (i) The function is not bounded above, so there is no upper integral.

(ii) $\displaystyle\int_{1/2}^{3/2} \lfloor x \rfloor = \frac{1}{2}$.

38 By qns 7.31, 7.32 and 7.34, the Maximum–Minimum Theorem.

39 (i) Yes, as in qn 38.

(ii) $s(x) = m_i$ when $x_{i-1} \le x < x_i$ gives a lower step function, and $S(x) = M_i$ when $x_{i-1} \le x < x_i$ gives an upper step function.

(iii) Choose n so that $(b-a)/n < \delta$; then

$$\frac{b-a}{n}\sum_{i=1}^{i=n}(M_i - m_i) < \frac{b-a}{n} \cdot \frac{n\varepsilon}{b-a} = \varepsilon.$$

40 No. Step functions and monotonic functions are integrable whether they are continuous or not.

41 Since $f(x) \ge 0$, $s(x) = 0$ is a lower step function and so 0 is a lower sum. Since $\int_a^b f = 0$, there must be arbitrarily small upper sums. However, if $f(c) > 0$ for some c, there is a neighbourhood of c on which $|f(x) - f(c)| < \frac{1}{2}f(c)$ as in qn 6.70. So for some δ, $|x - c| < \delta \Rightarrow f(x) > \frac{1}{2}f(c)$. So any upper sum $> 2\delta \cdot \frac{1}{2}f(c)$, which is not arbitrarily small. This contradiction shows that $f(x) = 0$ for all x. In qns 25 and 36, a discontinuous f satisfies the conditions.

42 The conclusion holds if both f and g are continuous functions, using qn 41. Otherwise we could take f as in qn 25 and $g(x) = 0$.

43 (i) $m_i = \inf\{f(x) | x_{i-1} \le x \le x_i\}$,
and $M_i = \sup\{f(x) | x_{i-1} \le x \le x_i\}$.
So $m_i \le f(x_i) \le M_i$, and
$m_i(x_i - x_{i-1}) \le f(x_i)(x_i - x_{i-1}) \le M_i(x_i - x_{i-1})$.
Thus

$$\text{lower sum } \le \sum_{i=1}^{i=n} f(x_i)(x_i - x_{i-1}) \le \text{upper sum.}$$

$x_i = a + (i/n)(b-a)$ and $x_i - x_{i-1} = (b-a)/n$.

(ii) We proved in qn 39 that the difference between these upper and lower sums may be made arbitrarily small for sufficiently large n, and it therefore follows that each of these tends to the integral as $n \to \infty$. Denoting the sum in question by Σ, lower sum $\leq \Sigma \leq$ upper sum $\Rightarrow 0 \leq \Sigma$ – lower sum $\leq$ upper sum – lower sum, and now, by a sandwich theorem, Σ tends to the integral as $n \to \infty$.

(iii) This summation gives a minimal upper sum for the given subdivision when the function is monotonic increasing.

44 The function f is continuous on $[0, 1]$ by qn 6.49 and is thus integrable. $s(x) = -1$ is a lower step function and $S(x) = 1$ is an upper step function for g. Consider step functions on $[0, 1/n]$ and $[1/n, 1]$. The function is continuous on $[1/n, 1]$ and therefore there are arbitrarily close upper and lower sums. On $[0, 1/n]$ the difference between upper and lower sums is $2/n$ which may be arbitrarily small. So the function is integrable by qn 24.

45 See qns 16 and 18. Since f is continuous, $(b - a)f$ is continuous. The minimum value of this function is $m(b - a)$ and the maximum value of this function is $M(b - a)$. Since the integral lies between these values, or at one of them, the integral is equal to $(b - a)f(c)$ for some $c \in [a, b]$ by the Intermediate Value Theorem.
Divide the result of qn 43(ii) by $(b - a)$.

46 Since f is integrable on $[a, b]$, there exist step functions s and S giving upper and lower sums which are arbitrarily close. If c and d were not part of the subdivisions for these step functions, introduce these points into the subdivision for each step function. Then the difference between the upper and lower sums on $[c, d] \leq$ the difference between the upper and lower sums on $[a, b]$, so f is integrable on $[c, d]$.

47 If f is a step function the result is obvious. To obtain the result for any integrable f, apply this result to upper and lower step functions for f.

48 To retain the equation for qn 47, we must define
$\int_a^a f = 0$ and $\int_b^a f = -\int_a^b f$.

49 The function f is monotonic.

$$F(x) = \begin{cases} 0 & \text{when} \quad 0 \leq x \leq 1, \\ x - 1 & \text{when} \quad 1 \leq x \leq 2, \\ 2x - 3 & \text{when} \quad 2 \leq x \leq 3. \end{cases}$$

F is continuous at $x = 1$ and $x = 2$, by considering limits from above and below at each of these points.

50 Since f is integrable, f is bounded. Let $m \leq f(x) \leq M$.
$F(c + h) - F(c) = \int_c^{c+h} f$, so for positive h,
$mh \leq F(c + h) - F(c) \leq Mh$.

Thus $\lim_{h\to 0^+}(F(c+h)-F(c))=0$, so $\lim_{h\to 0^+}F(c+h)=F(c)$.
Likewise for the limit from below. Then use qn 6.89, continuity by limits.
$L=\max\{|m|,|M|\}$.

51 (i) By the Mean Value Theorem applied to F.
(ii) Every summation of this kind lies between an upper sum and a lower sum evaluated on the given subdivision. Since the subdivision is arbitrary, the only number satisfying this condition for all possible upper step functions and all possible lower step functions is the integral itself.

52 $F'(x)=2x\sin(1/x)-\cos(1/x)$ when $x\neq 0$, and
$F'(0)=0$ from qn 8.22.
Note that F' is not continuous at 0. None the less, F' is integrable by qn 29, because $x\sin(1/x)$ gives a continuous function, which is necessarily integrable, and $\cos(1/x)$ gives an integrable function like g in qn 44. Thus $\int_0^x F'=x^2\sin(1/x)$ provided $x\neq 0$.

53 When $x\neq 0$, $F'(x)=2x\sin(1/x^2)-(2/x)\cos(1/x^2)$, which is unbounded near $x=0$.

$$\frac{F(x)-F(0)}{x-0}=x\sin\frac{1}{x^2},\ so\ -|x|\leq\frac{F(x)-F(0)}{x-0}\leq|x|,$$

and $F'(0)=0$. See qn 8.25(i).

54 (i) By the Mean Value Theorem for integrals, qn 45.
(ii) Since F is continuous at x, $f(x+\theta h)\to f(x)$ as $h\to 0$. So

$$\frac{F(x+h)-F(x)}{h}\to f(x)\text{ as }h\to 0,$$

and this implies that $F'(x)=f(x)$.

55 $F'(x)=G'(x)=f(x)\Rightarrow(F-G)'(x)=0\Rightarrow(F-G)(x)=\text{constant}$ by qn 9.17.
$\Rightarrow(F-G)(b)=(F-G)(a)\Rightarrow F(b)-F(a)=G(b)-G(a)$.
Since $(F-G)(x)=\text{constant}$, $G(x)=F(x)+c$.

56 Given $\varepsilon>0$, there exists δ such that

$$\begin{aligned}0<(x-a)=h<\delta&\\ &\Rightarrow|f(x)-L|<\varepsilon\\ &\Rightarrow L-\varepsilon<f(x)<L+\varepsilon\\ &\Rightarrow L-\varepsilon<\frac{1}{h}\int_a^{a+h}f<L+\varepsilon\\ &\Rightarrow L-\varepsilon<\frac{F(a+h)-F(a)}{h}<L+\varepsilon,\end{aligned}$$

so $\lim_{h \to 0^+} \dfrac{F(a+h) - F(a)}{h} = L$
using the Mean Value Theorem for integrals, qn 45. A jump discontinuity is illustrated by qn 49.

57 $(f \cdot g)' = f' \cdot g + f \cdot g'$, so, since $f' \cdot g + f \cdot g'$ is continuous, the Fundamental Theorem of Calculus gives
$[f \cdot g]_a^b = \int_a^b (f' \cdot g + f \cdot g') = \int_a^b f' \cdot g + \int_a^b f \cdot g'$.

58 $\int_a^b f''(x)(b-x)dx = [f'(x)(b-x)]_a^b - \int_a^b f'(x)(-1)dx$
$= -f'(a)(b-a) + \int_a^b f'(x)dx = -f'(a)(b-a) + f(b) - f(a)$.

59 Apply qn 57 to obtain

$$\int_a^b (b-x)^n f^{(n+1)}(x)dx = [(b-x)^n f^{(n)}(x)]_a^b - \int_a^b -n(b-x)^{n-1} f^{(n)}(x)dx$$

$$= -(b-a)^n f^{(n)}(a) + n\int_a^b (b-x)^{n-1} f^{(n)}(x)dx.$$

Question 57 must be applied n times in all.
If $f(x) = \sin x$, then $f'(x) = \cos x = \sin(x + \frac{1}{2}\pi)$,
so $f^{(n)}(x) = \sin(x + \frac{1}{2}n\pi)$.
So $\left|\frac{1}{n!}\int_0^b (b-x)^n f^{(n+1)}(x)dx\right| \le \frac{|b-c|^n}{n!}|\sin(c + \frac{1}{2}(n+1)\pi)| \cdot |b|$,
for some c between 0 and b, by qn 45,

$$\le \frac{|b|^{n+1}}{n!}.$$

The last term $\to 0$ as $n \to \infty$ from qn 3.74(ii).
So ($\sin x$ – partial sum of the first n terms) $\to 0$. So the series tends to $\sin x$.

60 $(F \circ g)' = (F' \circ g) \cdot g' = (f \circ g) \cdot g'$. The first equation comes from the application of the Fundamental Theorem to $F' = f$ on $[g(a), g(b)]$. The second equation comes from the application of the Fundamental Theorem to $(F \circ g)' = (f \circ g) \cdot g'$ on $[a, b]$.
All four expressions are equal provided g is an injection which follows, for example, if g' is positive on $[a, b]$.

61 Using the Fundamental Theorem, the integral $= -1/b + 1/a$. As $b \to \infty$, integral $\to 1/a$. Compare with qn 3.

62 If $I(a) = \int_a^b f$ exists for a negative and unbounded below, and $\lim_{a \to -\infty} I(a) = L$ then we write $\int_{-\infty}^b f = L$. Same example as qn 61.

63 Integral $= 2\sqrt{b} - 2\sqrt{a}$. $\lim_{a \to 0^+}(2\sqrt{b} - 2\sqrt{a}) = 2\sqrt{b}$. If $\int_0^1 \frac{dx}{\sqrt{x}}$ is reflected about $y = x$, its image is $\int_1^\infty \frac{dx}{x^2}$. The convergence of this integral is related to the convergence of $\sum \frac{1}{n^2}$ by qn 5.57.

64 $\lim_{a\to -1^+} [2\sqrt{(1+x)}]_a^0 = 2.$

$$-1 < x \le 0 \Rightarrow 1/\sqrt{2} < 1/\sqrt{(1-x)} \le 1$$
$$\Rightarrow 0 < 1/\sqrt{(1-x^2)} \le 1/\sqrt{(1+x)}.$$

Now $I(a) = \displaystyle\int_a^0 \frac{dx}{\sqrt{(1-x^2)}}$ increases as $a \to -1^+$, but is bounded above by 2.

Let $\sup\{I(a)|-1 < a < 0\} = L$, then as $a \to -1^+$, $I(a) \to L$ and $\displaystyle\int_{-1}^0 \frac{dx}{\sqrt{(1-x^2)}} = L.$

65 If $I(b) = \int_a^b f$ and $\lim_{b\to c^-} I(b) = L$, then we write $\int_a^c f = L$.

$$\int_{-1}^b \frac{dx}{\sqrt{(-x)}} = [-2\sqrt{(-x)}]_{-1}^b = -2\sqrt{(-b)} + 2 \to 2 \text{ as } b \to 0^-.$$

66 $[-2\sqrt{(1-x)}]_0^b = -2\sqrt{(1-b)} + 2 \to 2$ as $b \to 1^-$.
$0 \le x < 1 \Rightarrow 1/\sqrt{(1+x)} \le 1 \Rightarrow 1/\sqrt{(1-x^2)} \le 1/\sqrt{(1-x)}$.

Thus $0 < \displaystyle\int_0^b \frac{dx}{\sqrt{(1-x^2)}} < \int_0^b \frac{dx}{\sqrt{(1-x)}}$ for $0 < b < 1$.

Now $I(b) = \displaystyle\int_0^b \frac{dx}{\sqrt{(1-x^2)}}$ increases as $b \to 1^-$, and is bounded above by 2. So $\displaystyle\int_0^1 \frac{dx}{\sqrt{(1-x^2)}}$ exists, as in qn 64.

67 $\int_0^N f = 1 - \frac{1}{2} + \frac{1}{3} - \frac{1}{4} + \ldots + (-1)^{n-1}/N$. This series is convergent by the alternating series test (qn 5.62) to ln 2, by qn 9.41.
$\int_0^N |f| = 1 + \frac{1}{2} + \frac{1}{3} + \frac{1}{4} + \ldots + 1/N$. This harmonic series is divergent, qn 5.30.

68 The Fundamental Theorem must be used to obtain the integral as in the solution of qn 5.59. Divide the three parts of the inequality in qn 5.59 by $\sqrt{n}$, and then apply qn 3.54(iii).

11

Indices and circle functions

For an elegant treatment of exponential and logarithmic functions starting from the counter-intuitive definition $\log x = \int_1^x dt/t$ see Hardy, or for a presentation of Hardy's treatment using a problem sequence see Quadling. For an alternative treatment starting from the definition

$$\exp(x) = 1 + x + \frac{x^2}{2!} + \frac{x^3}{3!} + \dots$$

see Burkill.

Exponential and logarithmic functions

Although we have used logarithms and exponentials in this book, they have not played a formal rôle in defining any of the concepts of analysis.

We now have the results we need to define these functions and to establish their properties.

Positive integers as indices

DEFINITION
For all $x \in \mathbb{R}$, $x^1 = x$ and $x^{n+1} = x^n x$; $n \in \mathbb{N}$.
For $x \neq 0$, $x^0 = 1$ and $x^{-n} = 1/x^n$.

1 (*Dedekind*, 1887) For $m, n \in \mathbb{N}$, prove by induction on n that

(i) $x^{m+n} = x^m x^n$,
(ii) $x^{mn} = (x^m)^n$, and
(iii) $(xy)^n = x^n y^n$.

Properties (i) and (ii) are called the *laws of indices* though they are in fact theorems. It will be necessary to establish that these two laws of indices still hold as we extend the family of numbers which may be used as indices.

2 For $m, n \in \mathbb{N}$, prove that when $x > 1$, $m < n \Leftrightarrow x^m < x^n$, from order properties in qn 2.11.
Deduce that, when $0 < x < 1$, $m < n \Leftrightarrow x^m > x^n$, from qn 2.23.

3 From qn 6.28, show that, for $n \in \mathbb{N}$, the function $f: \mathbb{R}^+ \to \mathbb{R}^+$ given by $f(x) = x^n$ is strictly increasing, continuous and unbounded above.

Positive rationals as indices

4 With f as in qn 3, show that $f^{-1}: \mathbb{R}^+ \to \mathbb{R}^+$ exists by the Intermediate Value Theorem, and is strictly increasing, continuous and unbounded above.

DEFINITION
$f^{-1}(x)$ (as in qn 4) is denoted by $x^{1/n}$.

DEFINITION
For $x > 0$ and $n, m \in \mathbb{Z}^+$, $x^{n/m} = (x^n)^{1/m}$.

5 (i) For $x, y > 0$, prove that $(xy)^{1/n} = x^{1/n}y^{1/n}$.
(ii) By induction on n prove that $x^{n/m} = (x^{1/m})^n$.
(iii) Show that $x^{1/mn} = (x^{1/m})^{1/n}$. Use (ii) and the fact that the function $x \mapsto x^{mn}$ is a bijection of $\mathbb{R}^+$.

6 For $x > 0$ and $r, s \in \mathbb{Q}^+$, prove the two laws of indices

(i) $x^{r+s} = x^r x^s$ and (ii) $x^{rs} = (x^r)^s$.

Build on qns 1 and 5.

7 Prove that, when $x > 1$, and $r, s \in \mathbb{Q}^+$, $r < s \Leftrightarrow x^r < x^s$.
Deduce that, when $0 < x < 1$, and $r, s \in \mathbb{Q}^+$, $r < s \Leftrightarrow x^r > x^s$.

Rational numbers as indices

DEFINITION
For $x > 0$ and $r \in \mathbb{Q}^+$, $x^0 = 1$ and $x^{-r} = 1/x^r$.

8 For $x > 0$ and $r, s \in \mathbb{Q}$, prove the two laws of indices

(i) $x^{r+s} = x^r x^s$ and (ii) $x^{rs} = (x^r)^s$.

9 Prove that, when $x > 1$, and $r, s \in \mathbb{Q}$, $r < s \Leftrightarrow x^r < x^s$.
Deduce that, when $0 < x < 1$, and $r, s \in \mathbb{Q}$, $r < s \Leftrightarrow x^r > x^s$.

DEFINITION
For $a > 1$, define $A: \mathbb{Q} \to \mathbb{R}^+$ by $A(x) = a^x$.

Use computer graphics to examine the graph of A for $a = 1.5$, 2 and 3.
For qns 10–28, we will always assume that $a > 1$.

10 Verify from qn 3.57 that $(A(1/n)) \to 1 = A(0)$ as $n \to \infty$. Deduce that $(A(-1/n)) \to 1 = A(0)$ as $n \to \infty$. Use the fact that A is strictly increasing, established in qn 9, to show that A is continuous at 0.

11 Prove that A is continuous at $q \in \mathbb{Q}$ by considering that

$$A(x) - A(q) = A(q)(a^{x-q} - 1) \to 0 \text{ as } x \to q.$$

So far we have only given a meaning to rational indices, and found that for such indices the function $x \mapsto a^x$ is continuous. Because the rational numbers are dense on the real line we can complete the definition of this function on $\mathbb{R}$ simply by insisting that it shall be continuous.
Questions 12–18 provide the tools for showing that this extension may only be done in one way and for finding the derivative of the resulting function.

DEFINITION
Keeping $a > 1$, define $D\colon \mathbb{Q}\backslash\{0\} \to \mathbb{R}^+$ by $D(x) = \dfrac{a^x - 1}{x}$.

Use computer graphics to examine the graph of D for $a = 1.5$, 2 and 3.

12 Why is D continuous where it has been defined?

13 By considering qn 2.50(iii), show that, when $m, n \in \mathbb{Z}^+$,
$m < n \Leftrightarrow D(m) < D(n)$.
Use the idea of qn 2.50(iv) to deduce that,
when $r, s \in \mathbb{Q}^+, r < s \Leftrightarrow D(r) < D(s)$.
We proved in qn 4.41(i), and again in qn 10.9, that $(D(1/n))$ is convergent as $n \to \infty$.

DEFINITION
Let $(D(1/n)) \to L(a)$ as $n \to \infty$.

14 Use the fact that D is strictly increasing for $x > 0$ to show that

$$\lim_{x \to 0^+} D(x) = L(a).$$

15 Use the fact that, for non-zero x, $D(-x) = D(x)/A(x)$, to show that

$$\lim_{x \to 0^-} D(x) = L(a).$$

16 How should $D(0)$ be defined so that D is continuous on $\mathbb{Q}$?

17 Find a positive number M such that

$$\frac{a^x - a^y}{x - y} < M,$$

provided $0 \leq y < x \leq c$.

Real numbers as indices

18 How does the condition in qn 17 guarantee that A may be extended in a unique way to a continuous function on $[0, \infty)$?

19 How can you be sure that A may also be extended in a unique way to a continuous function on $(-\infty, 0]$?

This gives us a continuous function $A: \mathbb{R} \to \mathbb{R}^+$, and at last we have a well-defined meaning for irrational indices.

20 For $a > 1$ and $x, y \in \mathbb{R}$, prove the two laws of indices

$$\text{(i) } a^{x+y} = a^x a^y \text{ and (ii) } a^{xy} = (a^x)^y.$$

For (ii), first establish the result for $y = n \in \mathbb{N}$, then for $y = 1/n$, and then for $y = m/n \in \mathbb{Q}$, before attempting $y \in \mathbb{R}$.

21 Show that A is strictly increasing on $\mathbb{R}$.

Since $A: \mathbb{R} \to \mathbb{R}^+$ is strictly increasing it is a bijection and has a unique inverse $\mathbb{R}^+ \to \mathbb{R}$ called the logarithm to the base a.

DEFINITION
$A^{-1}(x) = \log_a x$.

22 Prove that, when $X, Y > 0$, $\log_a X + \log_a Y = \log_a XY$.
Deduce that $\log_a 1/X = -\log_a X$.

23
(i) For $x \neq 0$, check that D is continuous at x if and only if A is continuous at x. See qn 12.
(ii) Use the argument of qn 21 to show that D is strictly increasing on $\mathbb{R}^+$. Use qn 13.
(iii) If (x_n) is a null sequence of positive terms, and q_n is a rational number lying between x_n and $2x_n$, use the fact that $(D(q_n)) \to L(a)$ and $(D(\frac{1}{2}q_n)) \to L(a)$ to prove that $(D(x_n)) \to L(a)$ and deduce that $\lim_{x \to 0^+} D(x) = L(a)$.
(iv) By an argument like that of qn 15 show that $\lim_{x \to 0^-} D(x) = L(a)$.
(v) If $D(0) = L(a)$, must D be continuous on $\mathbb{R}$?

24 Prove that $A'(x) = A(x) \cdot L(a)$ for $x \in \mathbb{R}$.

Natural logarithms

25 From qn 10.9, we know that $D(1/n)$ is an upper sum and $D(-1/n)$ is a lower sum for $\int_1^a dx/x$.
Deduce that

$$L(a) = \int_1^a \frac{dx}{x}, \text{ for } a > 1.$$

26 Why must $L(a) \to 0$ as $a \to 1^+$, and $L(a) \to \infty$ as $a \to \infty$? Why is L strictly increasing, continuous and $L'(x) = 1/x$?

27 Why must there be a number $\mathrm{e} > 1$ such that $L(\mathrm{e}) = 1$? Check that $L(2) < 1$, so that $2 < \mathrm{e}$.

DEFINITION
When $a = \mathrm{e}$, write $A(x) = E(x) = \exp(x)$.

28 Prove that $E'(x) = E(x)$. What is the Taylor series for $E(x)$?

If we now define A for $0 < a < 1$, we can use the fact that $1 < 1/a$ to apply the results of qn 10 onwards. With this condition, in the analogues of qns 10 and 21, A is then strictly decreasing. The analogue of qn 13 holds for negative integers and negative rationals and the analogues of qns 14 and 15 transpose the originals. The analogue of qn 16 holds, but for the analogues of qns 17 and 18 an interval $[-c, 0]$ must be used and the result extended from the negative reals to the positive reals. The analogues of qns 20, 23 and 24 hold, and for qn 25 despite a plethora of minus signs; so we get $L(a)$ as before when $0 < a$ and $L(a) \to -\infty$ as $a \to 0^+$.

It is trivial to define $A(x) = 1$ when $a = 1$, and then $L(1) = 0$.

Now L is defined on the domain $\mathbb{R}^+$ and is strictly increasing and continuous.

29 By differentiating the function $f = L \circ E$, show that $L(E(x)) = x$ for $x \in \mathbb{R}$.

30 Since L and E are bijections, $E(L(x)) = x$ for $x \in \mathbb{R}^+$. Deduce that $L(x) = \log_\mathrm{e} x$.

The function L is called the natural logarithmic function and is always denoted on a pocket calculator by ln. When $\log x$ is written without an explicitly named base, in university texts, the logarithm to the base e is meant. When log appears on a pocket calculator, the logarithm to the base 10 is meant. The function A is an exponential function, but when *the* exponential function is referred to, it is E that is meant.

31 Provided that $a, b > 0$, show that

(i) $a^x = (\mathrm{e}^{\ln a})^x = \mathrm{e}^{x \ln a}$.
(ii) $\ln a^x = x \cdot \ln a$.
(iii) $\ln b = \log_a b \cdot \ln a$.

Exponential and logarithmic limits

32 (*Euler*, 1748, *Cauchy*, 1821) Use de l'Hôpital's rule to find the limit of

$$\frac{\ln(1+ax)}{x} \text{ as } x \to 0^+.$$

Deduce that $(n \ln(1 + a/n)) \to a$ as $n \to \infty$.
Use the continuity of E to show that $((1 + a/n)^n) \to \mathrm{e}^a$ as $n \to \infty$.

33 Construct an argument similar to that of qn 32 to show that

$$\lim_{n\to\infty} \left(1 - \frac{a}{n}\right)^{-n} = \mathrm{e}^a.$$

34 The function f is defined on $\mathbb{R}^+$ by $f(x) = x^a$ where a is a real number different from 0. Prove that $f'(x) = a \cdot x^{a-1}$.

35 Use the ratio test to prove that $(n^a \mathrm{e}^{-n}) \to 0$ as $n \to \infty$ (qn 3.74). Deduce that $x^a \mathrm{e}^{-x} \to 0$ as $x \to \infty$. Prove that $\mathrm{e}^x/x^a \to \infty$ as $x \to \infty$ and illustrate this with graph-drawing facilities on a computer.
Prove that, when $a > 0$, $\ln x/x^a \to 0$ as $x \to \infty$.

36 Investigate the function given by $f(x) = x^x$, defined for positive x. Show that f is continuous throughout its domain. Find the minimum value of f. Show that f is monotonic decreasing on the domain $0 < x < 1/\mathrm{e}$. Show that $(f(1/n)) \to 1$ as $n \to \infty$. Deduce that $\lim_{x\to 0^+} f(x) = 1$.

37 What is the Maclaurin series for $E(x)$? Does it converge to the value of the function for all values of x?

38 Use the equation

$$1 + (-t) + (-t)^2 + \ldots + (-t)^{n-1} = \frac{1 - (-t)^n}{1 - (-t)},$$

which is valid for all t unless $t = -1$, and the Fundamental Theorem of Calculus, to show that

$$\ln(1+x) = x - \frac{x^2}{2} + \frac{x^3}{3} - \ldots + (-1)^{n-1}\frac{x^n}{n} + (-1)^n \int_0^x \frac{t^n}{1+t}dt,$$

provided $-1 < x$. Check the terms of the Maclaurin expansion with qn 1.8(i).
For any positive number K, show that, when $-K/(K+1) < x$,

$$\left|(-1)^n \int_0^x \frac{t^n}{1+t}dt\right| \le (K+1)\int_0^x |t|^n dt = \frac{(K+1)|x|^{n+1}}{(n+1)}.$$

(The introduction of K, and the condition $-K/(K+1) < x$, is a device to establish the convergence for negative x, with $-1 < x$.) Check that this last expression gives a null sequence when $|x| \le 1$ and determine for what values of x the power series developed here is a valid expansion of $\ln(1+x)$.

Summary: Exponential and logarithmic functions

Definitions	$x^1 = x$ and $x^{n+1} = x^n x$ for $n \in \mathbb{N}$.
Theorem qns 3, 4	$f\colon \mathbb{R}^+ \to \mathbb{R}^+$ given by $f(x) = x^n$ is a continuous bijection and strictly increasing. $f^{-1}\colon \mathbb{R}^+ \to \mathbb{R}^+$ given by $f^{-1}(x) = \sqrt[n]{x} = x^{1/n}$ is a continuous bijection and strictly increasing.
Definition	$x^{n/m} = (x^n)^{1/m}$, for $m, n \in \mathbb{N}$.
Theorem qn 7	$f\colon \mathbb{R}^+ \to \mathbb{R}^+$ given by $f(x) = x^q$ for $q \in \mathbb{Q}^+$ is a continuous bijection and strictly increasing.
Definition	When $x \neq 0$, $x^0 = 1$ and $x^{-q} = 1/x^q$ for $q \in \mathbb{Q}$.
Theorem qns 10, 11	When $1 < a$, $A\colon \mathbb{Q} \to \mathbb{R}^+$ given by $A(x) = a^x$ is continuous and strictly increasing. When $0 < a < 1$, A is strictly decreasing.
Theorem qns 12, 13	$D\colon \mathbb{Q}\backslash\{0\} \to \mathbb{R}$ given by $D(x) = \dfrac{a^x - 1}{x}$ is continuous. If $1 < a$, D is increasing on $\mathbb{Q}^+$. If $0 < a < 1$, D is increasing on negative $\mathbb{Q}$.
Theorem qns 17, 18	If $c > 0$, $A\colon x \mapsto a^x$ is uniformly continuous on $[0, c] \cap \mathbb{Q}$ and so may be extended in a unique way to a continuous function on $[0, c]$ and so on $[0, \infty)$.
Theorem qn 19	$A\colon x \mapsto a^x$ may be extended in a unique way to a continuous bijection $\mathbb{R} \to \mathbb{R}^+$, which is increasing when $1 < a$ and decreasing when $0 < a < 1$.
Theorem qn 20	$a^{x+y} = a^x a^y$ and $a^{xy} = (a^x)^y$, when $0 < a$ and $x, y \in \mathbb{R}$.
Definition	If $0 < a$ and $a^x = y$, then $\log_a y = x$.
Theorem qn 22	For $x, y > 0$, $\log_a x + \log_a y = \log_a xy$.
Theorem qns 14, 15, 16, 23	$D\colon x \mapsto (a^x - 1)/x$ may be extended in a unique way to a continuous function on $\mathbb{R}$.
Theorem qns 14, 15, 25	$D(x) \to \int_1^a dy/y$ as $x \to 0$.
Theorem qns 24, 25	If $0 < a$ and $A(x) = a^x$ then $A'(x) = A(x) \cdot \int_1^a dy/y$.
Definition	$\int_1^{\mathrm{e}} dx/x = 1$, $E(x) = \mathrm{e}^x = \exp(x)$.
Theorems qns 28, 29, 30	$E'(x) = E(x)$, $E(\int_1^a dx/x) = a$, $\int_1^a dx/x = \log_{\mathrm{e}} a$.

Definition $\ln x = \log_e x$
Theorem When $a > 0$, $a^x = e^{x \ln a}$; $\ln a^x = x \ln a$.
qn 31
Theorem
qn 32 $\lim_{n\to\infty} \left(1 + \frac{a}{n}\right)^n = e^a$.

Circular or trigonometric functions

The origins of the functions sine, cosine, tangent, cotangent, secant and cosecant are geometric. A particle, P, moves around the circumference of a circle with centre O and radius 1. The *angle* through which the radius turns is the length of arc which is traced out by P on the circumference. If the particle moves anti-clockwise from the point A to the point B, and the arc length from A to B is x, then the perpendicular from B to OA has length 'sine of x'. The cosine of x is the sine of the complementary angle. The tangent of x is the length along the tangent at B from B to the point where it meets OA produced. The cotangent of x is the tangent of the complementary angle.

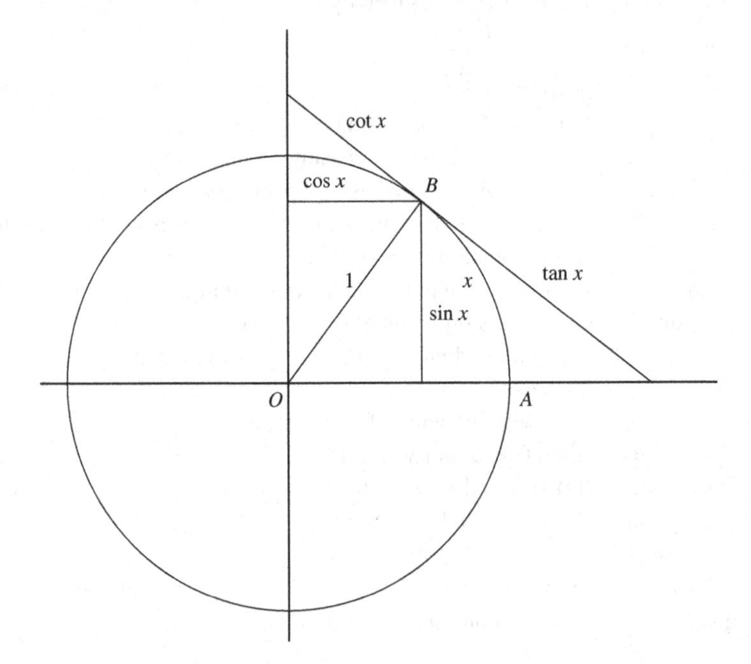

Figure 11.1

If we only have the established properties of the real numbers, how do we define these functions? If logic was all that mattered, we could ignore the geometric origins of the functions and define sine and cosine by their power series: this is the procedure adopted by Burkill. It is also possible to define the sine function by an infinite product or to develop the circular functions from the definition

$$\arctan x = \int_0^x \frac{dt}{1+t^2}:$$

this is the procedure adopted by Hardy.

But if the geometric origins of these functions are to be respected we must develop a formal definition of angle either from the notion of the area of a sector of a unit circle (the procedure adopted by Spivak) or from a formal definition of arc length. This is what is done in qns 39–48. You may, if you wish, skip to the definition of circular arc length following qn 48 and explore the intervening problems when your curiosity is aroused.

Length of a line segment

39 Give an algebraic formula for the non-negative function $f: [-1, 1] \to \mathbb{R}$ whose graph will appear as a semicircle with centre at the origin and radius 1.

40 If the distance between two points (x, y) and (a, b) is defined to be $\sqrt{((x-a)^2 + (y-b)^2)}$, show that this distance $\leq |x-a| + |y-b|$. What is the distance between the two points $(a, f(a))$ and $(b, f(b))$?

41 Why is the function f, of qn 39, continuous on $[-1, 1]$ and differentiable on $(-1, 1)$?

42 Apply the Mean Value Theorem to f to show that the distance between the points of qn 40 is equal to

$$\frac{b-a}{\sqrt{(1-c^2)}}$$

for some c with $-1 \leq a < c < b \leq 1$.

Arc length

43 For any subdivision $a = x_0 < x_1 < x_2 < \ldots < x_n = b$, show that the *polygonal arc length* of the function f on $[a, b]$, defined by

$$\sum_{i=1}^{i=n} \sqrt{((x_i - x_{i-1})^2 + (f(x_i) - f(x_{i-1}))^2)}$$

can only increase if an additional point, or finite set of points, is added to the subdivision. Use qn 2.64, the triangle inequality in the plane.

44 Use qn 40 to show that any polygonal arc length for all or part of the function of qn 39 is less than or equal to 4.

Deduce that the polygonal arc length of this function has a supremum. The supremum for polygonal arc length on the interval $[a, b]$ is called the *arc length* of the function on $[a, b]$.

45 By applying qn 42, show that any polygonal arc length on the interval $[a, b]$ for the function of qn 39 has the value

$$\sum_{i=1}^{i=n} \frac{x_i - x_{i-1}}{\sqrt{(1 - c_i^2)}},$$

with x_i as in qn 43, for some c_is, with $x_{i-1} \le c_i \le x_i$.

46 We define the function $g: (-1, 1) \to \mathbb{R}$ by

$$g(x) = \frac{1}{\sqrt{(1 - x^2)}}.$$

Show that any polygonal arc length on the interval $[a, b]$ for the function f of qn 39 is greater than a lower sum for the function g on the interval $[a, b]$, and less than an upper sum for this function provided $-1 < a < b < 1$. Use qn 10.39.
Deduce that the lower integral $\underline{\int}_a^b g \le$ arc length on $[a, b]$.

47 Let $-1 < a = x_0 < x_1 < x_2 < \ldots < x_n = b < 1$ and $a = y_0 < y_1 < y_2 < \ldots < y_m = b$ be two subdivisions of the interval $[a, b]$, and let the union of these two subdivisions be $a = z_0 < z_1 < z_2 < \ldots < z_l = b$.
Use qn 43 to show that

the polygonal arc length of f with the 'x' subdivision
$\le$ the polygonal arc length of f with the 'z' subdivision.

Explain why

the upper sum for g on the 'y' subdivision
$\ge$ the upper sum for g on the 'z' subdivision.

Deduce from qn 46 that

any polygonal arc length of f on $[a, b] \le$ any upper sum for g on $[a, b]$.

Deduce that

the arc length on $[a, b] \le$ the upper integral $\overline{\int}_a^b g$.

48 From qns 46 and 47 and the definition of the Riemann integral we know that the arc length of f on $[a, b]$ is equal to $\int_a^b g$.
Use qns 10.64 and 10.66 to show that the integral exists as an improper integral even when $a = -1$ or $b = 1$ or both.

Arc cosine

We now *define* an angle function (or arc length function) $A: [-1, 1] \to \mathbb{R}$ by

$$A(y) = \int_y^1 \frac{dx}{\sqrt{(1-x^2)}}.$$

49 Say why the function A is

(i) continuous,
(ii) monotonic decreasing,
(iii) differentiable on $(-1, 1)$ with $A'(y) = \dfrac{-1}{\sqrt{(1-y^2)}}$.

50 Say why

(i) $A(1) = 0$,
(ii) $A(-1) = 2A(0)$.
(iii) A is a bijection $[-1, 1] \to [0, A(-1)]$.

The function A is usually called arccos, and sometimes $\cos^{-1}$. We may now *define* $A(-1) = \pi$ and A^{-1} as the cosine function, so that cosine is a continuous and monotonic decreasing bijection with domain $[0, \pi]$ and range $[-1, 1]$. On the same domain we define the sine function by

$$\sin x = \sqrt{(1 - \cos^2 x)}.$$

Cosine and sine

51 Find the values of $\cos x$ and $\sin x$ when $x = 0$, $\frac{1}{2}\pi$ and π.

52 Use the equation $A(\cos x) = x$ to prove that $\cos' x = -\sin x$ for $0 < x < \pi$. Use the definition of sine to prove that $\sin' x = \cos x$ for $0 < x < \pi$.

53 Sketch the graphs of cosine and sine on the domain $[0, \pi]$. For $\pi < x \leq 2\pi$, *define* $\cos x = \cos(2\pi - x)$ and $\sin x = -\sin(2\pi - x)$. Sketch the graphs of cosine and sine on the domain $[0, 2\pi]$.
Verify that $\sin^2 x + \cos^2 x = 1$ on $[0, 2\pi]$, and that $\cos' x = -\sin x$ on $(\pi, 2\pi)$.

54 Prove that $\cos' \pi = -\sin \pi = 0$, by applying the Mean Value Theorem to $(\cos x - \cos \pi)/(x - \pi)$, so that $\cos' x = -\sin x$ on $(0, 2\pi)$. Prove likewise that $\sin' \pi = \cos \pi = -1$.

For any integer k, we now *define* sine and cosine for $2k\pi \le x < 2(k+1)\pi$ by

$\cos x = \cos(x - 2k\pi)$ and

$\sin x = \sin(x - 2k\pi)$.

55 Prove that $\sin^2 x + \cos^2 x = 1$ for all real x.
Prove that $\cos' x = -\sin x$ and $\sin' x = \cos x$, except possibly when $x = 2k\pi$.
Use the method of qn 54 to prove that $\cos' 0 = -\sin 0 = 0$ and $\sin' 0 = \cos 0 = 1$, so that the formulae for the derived functions hold for all values of x.

56 Define $f(x) = \sin(a - x) \cdot \cos x + \cos(a - x) \cdot \sin x$. Prove that $f'(x) = 0$ for all x. Use the Mean Value Theorem to prove that $f(x) = \sin a$. Deduce the formula for $\sin(x + y)$.

57 Use the method of qn 56 to prove that

$$\cos(x + y) = \cos x \cdot \cos y - \sin x \cdot \sin y.$$

58 Prove that

$$\begin{aligned}\cos 2x &= \cos^2 x - \sin^2 x \\ &= 2\cos^2 x - 1 \\ &= 1 - 2\sin^2 x.\end{aligned}$$

Prove also that

$$\sin 2x = 2\sin x \cdot \cos x.$$

Tangent

59 Define $\tan x = \sin x / \cos x$, except where $x = (k + \frac{1}{2})\pi$, and prove that $\tan' x = 1/\cos^2 x = \tan^2 x + 1$. Show that the tangent function is a monotonic increasing bijection $(-\frac{1}{2}\pi, \frac{1}{2}\pi) \to \mathbb{R}$.

60 Define arctan: $\mathbb{R} \to (-\frac{1}{2}\pi, \frac{1}{2}\pi)$ as the function inverse to tan, so that $\arctan(\tan x) = x$. Explain why arctan is differentiable and why

$$\arctan' x = 1/(1 + x^2).$$

61 Use the equation

$$1 + (-t^2) + (-t^2)^2 + \ldots + (-t^2)^{n-1} = \frac{1 - (-t^2)^n}{1 - (-t^2)},$$

which is valid for all t, and the Fundamental Theorem of Calculus, to show that

$$\arctan x = x - \frac{x^3}{3} + \frac{x^5}{5} - \ldots + (-1)^{n-1}\frac{x^{2n-1}}{2n-1} + (-1)^n \int_0^x \frac{t^{2n}}{1+t^2}dt.$$

Check the terms of this series with the Maclaurin expansion derived from qn 1.8(ii).
Prove that

$$\left|(-1)^n \int_0^x \frac{t^{2n}}{1+t^2} dt\right| \le \int_0^x |t|^{2n} dt = \frac{|x|^{2n+1}}{2n+1}.$$

Check that this last expression gives a null sequence when $|x| \le 1$ and determine for what values of x the power series developed here is a valid expansion of $\arctan x$.

62 (*Leibniz*, 1674) Prove that

$$\frac{1}{4}\pi = 1 - \frac{1}{3} + \frac{1}{5} - \frac{1}{7} + \ldots + (-1)^{n-1}\frac{1}{2n-1} + \ldots.$$

Summary: Circular or trigonometric functions

Theorem qn 48	The arc length on the unit circle from $(1, 0)$ to $(x, \sqrt{(1-x^2)})$, $-1 \le x \le 1$, is $A(x) = \int_x^1 \frac{dy}{\sqrt{(1-y^2)}}$.
Theorem qn 49	The function $A\colon [-1, 1] \to [0, A(-1)]$ is continuous, decreasing and $A'(x) = \frac{-1}{\sqrt{(1-x^2)}}$ on $(-1, 1)$.
qn 50	$A(-1) = 2A(0)$.
Definition	$A(-1) = \pi$. $A^{-1} =$ cosine.
Theorem	$\cos : [0, \pi] \to [-1, 1]$
qn 51	$\cos 0 = 1$, $\cos \frac{1}{2}\pi = 0$, $\cos \pi = -1$.
Definition	$\sin x = \sqrt{(1 - \cos^2 x)}$ on $[0, \pi]$.
	$\cos x = \cos(2\pi - x)$ on $(\pi, 2\pi]$.
	$\sin x = -\sin(2\pi - x)$ on $(\pi, 2\pi]$.
	$\cos x = \cos(x - 2k\pi)$ on $[2k\pi, 2(k+1)\pi]$.
	$\sin x = \sin(x - 2k\pi)$ on $[2k\pi, 2(k+1)\pi]$.
Theorem	$\sin 0 = 0$, $\sin \frac{1}{2}\pi = 1$, $\sin \pi = 0$.
qns 52, 55,	$\cos' x = -\sin x$, $\sin' x = \cos x$.
56, 57	$\sin^2 x + \cos^2 x = 1$.
	$\sin(x + y) = \sin x \cdot \cos y + \cos x \cdot \sin y$.
	$\cos(x + y) = \cos x \cdot \cos y - \sin x \cdot \sin y$.
Definition	$\tan x = \sin x / \cos x$ provided $x \ne (k + \frac{1}{2})\pi$.
Theorem	$\tan' x = 1 + \tan^2 x$.
qn 59	$\tan\colon (-\frac{1}{2}\pi, \frac{1}{2}\pi) \to \mathbb{R}$ is a continuous bijection and strictly increasing.

Historical Note

In 1614, John Napier published tables matching $10^7(1-10^{-7})^n$ with n, for $n = 0, 1, 2, \ldots, 100$ and for a host of larger n, interpolating to build up a table of log sines to seven significant figures. Such 'logarithms', matching the terms of a geometric progression with those of an arithmetic progression, satisfy $\log a + \log b = \log c + \log d \Leftrightarrow ab = cd$, and provide some of the advantages of modern logarithms for computational purposes. Napier's achievement was the more remarkable when one realises that neither the notion nor the notation of exponents was developed until some 20 years later. Briggs had discussions with Napier and then constructed tables (of common logarithms) in which log 1 was 0 and log 10 was 1. These tables were first published in 1617.

In 1647, work by Gregory of St Vincent on the hyperbola showed that a geometric progression along one asymptote produced strips parallel to the other asymptote whose areas were equal. In 1649, Anthony de Sarasa claimed that this relationship was logarithmic.

In his *De Analysi* (written in 1669 and circulated among friends, but not published until 1711) Newton had used term-by-term integration of the series for $1/(1+x)$ (which he obtained by long division) to find the series for $\ln(1+x)$, and then developed a method of repeated successive approximations to calculate the inverse function of a power series. This enabled him to construct the series for $\exp(x) - 1$. Newton did not use the terminology of logarithms or exponentials. In 1668, Mercator published the series for $\ln(1+x)$ in his *Logarithmotechnia* and referred to areas under the rectangular hyperbola as *natural* logarithms. Leibniz described the integral $\int dx/x$ as a logarithm in 1676.

Negative and fractional exponents were first proposed by Wallis in 1656 and first used by Newton in his discussion of the Binomial Theorem in 1669. The first overt claim of the connection between exponents and logarithms was made by Wallis in 1685 in relation to the terms of a geometric progression.

The first claim that the common logarithm of x was the power to which 10 had to be raised to obtain x was made in tables published in England in 1742. In his *Introductio in Analysin Infinitorum* published in 1748, Euler explicitly defined $\log_a x = y$ where $a^y = x$. He used infinitesimals and infinite numbers to great effect. For an infinitesimal ε, he let $(a^\varepsilon - 1)/\varepsilon = k$ (which depends on a) and called the value of a which makes $k = 1$, e (the first letter of his name). By combining infinitesimal and infinite numbers $\varepsilon N = x$, he obtained a^x as the exponential series $\exp(kx)$. The conventional exponential series follows. The original equation giving k, was rearranged to give $(1+x/N)^N = \mathrm{e}^x$. Further ingenious manipulation led to the result that $\log_a x = (N/k)(x^{1/N} - 1)$, and he then used the binomial series to obtain the series for $\ln(1+x)$.

In 1821, Cauchy, presuming that a^x was well defined for positive a, proved that every continuous function f satisfying $f(x) \cdot f(y) = f(x+y)$ was of the form $f(x) = a^x$ for some a. He used the Cauchy product to show that the exponential power series belonged to this family of functions, and defining $\mathrm{e} = \Sigma 1/n!$ had thereby proved that the exponential power series was equal to e^x. Cauchy also used the Binomial Theorem to prove that, as $a \to 0$, $(1+ax)^{1/a} \to \mathrm{e}^x$. In 1881, A. Harnack defined a^x for rational x and extended his definition to irrational x by affirming the continuity of the new function.

In order to extend exponentials to complex numbers, Euler (1748) had defined $\exp(z)$ by means of the exponential series $1 + z + z^2/2! + \ldots$ and in the late nineteenth century it was recognised that, if this definition was used for a real variable, a formal definition of a^x as $\exp(x \ln a)$ avoids some of the difficulties with which Cauchy had struggled.

Newton's discovery of the sine series depended on his previous discovery of the Binomial Theorem for rational index. He derived the series for arcsine from the term-by-term integration of the power series for $\surd(1 - x^2)$ and then used his repeated approximation method to construct the series for $\sin x$. This method is based on the formula $\frac{1}{2}r^2\theta$ for the area of a circular sector and appears in Newton's notes from 1666. In *De Analysi* Newton's method for these series was based on a calculation of arc length, which is closely related to the treatment in this chapter. Again, it was Euler who recognised that an extension of the trigonometric functions to complex variables required a formal definition of these functions by series, and he obtained the equation $\mathrm{e}^{\mathrm{i}x} = \cos x + \mathrm{i} \sin x$ in 1748.

In 1821, Cauchy proved that a continuous solution of the functional equation $f(x+y) + f(x-y) = 2f(x)f(y)$ must take the form $\cos ax$ or $\cosh ax$ with the cosine occurring when $f(x) < 1$ for some x.

In the late nineteenth century it was recognised that, if Weierstrass' arithmetisation of analysis was to be carried through, a non-geometric definition of these functions was required for real variables. In 1880, Thomae gave an analytic definition of cosine using Cauchy's equation above. The series definition (as for complex variables) was also available for this purpose, and the alternative definition with an integral (which we have used) was commended by Felix Klein in 1908.

Answers

1 (i) The definition gives the basis for the induction.
Then $x^{m+n+1} = x^{m+n}x = x^m x^n x = x^m x^{n+1}$.
(ii) The definition gives the basis for the induction.
$x^{m(n+1)} = x^{mn+m} = x^{mn}x^m = (x^m)^n x^m = (x^m)^{(n+1)}$.
(iii) $(xy)^{n+1} = (xy)^n(xy) = x^n y^n xy = x^n x y^n y = x^{n+1}y^{n+1}$.

2 $m < n \Leftrightarrow 0 < n - m \Leftrightarrow 1 < x^{n-m} \Leftrightarrow x^m < x^n \Leftrightarrow 1/x^n < 1/x^m$.

3 See qn 2.20 for strict increase, qn 6.28 for continuity.
$1 < x \le x^n$ shows unboundedness above.

4 See qn 7.27, continuity of nth root function. For $c > 1$, $\sqrt[n]{c^{n+1}} > c$.

5 (i) Use qn 1(iii) and the fact that $x \mapsto x^n$ is a bijection.
(ii) $(x^{1/m})^{(n+1)} = (x^{1/m})^n x^{1/m} = (x^{n/m})x^{1/m}$

$$= (x^n)^{1/m}x^{1/m} = (x^n x)^{1/m} = (x^{n+1})^{1/m}.$$

6 Let $r = p/m$ and $s = q/n$, where $m, n, p, q \in \mathbb{Z}^+$.

(i) $x^{r+s} = x^{(pn+qm)/mn} = (x^{1/mn})^{(pn+qm)} = (x^{1/mn})^{pn}(x^{1/mn})^{qm}$
$= x^{pn/mn}x^{qm/mn} = x^r x^s$
(ii) $x^{rs} = x^{pq/mn} = (x^{pq})^{1/mn} = (x^{p/m})^{q/n}$, using qn 5(ii) and (iii).

7 $r < s \Leftrightarrow 0 < s - r \Leftrightarrow 1 < x^{s-r} \Leftrightarrow x^r < x^s \Leftrightarrow 1/x^s < 1/x^r$.
Use qns 3.57 and 2.23, reciprocal inequalities.

8 The argument of qn 6 applies. For negative r and s, work with reciprocals.

9 The argument of qn 7 applies.

10 $A(-x) = 1/A(x)$. For some N, $m > N \Rightarrow |A(\pm 1/m) - 1| < \varepsilon$, and because A is strictly increasing, $-1/(N+1) < x < 1/(N+1) \Rightarrow |A(x) - 1| < \varepsilon$.
Thus $A(x) \to 1 = A(0)$ as $x \to 0$. A is continuous at 0 by qn 6.89.

11 $x \to q \Rightarrow x - q \to 0 \Rightarrow A(x-q) \to 1 \Rightarrow A(x-q) - 1 \to 0$
$\Rightarrow A(x) - A(q) \to 0 \Rightarrow A(x) \to A(q)$.

12 By qn 6.23, continuity of sums of functions, and qn 6.54, continuity of quotients of functions.

13 Let $r = p/m$ and $s = q/n$, where $m, n, p, q \in \mathbb{Z}^+$.
$r < s \Leftrightarrow pn < qm \Leftrightarrow D(pn) < D(qm)$. Now put $b = a^{mn}$, and we get $D(r) < D(s)$ but with b for a in the definition of D.

14 If $m > N \Rightarrow |D(1/m) - L(a)| < \varepsilon$, then because D is strictly increasing on $\mathbb{Q}^+$, $0 < x < 1/(N+1) \Rightarrow |D(x) - L(a)| < \varepsilon$.

15 Apply qn 6.93, the algebra of limits, from above.

16 $D(0) = L(a)$ by qn 6.89, the neighbourhood definition of continuity.

17 Expression $= A(y)D(x-y) < A(c) \cdot D(c) = M$, since both A and D are strictly increasing on $\mathbb{Q}^+ \cup \{0\}$.

18 A is uniformly continuous on $\mathbb{Q} \cap [0, c]$ by qn 7.42, the Lipschitz condition, and therefore may be extended to a continuous function in a unique way by qn 7.47. This extension may be made at any point $x > 0$ by selecting $c > x$.

19 $A(-x) = 1/A(x)$. Since $A(x) > 0$ for all x, the continuity of A at x is equivalent to the continuity of A at $-x$ from qn 6.52.

20 If x and y are rational numbers, these results were obtained in qn 8. Let (x_n) be a sequence of rational numbers tending to x (possibly irrational) and let (y_n) be a sequence of rational numbers tending to y (also possibly irrational), then $(x_n + y_n) \to x + y$ and $(x_n y_n) \to xy$ by qn 3.54(iii) and (vi). By the continuity of A, $(A(x_n)) \to A(x)$, $(A(y_n)) \to A(y)$, $(A(x_n + y_n)) \to A(x + y)$, $(A(xy_n)) \to A(xy)$ and, by qn 8(i), $A(x_n + y_n) = A(x_n) \cdot A(y_n)$, so by qn 3.54(vi), the product rule, $A(x + y) = A(x) \cdot A(y)$. So we have the first law. For the second law, first establish that it holds for $y = n \in \mathbb{N}$ by induction. Then establish that it holds for $y = 1/n$ using the bijection of qn 4. Consider a product of n terms each equal to $a^{x/n}$. Then establish that it holds for $y = m/n$, a positive rational, using the two previous cases and extend this to negative rational y from the definitions. We now have the equipment to claim that $A(xy_n) = (A(x))^{y_n}$. Now $(A(xy_n)) \to A(xy)$ as we saw before and $((A(x))^{y_n}) \to (A(x))^y$, by the continuity of A, substituting a^x for a. So $A(xy) = (A(x))^y$.

21 We know that A is strictly increasing on $\mathbb{Q}$ and that A is continuous on $\mathbb{R}$. If (p_n) is an increasing sequence of rationals tending to an irrational r, then by the continuity of A, $(A(p_n)) \to A(r)$, and since A is increasing, $(A(p_n))$ is increasing, so by qn 4.66, $A(r) = \sup\{A(p_n) | \, n \in \mathrm{N}\}$, and since every rational $p < r$ can belong to a possible (p_n) sequence, $A(r) = \sup\{A(p) | \, p \in \mathbb{Q}, p < r\}$. Likewise $A(r) = \inf\{A(q) | \, q \in \mathbb{Q}, r < q\}$. If $p, q \in \mathbb{Q}$ and $p < r < q$, then $A(p) < A(r) < A(q)$. Further, for any two distinct real numbers r and s, with $r < s$, there exists a rational number q such that $r < q < s$, and so $A(r) < A(q) < A(s)$, and A is strictly increasing on $\mathbb{R}$.

22 Since A is a bijection $\mathbb{R} \to \mathbb{R}^+$, there is a unique x such that $A(x) = X$ and a unique y such that $A(y) = Y$. From qn 20(i), $A(x + y) = XY$, so $A^{-1}(XY) = x + y$, and $\log_a XY = \log_a X + \log_a Y$.

23 (iii) $0 < \frac{1}{2}q_n < x_n < q_n$, and D strictly increasing, implies $D(\frac{1}{2}q_n) < D(x_n) < D(q_n)$. From qn 14, $(D(\frac{1}{2}q_n)) \to L(a)$ and $(D(q_n)) \to L(a)$. The result comes from qn 3.54(viii), the squeeze rule.

(v) From (i) and qn 6.89, continuity by limits.

24 $(A(x') - A(x))/(x' - x) = A(x) \cdot D(x' - x) \to A(x) \cdot L(a)$ as $x' \to x$.

25 Both $(D(1/n))$ and $(D(-1/n)) \to L(a)$ as $n \to \infty$.

26 $0 < L(a) < a - 1$, so $L(a) \to 0$ as $a \to 1^+$.
For $a \geq 2$, $L(a) > \sum_{n=2}^{n=\lfloor a \rfloor} \frac{1}{n} \to \infty$ as $a \to \infty$, by qns 5.57 and 5.30, the harmonic series.
L is strictly increasing because $x > 0 \Rightarrow 1/x > 0$.
L is continuous by qn 10.50, the continuity of integrals.
$L'(x) = 1/x$ by the Fundamental Theorem of Calculus.

27 Since L is continuous and takes values from small positive to large positive on $(1, \infty)$, L takes the value 1 by the Intermediate Value Theorem for some value of a in this domain. On $(1, 2)$, $1/x < 1$, so $L(2) < 1$.

28 Follows from $E'(x) = E(x) \cdot L(e)$ from qn 24 and $L(\mathrm{e}) = 1$. Taylor series from qn 9.38.

29 $f'(x) = L'(E(x)) \cdot E'(x) = (1/E(x)) \cdot E(x) = 1$. Thus $f(x) = x + c$. But $f(0) = 0$, so $c = 0$.

30 Since $E = L^{-1}$, $E(L(x)) = x$, and $L(x) = \log_{\mathrm{e}} x$.

31 (i) First equality by definition of ln. Second equality from second law of indices.
(ii) The logarithm of the equality in (i). Use qn 29.
(iii) $E(\ln b) = b$. $E(\log_a b \cdot \ln a) = A(\log_a b)$ by (i), $= b$ by definition.

32 The function is well defined provided $|x| < |1/a|$.
Limit $= \lim_{x \to 0^+} \frac{a}{1+ax} = a$. Then consider $x = 1/n$.

34 $f(x) = E(a \ln x)$. $f'(x) = E'(a \ln x) \cdot (a/x) = f(x)(a/x)$.

35 Let $f(x) = x^a \mathrm{e}^{-x}$. $f(n+1)/f(n) = (1 + 1/n)^a/\mathrm{e} \to 1/\mathrm{e}$ as $n \to \infty$. By qn 3.71, d'Alembert, $(f(n)) \to 0$ as $n \to \infty$.
$f'(x) = f(x)(a/x - 1) < 0$ when $x > a$. So f is strictly decreasing when $x > a$. Since the sequence $\to 0$, the function $\to 0$ as $x \to \infty$. So $1/f(x) \to \infty$.
Let $g(x) = (\ln x)/x^a$, then $g'(x) = \dfrac{1 - a \ln x}{x^{a+1}} < 0$ when $x > \mathrm{e}^{1/a}$.
Let $h(n) = g(\mathrm{e}^n)$, then $h(n+1)/h(n) = (1 + 1/n)/\mathrm{e}^a \to 1/\mathrm{e}^a$ as $n \to \infty$.
So, by qn 3.71, $(h(n)) \to 0$ and since g is strictly decreasing, $g(x) \to 0$ as $x \to \infty$.

36 $f(x) = E(x \ln x)$. f is continuous since ln and E are continuous.
$f'(x) = E'(x \ln x) \cdot (\ln x + x \cdot 1/x) = 0$ when $\ln x = -1$ or $x = 1/\mathrm{e}$. On $(0, 1/\mathrm{e})$, $f'(x) < 0$. $f(1/n) = \sqrt[n]{(1/n)} = 1/\sqrt[n]{n} \to 1$ by qn 3.59. Since f is strictly decreasing on this domain, $f(x) \to 1$ as $x \to 0^+$. Minimum is $f(1/\mathrm{e})$.

37 See qn 9.38.

38 Use qn 10.60 (integration by substitution) to show that

$$\int_1^{1+x} \frac{dt}{t} = \int_0^x \frac{dt}{1+t}.$$

$$\int_0^x 1 + (-t) + \ldots + (-t)^{n-1} dt = \int_0^x \frac{dt}{1-(-t)} - \int_0^x \frac{(-t)^n dt}{1-(-t)}.$$

$$-K/(K+1) < t \Rightarrow 1/(K+1) < t+1 \Rightarrow 1/(t+1) < K+1$$

$$\Rightarrow |t|^n/(t+1) < |t|^n(K+1).$$

Also $\dfrac{(K+1)|x|^n}{n+1} \le \dfrac{K+1}{n+1}$ if $|x| \le 1$.
But K is constant, so $((K+1)/(n+1)) \to 0$ as $n \to \infty$.
Expansion is valid when $-1 < x \le 1$.

39 $f(x) = \sqrt{(1-x^2)}$.

40 Squaring both sides of the inequality makes the comparison trivial.
$\sqrt{((a-b)^2 + (f(a) - f(b))^2)}$.

41 f is continuous by qn 6.29, the continuity of polynomials, qn 7.27, the continuity of root functions, and qn 6.40, the continuity of composite functions.
$f'(x) = -x/\sqrt{(1-x^2)}$ on $(-1, 1)$.

42 $f(b) - f(a) = f'(c) \cdot (b-a)$, so

$$\sqrt{((b-a)^2 + (f(b) - f(a))^2)} = (b-a)\sqrt{(1 + (f'(c))^2)}$$

$$= \frac{b-a}{\sqrt{(1-c^2)}}.$$

43 Adding a point to the subdivision replaces one part
of the polygonal arc length by two new ones using the original end points;
this increases the polygonal arc length by the triangle inequality, qn 2.64.

44 On $[-1, 0]$ f is strictly increasing, so the polygonal arc length
$\le \Sigma|x_i - x_{i-1}| + \Sigma|f(x_i) - f(x_{i-1})| \le 1 + 1$.
Likewise on
$[0, 1]$ f is strictly decreasing and again the polygonal arc length is bounded above by 2. So all polygonal arc lengths, by qn 43, are bounded above by 4.

46 The function g is bounded on $[a, b]$ by $1/\sqrt{(1-a^2)}$ or $1/\sqrt{(1-b^2)}$
and continuous and so integrable. Working on the subdivision
$-1 < a = x_0 < x_1 < x_2 < \ldots < x_n = b < 1$ and letting
$m_i = \inf\{g(x)|x_{i-1} \le x \le x_i\}$ and $M_i = \sup\{g(x)|x_{i-1} \le x \le x_i\}$,
from qn 45, $\Sigma m_i(x_i - x_{i-1}) \le$ polygonal arc length of f for this subdivision
$\le \Sigma M_i(x_i - x_{i-1})$.
So sup (lower sums) $\le$ sup (polygonal arc lengths) $=$ arc length.
So lower integral of g on $[a, b] \le$ arc length of f on $[a, b]$.

47 The z-subdivision
is the same as the x-subdivision but with some additional intervening points. The additional points increase the polygonal arc length by qn 43.
The z-subdivision is the same as the y-subdivision
but with some additional intervening points, so if $y_{j-1} = z_k$, then
$y_j = z_{k+p}$ for some p, so $[z_{s-1}, z_s] \subseteq [y_{j-1}, y_j]$ for $s = k+1, \ldots, k+p$.
So $\sup\{g(x) | z_{s-1} \leq x \leq z_s\} \leq \sup\{g(x) | y_{j-1} \leq x \leq y_j\}$
for $s = k+1, \ldots, k+p$.
Polygonal arc length on x-subdivision
$\leq$ polygonal arc length on z-subdivision
$\leq$ upper sum on z-subdivision (using qn 45)
$\leq$ upper sum on y-subdivision.
So every polygonal arc length $\leq$ every upper sum.
So sup (polygonal arc lengths) $\leq$ every upper sum.
So arc length is a lower bound to the upper sums.
So arc length $\leq$ inf (upper sums) = upper integral.

48 Lower integral $\leq$ arc length $\leq$ upper integral (from qns 46 and 47). But g is bounded and continuous on $[a, b]$, so g is integrable on $[a, b]$, and its upper integral and lower integral are equal. So arc length = integral.

49 (i) A is continuous by qn 10.50, the continuity of integrals.
(ii) If $-1 \leq t < s \leq 1$, then $A(s) - A(t) = \int_s^1 - \int_t^1 = \int_s^t = -\int_t^s < 0$.
(iii) A is differentiable by the Fundamental Theorem, qn 10.54.

50 (i) By qn 10.48, definition.
(ii) $A(0) = \int_0^1 \frac{dx}{\sqrt{(1-x^2)}}$.
Putting $u = -x$,

$$\int_{-1}^{0} \frac{dx}{\sqrt{(1-x^2)}} = \int_{1}^{0} \frac{-du}{\sqrt{(1-u^2)}} = A(0).$$

(iii) A is monotonic decreasing and so a bijection. Domain is given. The range is bounded by the values of the function at the extremities of the domain.

51 $A(1) = 0$, so $\cos 0 = 1$ and $\sin 0 = 0$.
$A(-1) = \pi$, so $\cos \pi = -1$ and $\sin \pi = 0$.
$A(0) = \frac{1}{2}A(-1) = \frac{1}{2}\pi$, so $\cos \frac{1}{2}\pi = 0$ and $\sin \frac{1}{2}\pi = 1$.

52 $A(\cos x) = x$
and cos is differentiable by qn 8.42, inverse functions. By qn 8.40,
$A'(\cos x) \cdot \cos' x = 1$, so $(-1/\sqrt{(1-\cos^2 x)}) \cdot \cos' x = 1$. $\cos' x = -\sin x$.
$\sin x = \sqrt{(1-\cos^2 x)}$, so $\sin' x = \dfrac{\frac{1}{2}(-2\cos x)(-\sin x)}{\sqrt{(1-\cos^2 x)}} = \dfrac{\cos x \cdot \sin x}{\sin x}$.

54 The function cos is continuous on $[0, \pi]$ and differentiable on $(0, \pi)$ so we may apply the Mean Value Theorem and get

$$\frac{\cos x - \cos \pi}{x - \pi} = \cos' c = -\sin c \text{ for some } c,\ x < c < \pi.$$

As $x \to \pi$, $c \to \pi$, so $-\sin c \to 0$ and $\cos' \pi = 0$.
So we have $\cos' x = -\sin x$ on $(0, 2\pi)$.
Similarly $\sin' \pi = \cos \pi = -1$.

56 $$\begin{aligned} f'(x) &= -\cos(a-x)\cos x - \sin(a-x)\sin x \\ &\quad + \sin(a-x)\sin x + \cos(a-x)\cos x \\ &= 0. \end{aligned}$$

From qn 9.17, f is constant. But $f(0) = \sin a$. Now put $x + y = a$ to obtain $\sin(x + y) = \sin x \cdot \cos y + \cos x \cdot \sin y$.

59 Apply qns 56 and 57 putting $y = x$, and use $\sin^2 x + \cos^2 x = 1$.

$$\tan' x = \frac{\cos x \cdot \sin' x - \sin x \cdot \cos' x}{\cos^2 x} = \frac{\cos^2 x + \sin^2 x}{\cos^2 x} = 1 + \tan^2 x.$$

The tan function is continuous on $(-\frac{1}{2}\pi, \frac{1}{2}\pi)$ and has positive derivative and is therefore strictly increasing, and so a bijection.
As $x \to \frac{1}{2}\pi^-$, $\cos x \to 0$, so $1/\cos x \to +\infty$; also $\sin x \to 1$, so $\tan x \to \infty$.
Similarly as $x \to -\frac{1}{2}\pi^+$, $\tan x \to -\infty$.

60 arctan is differentiable by qn 8.42, inverse functions.

$$\begin{aligned} \arctan(\tan x) = x &\Rightarrow \arctan'(\tan x)\tan' x = 1 \\ &\Rightarrow \arctan'(\tan x) = \frac{1}{1 + \tan^2 x} \end{aligned}$$

61 First equation from qn 1.3(vi). Then integrate from 0 to x, using qn 60. Power series valid for $-1 \leqslant x \leqslant 1$. Compare with qn 5.8.

62 Put $x = 1$ in the series.
Let $\arctan 1 = X$, then $\tan X = 1$,
so $\sin X = \cos X$ and $\cos 2X = 0$ from qn 58. Thus $2X = \frac{1}{2}\pi$ and $X = \frac{1}{4}\pi$.

12

Sequences of functions

For this chapter, more than any other, the availability of a computer-assisted graph-drawing package is essential, and the preliminary exercise (i)–(x) is dependent on this.

Concurrent reading: Burkill and Burkill ch. 5, Spivak ch. 24.
Further reading: Rudin ch. 7.

Sequences of functions are often considered in a second course of analysis. The difficulties encountered are not like those in earlier analysis since the apparatus of limits has already been developed. Difficulties arise from having two variables of unsymmetric status and having to think of keeping one constant and letting the other vary according to context. Before starting the questions in this chapter it is necessary to have some intuitive feel for the subject matter, and to have observed, in graphical form, how a sequence of functions may behave. To do this, use graph-drawing facilities on a computer to illustrate the functions f_n for $n = 1,\ 2,\ 5$ and 10, where

(i) $f_n(x) = x + 1/n,$

(ii) $f_n(x) = x/n,$

(iii) $f_n(x) = \dfrac{nx}{x+n}$ for $x \in \mathbb{R}^+,$

(iv) $f_n(x) = \dfrac{1}{x^2+n^2},$

(v) $f_n(x) = \dfrac{1}{1+n^2x^2},$

(vi) $f_n(x) = \dfrac{nx}{1+n^2x^2},$

(vii) $f_n(x) = \dfrac{x}{1+n^2x^2},$

(viii) $f_n(x) = \dfrac{1}{(1+x^2)^n},$

(ix) $f_n(x) = x^n$,

(x) $f_n(x) = \dfrac{x^n}{1 + x^{2n}}$.

The first thing to look for in these graphs are the values of x for which the sequence of real numbers $(f_n(x))$ is convergent. One does this by thinking of x as constant and letting $n \to \infty$. Having obtained a limit function f in this way, one then compares the graphs of f_n and f as a whole, looking for their greatest distance apart and thinking of n as constant.

Pointwise limit functions

1 Sketch the graphs of the functions given by

$$f_1(x) = \frac{1}{1 + x^2}, \quad f_2(x) = \frac{1}{4 + x^2}, \quad f_3(x) = \frac{1}{9 + x^2},$$

superimposing the diagrams on the same axes, using appropriate computer software.

For a constant real number x, find $\lim_{n \to \infty} \dfrac{1}{n^2 + x^2}$.

Because this limit is well defined for each real number x, we have a *limit function* $f\colon \mathbb{R} \to \mathbb{R}$ for the sequence of functions (f_n) defined by

$$f_n(x) = \frac{1}{n^2 + x^2}.$$

The limit function in this case is given by $f(x) = 0$, for all x.

2 Sketch the graphs of the functions given by

$$f_n(x) = (\sin x)/n, \text{ for } n = 1, 2 \text{ and } 3,$$

using computer software.

Is the limit $\lim_{n \to \infty} \sin x/n$ well defined, for each real number x?

Since $\lim_{n \to \infty} f_n(x) = 0$ for all x in this case, the function f defined by $f(x) = 0$ is the limit function for the sequence (f_n).

In qns 1 and 2, the limit functions were constant, giving the same value for each x. In qns 3, 4 and 5, the values of the limit are not the same for all values of x.

3 If $f_n(x) = \dfrac{1}{1 + n^2x^2}$, find the limit function for the sequence (f_n) on the domain $\mathbb{R}$.

4 If $f_n(x) = x^n$, find the limit function for the sequence (f_n) on the domain $[0, 1]$.

5 If $f_n(x) = \dfrac{x^n - 1}{x^n + 1}$, find the limit function for the sequence (f_n) on the domain $[0, \infty)$.

Because these limit functions are determined point by point for each x of the domain of the functions, they are called *pointwise limit functions*.

A formal definition runs like this: if a sequence of real functions (f_n), $f_n \colon A \to \mathbb{R}$, have the same domain A and, for each $x \in A$, the limit $\lim_{n\to\infty} f_n(x)$ exists, then a function $f \colon A \to \mathbb{R}$ may be defined by $f(x) = \lim_{n\to\infty} f_n(x)$. This f is the *pointwise limit function* for the sequence (f_n). It is important to stress that pointwise limit functions are found by keeping x constant and letting $n \to \infty$.

It is somewhat disconcerting to find that even though the functions in the sequence are continuous it is possible for the pointwise limit function to be discontinuous as in qns 3, 4 and 5.

This chapter investigates the context in which a property which is common to all the functions of a sequence is retained by the limit function. The possibility that continuity may be lost in the limit shows how significant this question can be. We will distinguish carefully between the cases where the pointwise limit function is continuous, and the cases where it is not.

6 In each of qns 1, 2, 3, 4 and 5, attempt to find a value of x such that

(i) $|f_{10}(x) - f(x)| > \frac{1}{2}$,

(ii) $|f_{100}(x) - f(x)| > \frac{1}{2}$.

7 Continuing the investigation started in qn 6, show that, for the functions of qn 1, $|f_n(x) - f(x)| \leq 1/n^2$ for all x in the domain.

8 Show that, for the functions of qn 2, $|f_n(x) - f(x)| \leq 1/n$ for all x in the domain.

9 For the functions of qn 3, find values of x for which $|f_n(x) - f(x)| > \frac{1}{2}$.

10 For the functions of qn 4, find values of x for which $|f_n(x) - f(x)| > 0.99$.

11 For the functions of qn 5, find values for x for which $|f_n(x) - f(x)| > \frac{1}{2}$.

One way of distinguishing between the cases in qns 1 and 2 on the one hand and qns 3, 4 and 5 on the other is to imagine the graph of the pointwise limit functions f as an arm onto which sleeves of various diameters are pulled. In qns 1 and 2, however narrow the sleeves get, the sequence of functions is eventually wholly within them. But in qns 3, 4 and 5 there are sleeves which the functions in the sequence never get wholly inside.

12 For the functions in qn 3, find $\sup\{|f_n(x) - f(x)| : x \in \mathbb{R}\}$.

13 For the functions in qn 4, find $\sup\{|f_n(x) - f(x)| : x \in [0, 1]\}$.

14 For the functions in qn 5, find $\sup\{|f_n(x) - f(x)| : x \in [0, \infty)\}$.

Uniform convergence

When a sequence of functions, (f_n), with each $f_n : A \to \mathbb{R}$, has a pointwise limit function $f : A \to \mathbb{R}$, and, for any $\varepsilon > 0$, the graphs of the functions f_n are eventually inside a sleeve of radius ε about the graph of the limit function f, then, for sufficiently large n,

$f(x) - \varepsilon < f_n(x) < f(x) + \varepsilon$ for all values of x.

In this case we say that the sequence (f_n) *converges uniformly* to f. [Think of a large fixed n, and check on the sleeve property by letting x vary right across the domain.]

We can also express the uniform convergence of (f_n) to f by saying that

$(\sup\{|f_n(x) - f(x)| : x \in A\}) \to 0$ as $n \to \infty$,

or by saying that, given $\varepsilon > 0$, there exists an N such that

$n > N \Rightarrow \sup\{|f_n(x) - f(x)| : x \in A\} < \varepsilon$.

15 Use qns 7 and 8 to prove that the sequences of functions (f_n) in qns 1 and 2 converge uniformly to their respective pointwise limit functions f.

16 Use qns 12, 13 and 14 to prove that the sequences of functions (f_n) in qns 3, 4 and 5 do not converge uniformly to their respective pointwise limit functions.

17 The convergence of a sequence of functions, uniform or otherwise, may depend on the domain of those functions. Examine the convergence of the sequence of functions given by

$f_n(x) = x/n$

(i) on the domain $[-a, a]$,
(ii) on the domain $\mathbb{R}$.

18 Examine the convergence of the sequence of functions in qn 4

(i) on the domain $[0, a]$, where $0 < a < 1$, and
(ii) on the domain $[0, 1)$.

19 The convergence of a seemingly well-behaved sequence of functions may fail to be uniform. Examine the convergence of the sequence of functions given by $f_n(x) = nx/(1 + n^2x^2)$ on the domain $\mathbb{R}$.
Draw the graphs of f_n for $n = 1, 2, 3, 5, 10$.

This particular example is a useful corrective to a wrong impression that might be gleaned from qns 1, 2, 3, 4 and 5. For in those questions there was uniform convergence to constant functions, and non-uniform convergence to non-constant functions. Question 19 shows that this coincidence was accidental.

Uniform convergence and continuity

Having defined uniform convergence with the intention of giving a condition that would make a sequence of continuous functions converge to a continuous limit, we need to see whether the condition we have given is sufficient to guarantee this in every case.

20 We suppose that the sequence of functions (f_n) converges uniformly to the function f and that each of the functions f_n is continuous on its domain A. We seek to prove that f is continuous on A or, in other words, that, for each $a \in A$, $\lim_{x \to a} f(x) = f(a)$.
Or, again, given $\varepsilon > 0$, there exists a δ such that

$$|x - a| < \delta \Rightarrow |f(x) - f(a)| < \varepsilon.$$

We break down $f(x) - f(a)$ into manageable pieces.

$$f(x) - f(a) = f(x) - f_n(x) + f_n(x) - f_n(a) + f_n(a) - f(a)$$

so

$$|f(x) - f(a)| \leq |f(x) - f_n(x)| + |f_n(x) - f_n(a)| + |f_n(a) - f(a)|.$$

For what reason must both $|f(x) - f_n(x)|$ and $|f_n(a) - f(a)|$ be less than $\frac{1}{3}\varepsilon$ for sufficiently large n?
Keeping to one of these sufficiently large ns, for what reason is it possible to find a δ such that $|x - a| < \delta \Rightarrow |f_n(x) - f_n(a)| < \frac{1}{3}\varepsilon$?
Now complete the proof.

21 We have proved that the uniform limit of a sequence of continuous functions is continuous. By considering qns 17(ii) and 19, show that the converse of this theorem is false: namely that a non-uniform limit of a sequence of continuous functions may also be continuous.

22 If (f_n) is a sequence of continuous functions which converges uniformly to the function f, and a is a point in the domain of these functions, justify each step of the following argument.

$$\lim_{n\to\infty}\lim_{x\to a} f_n(x) = \lim_{n\to\infty} f_n(a)$$
$$= f(a)$$
$$= \lim_{x\to a} f(x)$$
$$= \lim_{x\to a}\lim_{n\to\infty} f_n(x).$$

23 Illustrate the dependence of the argument in qn 22 on uniform convergence by showing how it would fail for the sequence of functions in qn 3 if we were to take $a = 0$.

Uniform convergence and integration

We have shown that the uniform limit of a sequence of continuous functions is continuous. Because of qn 10.39 this allows us to claim in certain cases that the uniform limit of a sequence of integrable functions is integrable. We seek to show this in general.

24 Let (f_n) be a sequence of continuous functions on the domain $[a, b]$ which converges uniformly to f. The function f is continuous from qn 20 and so must be integrable from qn 10.39. Show that the limit of the integrals $(\int f_n)$ is equal to the integral of the limit function $\int f$.
Use the inequality $|\int f_n - \int f| = |\int (f_n - f)| \leq \int |f_n - f|$ (from qn 10.35), to prove that

$$\lim_{n\to\infty} \int f_n = \int f, \text{ or } \lim_{n\to\infty} \int f_n = \int (\lim_{n\to\infty} f_n).$$

We have shown that the limit of an integral is equal to the integral of the limit in a context of uniform convergence but have not shown that there was any need for uniform convergence for this result.

25 Let $f_n(x) = nx(1 - x^2)^n$ on the domain $[0, 1]$. Verify that the pointwise limit function is the zero function throughout this domain.
Prove that $\int f_n = n/(2n + 2)$ on $[0, 1]$ and deduce that

$$\lim_{n\to\infty} \int f_n \neq \int \lim_{n\to\infty} f_n.$$

Is the convergence of the sequence (f_n) uniform? Use qn 24. Use computer software to draw the graphs of some functions from this sequence.
What is $\lim_{n\to\infty} \sup\{|f_n(x) - f(x)| : 0 \leq x \leq 1\}$?

26 Show that the result of qn 24 may *not* be extended to improper integrals by attempting to apply it to the sequence of functions defined by

$$f_n(x) = \begin{cases} (n - |x|)/n^2 & \text{when } x \in [-n, n], \\ 0 & \text{otherwise.} \end{cases}$$

Determine the pointwise limit function of the sequence (f_n). Prove that the convergence is uniform, and evaluate $\int f_n$ and $\int f$ on the domain $\mathbb{R}$.

If we ask whether convergent sequences of integrable functions which are not necessarily continuous must converge to an integrable limit we would have part of an answer if we could construct a sequence of integrable functions which converge to Dirichlet's function (qn 10.11) given by

$$f(x) = \begin{cases} 1 & \text{when } x \text{ is rational, and} \\ 0 & \text{when } x \text{ is irrational,} \end{cases}$$

since this is an example of a function which is not Riemann integrable on any interval.

27 We construct a sequence of functions on the closed interval $[a, b]$. There is a countable infinity of rational numbers in this interval so there is a sequence in $[a, b]$, which we denote by $x_1, x_2, x_3, \ldots, x_n, \ldots$, which contains them all.
On the closed interval $[a, b]$ we define the function f_n by

$$f_n(x) = \begin{cases} 1 & \text{when } x = x_1, x_2, \ldots, x_n, \text{ and} \\ 0 & \text{otherwise.} \end{cases}$$

Illustrate f_1, f_2 and f_3 on a graph. Are these functions integrable? Is every member of the sequence (f_n) integrable?
What is the pointwise limit function?

28 In qn 27, evaluate $|f_n(x_{n+1}) - f(x_{n+1})|$. Determine whether the convergence of the sequence in qn 27 is uniform.

Having found that the pointwise limit of integrable functions may not be integrable, we now ask whether the uniform limit of a sequence of integrable functions must be integrable.

29 On the domain $[a, b]$ we let (f_n) be a sequence of integrable functions which tends uniformly to the function f. The condition of uniform convergence, that for sufficiently large n, $|f_n(x) - f(x)| < \varepsilon_1$, for any positive ε_1, allows us to set an upper bound and a lower bound on the function f, since $f_n(x) - \varepsilon_1 < f(x) < f_n(x) + \varepsilon_1$, for all x. Now for

any subdivision of the domain whatsoever, show how an upper sum for f_n may be used to construct an upper sum for f.
Likewise show how, for the same subdivision, the lower sum for f_n may be used to construct a lower sum for f.
How close together can you be sure the upper and lower sums for f must be? For given $\varepsilon > 0$, can you choose n sufficiently large to make $2\varepsilon_1(b-a) < \frac{1}{2}\varepsilon$?
And can you then choose a subdivision for which the difference between the upper and lower sums for $f_n < \frac{1}{2}\varepsilon$? What then follows about the difference between the upper and lower sums for f for this subdivision of the domain $[a, b]$? Deduce that f is integrable on $[a, b]$.

30 If, on the domain $[a, b]$, the sequence of integrable functions (f_n) converges uniformly to the function f, use the argument of qn 24 to prove that the limit as $n \to \infty$ of $\int f_n$ is equal to $\int f$.

31 Let (f_n) be a sequence of functions which are integrable on the domain $[a, b]$ and which converge uniformly. Use an argument like that of qn 24 to prove that the sequence of functions (F_n) defined by $F_n(x) = \int_a^x f_n$ is uniformly convergent on the domain $[a, b]$.

Summary: Uniform convergence, continuity and integration

Definition qns 5, 6 — (f_n) is a sequence of real functions, $f_n : A \to \mathbb{R}$, all of which have the same domain A. If for each $x \in A$, the limit $\lim_{n\to\infty} f_n(x)$ exists, then a function $f : A \to \mathbb{R}$ may be defined by $f(x) = \lim_{n\to\infty} f_n(x)$. This f is called the *pointwise limit function* of the sequence (f_n).

Definition qns 14, 15 — The sequence of functions, (f_n), has the pointwise limit function $f : A \to \mathbb{R}$.
If $(\sup\{|f_n(x) - f(x)| : x \in A\}) \to 0$ as $n \to \infty$, we say that the sequence (f_n) *converges uniformly* to f.

Theorem qn 20 — If the sequence of functions (f_n) converges uniformly to the function $f : A \to \mathbb{R}$ and each of the functions f_n is continuous on its domain A, then f is continuous on A.

Theorem qns 29, 30 — If the sequence of functions (f_n) converges uniformly to the function $f : [a, b] \to \mathbb{R}$ and each of the functions f_n is integrable on $[a, b]$ then f is integrable on $[a, b]$ and $\lim_{n\to\infty} \int_a^b f_n = \int_a^b f$.

Uniform convergence and differentiation

32 Investigate the convergence of the sequence of functions defined by $f_n(x) = x/(1+nx^2)$ on $\mathbb{R}$.
Prove that $f_n'(x) = \dfrac{1-nx^2}{(1+nx^2)^2}$.
If f is the pointwise limit function, show that

$$\sup\{|f_n(x) - f(x)| x \in \mathbb{R}\} = 1/2\sqrt{n},$$

and deduce that the convergence is uniform.
Deduce that $\lim_{n\to\infty} f_n'(0) \neq f'(0)$.
Is $\lim_{n\to\infty} \int_0^1 f_n = \int_0^1 f$? Use qn 24.

Question 32 shows that differentiability is less well behaved under uniform convergence than is integrability.

33 Let $f_n(x) = (\cos nx)/n$.
Find the pointwise limit function for the sequence (f_n) and determine whether the convergence is uniform on $\mathbb{R}$.
Determine the function $f_n'(x)$. Is there a pointwise limit function for the sequence (f_n')? Consider values of $x \neq k\pi$. Determine the value of $\int_0^x f_n$. Is $\lim_{n\to\infty} \int_0^x f_n = \int_0^x \lim_{n\to\infty} f_n$?

From qns 32 and 33 it is clear that, even when the sequence (f_n) converges uniformly to f and every function f_n is differentiable, the sequence (f_n') need not be convergent and when it is convergent it may not be uniformly convergent. If we are to formulate a theorem about the uniform convergence of a sequence of derivatives, the uniform convergence of the sequence of derivatives will have to be assumed. Then, if the functions under discussion are all continuous, the Fundamental Theorem of Calculus may enable us to use the good behaviour of integrals under uniform convergence to claim results about derivatives.

34 We suppose that we have a sequence of differentiable functions (f_n) on an interval $[a, b]$ converging to the pointwise limit function f. We further suppose

(a) that the sequence (f_n') consists of continuous functions, and
(b) that the sequence $(f_n') \to \phi$ uniformly.

Give reasons for each of the following propositions.

(i) The function ϕ is continuous.
(ii) Each function f_n' and the function ϕ are integrable on $[a, b]$.
(iii) $\int_a^x \phi = \lim_{n\to\infty} \int_a^x f_n'$.

(iv) $\quad = \lim_{n\to\infty} (f_n(x) - f_n(a))$
(v) $\quad = f(x) - f(a)$.
(vi) $\quad \phi = f'$
(vii) $(f_n) \to f$ uniformly on $[a, b]$.

The conditions of qn 34 appear to be quite restrictive. But we need to apply it, typically, to prove theorems about power series and, in this case, when the functions concerned are polynomials, some of the conditions we have used are obvious. One of the issues at stake is when we may integrate or differentiate a power series term by term while still guaranteeing its continuity or differentiability. Our theorems relating integration and differentiation to uniform convergence will enable us to decide this matter.

Uniform convergence of power series

A series of functions is said to be uniformly convergent when its sequence of partial sums is uniformly convergent.

35 Let $e_n(x) = 1 + x + x^2/2! + \ldots + x^n/n!$.

(i) Why is the sequence $(e_n(x))$ convergent for each x?
We define $e(x) = \lim_{n\to\infty} e_n(x)$.
If $-a \le x \le a$, then $|x^n/n!| \le a^n/n!$.
(ii) Prove that $|e_{n+m}(x) - e_n(x)| \le e_{n+m}(a) - e_n(a)$.
(iii) Let $m \to \infty$ in (ii) and prove that $|e(x) - e_n(x)| \le e(a) - e_n(a)$ for all x in $[-a, a]$.
(iv) Prove that $(e_n) \to e$ uniformly on $[-a, a]$.
(v) Use qn 20 to prove that the function e is continuous on $[-a, a]$.
(vi) Use qn 34 to prove that $e'(x) = e(x)$.

The earlier results of this chapter led to the proof of the uniform convergence of the sequence (e_n) in qn 35 on the basis of the inequality $|x^n/n!| \le a^n/n!$ and the convergence of $\Sigma a^n/n!$. We isolate this claim in the basic theorem about the uniform convergence of series.

36 *The Weierstrass M-test*
We let $f_N(x) = \Sigma_{n=1}^{n=N} u_n(x)$ for $x \in A \subseteq \mathbb{R}$, and assume

(i) for each n, and for all $x \in A$, there exists M_n such that $|u_n(x)| \le M_n$, and
(ii) ΣM_n is convergent.

We investigate the convergence of the sequence (f_n).

Explain why the series $\Sigma u_n(x)$ is absolutely convergent for all $x \in A$. Deduce that the sequence (f_n) is pointwise convergent to a function f, say.

Prove that

$$\left|\sum_{n=N+1}^{n=N+m} u_n(x)\right| \le \sum_{n=N+1}^{n=N+m} M_n,$$

and deduce that

$$|f_{N+m}(x) - f_N(x)| \le \sum_{n=1}^{n=N+m} M_n - \sum_{n=1}^{n=N} M_n,$$

so that, letting $m \to \infty$,

$$|f(x) - f_N(x)| \le \sum_{n=1}^{\infty} M_n - \sum_{n=1}^{n=N} M_n.$$

Now deduce that (f_n) converges uniformly.

37 Use the Weierstrass M-test to prove that each of the following series is uniformly convergent on the domain given.

(i) $\Sigma\, x^n/n^2$ on the domain $[-1, 1]$,
(ii) $\Sigma\, (\sin nx)/2^n$ on the domain $\mathbb{R}$,
(iii) $\Sigma\, 1/(n^2 + x^2)$ on the domain $\mathbb{R}$,
(iv) $\Sigma\, x/(x^2 + n^2)$ on the domain $[-a, a]$,
(v) $\Sigma\, (x^2 + n)/(x^2 + n^3)$ on the domain $[-a, a]$,
(vi) $\Sigma\, (n + 1)x^n$ on the domain $[-a, a]$, where $0 < a < 1$,
(vii) $\Sigma\, x^n(1 - x)/n$ on the domain $[0, 1]$.

38 Let $f_n(x) = 1 + x + x^2 + \ldots + x^{n-1}$.
For $-1 < x < 1$, find the pointwise limit function f of the sequence (f_n).
For x in this range, prove that $|f_n(x) - f(x)| = |x|^n/(1 - x)$.
Deduce that the convergence of the sequence (f_n) is not uniform on the domain $(-1, 1)$.
Prove that the convergence of the sequence (f_n) is uniform on the domain $[-a, a]$, provided $0 < a < 1$.

39 Prove that each of the following series is not uniformly convergent on the domain given.

(i) $\Sigma\, (x^2 + n)/(x^2 + n^3)$ on the domain $\mathbb{R}$,
(ii) $\Sigma\, (n + 1)x^n$ on the domain $(-1, 1)$.

The heart of the proof lies in finding a value of x which shows that the sum of the series from n to ∞ cannot be made arbitrarily small by choice of n, however large.
Compare these results with qn 37(v) and 37(vi).

It is in fact quite straightforward to show that any power series is uniformly convergent within a closed interval inside its circle of convergence (*Abel*, 1827).

40 Suppose R is the radius of convergence of the power series $\Sigma\, a_n x^n$. Show that if $0 < a < R$ then the series is uniformly convergent on the interval $[-a, a]$ by choosing a real number r such that $a < r < R$, and using the Weierstrass M-test with $M_n = |a_n r^n|$. Deduce that the limit function must be continuous on this domain.

41 Use qn 24 to show that if a power series is integrated term by term then the integral of the limit function is equal to the limit of the power series obtained by term-by-term integration, on an interval $[0, x]$ inside its circle of convergence.
Is the radius of convergence of the integrated series
the same as the radius of convergence of the original power series?
See qn 5.107.

42 If a power series is differentiated term by term show that the radius of convergence of the differentiated series is the same as the radius of convergence of the original series. See qn 5.107. Use qn 34 to show that the limit function is differentiable and that the derivative of the limit function is equal to the limit of the power series obtained by term-by-term differentiation, within a closed interval inside its circle of convergence.

43 By differentiating the series for $1/(1-x)$ m times, establish the Binomial Theorem for negative integral index.

44 By integrating the power series

$$1 + x + x^2 + \ldots + x^n + \ldots$$

within its circle of convergence, prove that

$$\ln(1-x) = -x - x^2/2 - \ldots - x^n/n - \ldots$$

and determine the radius of convergence of this power series.

The Binomial Theorem for any real index

45 (*Cauchy*, 1821) A function f is defined for $-1 < x < 1$, by

$$f(x) = \sum_{n=0}^{\infty} \binom{a}{n} x^n.$$

Here a is a real number which is neither zero nor a positive integer. See qn 5.98. Prove that $(1 + x)f'(x) = af(x)$.
Use the Mean Value Theorem to prove that the function g defined on $(-1, 1)$ by $g(x) = (1 + x)^{-a} f(x)$ is constant. Since $g(0) = 1$, prove that $f(x) = (1 + x)^a$.

The blancmange function

A function which is continuous everywhere and differentiable nowhere.

46 (i) *The saw-tooth function*
Let $f(x) = x - \lfloor x \rfloor$.
Define a real function s (the saw-tooth function) by

$$\begin{aligned} s(x) &= f(x), && \text{when } f(x) \le \tfrac{1}{2}, \text{ and} \\ s(x) &= 1 - f(x), && \text{when } f(x) > \tfrac{1}{2}. \end{aligned}$$

The graph of s is shown on the interval $[0, 2]$.

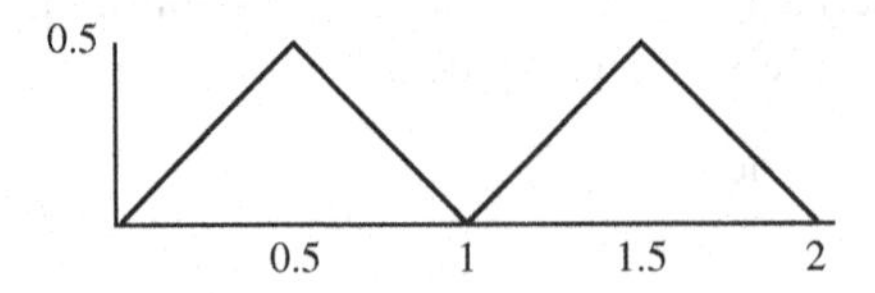

Figure 12.1

Check that $f(x + 1) = f(x)$, and that $f(m) = 0$ for any integer m.
Deduce that $s(x + 1) = s(x)$, and that $s(m) = 0$ for any integer m.

(ii) The real functions s_1, s_2, s_3 are defined by $s_1(x) = s(x)$, $s_2(x) = \frac{1}{2}s(2x)$ and $s_3(x) = \frac{1}{4}s(4x)$. The graphs of the functions s_1, s_2 and s_3 have been illustrated on $[0, 2]$.

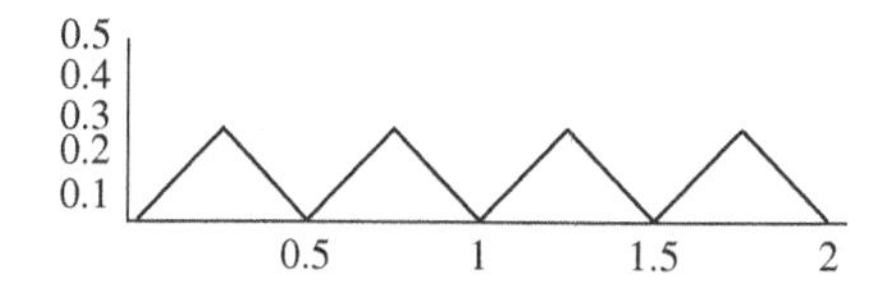

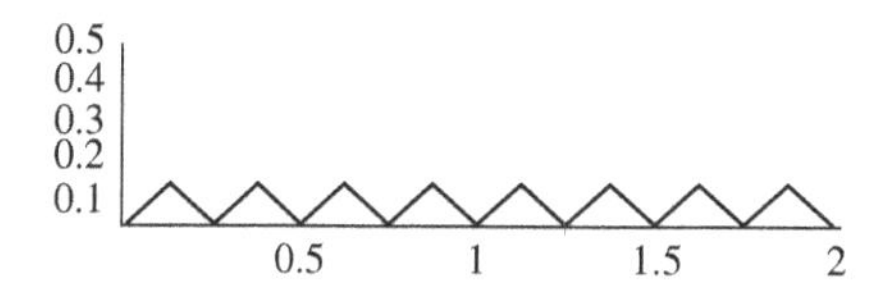

Figure 12.2

Notice that the graph of s_2 is a version of s, scaled down by a factor 2 in both the x- and y-directions. Scaling down by a factor k turns $y = f(x)$ into $ky = f(kx)$ or $y = \dfrac{1}{k} f(kx)$.

(iii) The graphs of $s_1 + s_2$ and $s_1 + s_2 + s_3$ have been illustrated on $[0, 2]$.

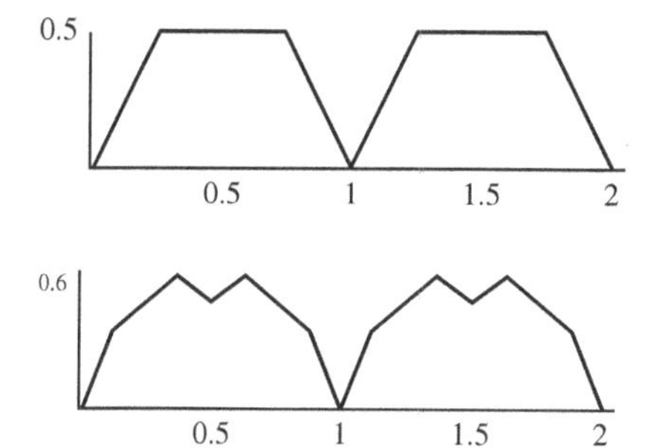

Figure 12.3

Since $0 \leq f(x) < 1$ for all x, deduce that $0 \leq s(x) \leq \frac{1}{2}$ for all x. Give an upper bound for $s_1 + s_2$ and for $s_1 + s_2 + s_3$.

(iv) Is s continuous for all x? See qn 6.96, the continuity of contiguous continuous functions. Define

$$s_n(x) = \frac{1}{2^{n-1}} s(2^{n-1}x).$$

Is s_n continuous for all n?

(v) *The blancmange function*
Show that $|s_n(x)| \leq \frac{1}{2}^n$ for all x, and deduce that the function b defined by

$$b(x) = \sum_{n=1}^{\infty} s_n(x)$$

is continuous for all x.

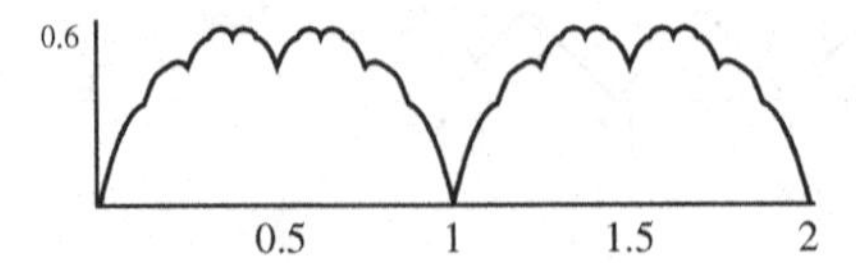

Figure 12.4

(vi) Verify that

$$b(x) = \sum_{n=1}^{n=N} s_n(x) + \frac{1}{2^N} b(2^N x).$$

(Notice that $\frac{1}{2^N} b(2^N x)$ is b transformed by scaling down by the same proportion in both the x- and y-directions.)

(vii) Since s_1 is linear (a straight line) on an interval of the type $[m/2, (m+1)/2]$, where m is an integer, state an interval on which s_2 is linear. Must $s_1 + s_2$ be linear on this second interval? Check that the function L_N defined by

$$L_N(x) = \sum_{n=1}^{n=N} s_n(x)$$

is linear on the interval

$$\left[\frac{m}{2^N}, \frac{m+1}{2^N}\right]$$

for any integer m. Since $b(m) = 0$ for any integer m, deduce that $b(m/2^N) = L_N(m/2^N)$. (On this interval, the function b is a scaled-down version of the function on $[m/2, (m+1)/2]$ with a linear factor $L_N(x)$ added, so the bending of b on the smaller interval is like the bending of b on the wider interval.)

(viii) Prove that $b(1/3) = 2/3$.

(ix) To show that b is *not* differentiable at any point a, we must show that

$$\frac{b(x) - b(a)}{x - a}$$

does *not* have a limit as x tends to a, and we do this by finding values of x inside any neighbourhood of a for which this gradient takes values separated by at least $\frac{1}{2}$.

Let c be the least number in the interval $(a, a + 1/2^{N-1}]$ of the form $m/2^N$, where m is an integer. Then $d = (m+1)/2^N$ and

$k = (m + 1/3)/2^N$ are also in this interval.

$a < c = m/2^N < k < d = (m+1)/2^N \leq a + 1/2^{N-1}$.

Check that $b(c) = L_N(c)$, $b(d) = L_N(d)$ and
$b(k) = L_N(k) + 2/(3 \cdot 2^N)$, from (vii) and (viii).

(x) Let L be the linear (straight line) function which coincides with L_N on the interval

$$[c, d] = \left[\frac{m}{2^N}, \frac{m+1}{2^N}\right].$$

By definition

$$\frac{L(k) - L(a)}{k - a} = \frac{L(c) - L(a)}{c - a}.$$

Use this to show that

$$\frac{b(k) - b(a)}{k - a} - \frac{b(c) - b(a)}{c - a} \geq \frac{b(k) - L(k)}{k - a},$$

provided $b(a) \geq L(a)$. Show further that

$$\frac{b(k) - L(k)}{k - a} \geq \frac{(2/3)2^{-N}}{(4/3)2^{-N}} = \tfrac{1}{2},$$

using (ix).

(xi) Use the notation in (x) to construct a similar proof in the case $L(a) \geq b(a)$.

Start with $\dfrac{L(k) - L(a)}{k - a} = \dfrac{L(d) - L(a)}{d - a}$.

Show that $\dfrac{b(k) - b(a)}{k - a} - \dfrac{b(d) - b(a)}{c - a} \geq \dfrac{b(k) - L(k)}{k - a}$,

and as before show that

$$\frac{b(k) - L(k)}{k - a} \geq \frac{(2/3)2^{-N}}{(4/3)2^{-N}} = \tfrac{1}{2}.$$

(xii) Use (x) and (xi) to show that in every neighbourhood of a,

$$\frac{b(x) - b(a)}{x - a}$$

takes values which differ by $\frac{1}{2}$ or more and therefore

$$\frac{b(x) - b(a)}{x - a}$$

cannot have a limit as $x \to a$.

The blancmange function is continuous everywhere and differentiable nowhere. A self-similar function like the blancmange function is equally bumpy on every interval.

Summary: Differentiation and the M-test

Theorem If
qn 34 (i) the sequence of functions (f_n) converges pointwise to the function $f: A \to \mathbb{R}$;
(ii) each of the functions f_n has a continuous derivative on its domain A;
(iii) the sequence (f_n') converges uniformly to $\phi : A \to \mathbb{R}$;
then $\phi = f'$ and (f_n) converges uniformly to f on A.

The Weierstrass M-test
qn 36 A sequence of functions (f_n) is defined by the partial sums of a series:

$$f_N(x) = \sum_{n=1}^{n=N} u_n(x).$$

If there exist real numbers M_n such that
(i) $|u_n(x)| \leq M_n$, and
(ii) $\Sigma\, M_n$ is convergent,
then (f_n) is uniformly convergent.

Theorem qn 40 A power series is uniformly convergent on any closed interval inside its circle of convergence.

Theorem qns 41, 42 If $f(x) = \Sigma_{n=0}^{\infty} a_n x^n$ has radius of convergence R, then

$$f'(x) = \sum_{n=1}^{\infty} n a_n x^{n-1} \text{ and } \int_0^x f = \sum_{n=0}^{\infty} \frac{a_n x^{n+1}}{n+1}$$

when $|x| < R$.

The Binomial Theorem

qn 45 $$(1+x)^a = \sum_{n=0}^{\infty} \binom{a}{n} x^n, \text{ for all real } a,$$

provided $-1 < x < 1$.

Historical Note

Newton and his contemporaries differentiated and integrated power series, term by term, without reference to their circle of convergence. During the eighteenth century it was presumed that the limit of a sequence of continuous functions was continuous. In 1821, Cauchy believed that he had a proof of this and in 1823 a proof that a sequence of integrable functions was integrable. However, he had not identified the distinctive features of uniform convergence and Cauchy's view of continuity differed from that of later mathematicians. Abel pointed out (1826) that Cauchy's theorem about the convergence of a sequence of continuous functions 'admitted exceptions' using the example of the series

$$\sin x - \frac{1}{2}\sin 2x + \frac{1}{3}\sin 3x - \cdots$$

which converges to $\frac{1}{2}x$ on $(-\pi, \pi)$ but which is discontinuous at $x = (2n+1)\pi$. Abel showed that a power series was uniformly convergent inside its radius of convergence and showed that this convergence implied the continuity of the limit. When Abel reflected on why the results in analysis obtained before Cauchy were generally sound, despite the lack of precision relating to limiting processes, he suggested that the reason might be that the functions considered were expressible as power series. In the same paper, Abel proved the Binomial Theorem for complex numbers, stimulated by a paper by Bolzano (1816) which exposed the flaws in earlier treatments and by Cauchy's treatment for real index (1821). In 1847 Seidel investigated Cauchy's 1821 theorem by examining the convergence of the series $\sum f_n$ where $f_n(x) \int_n^{n+1} \frac{\sin xt}{t} dt$. When x is positive, $\sum_0^\infty f_n(x) = \frac{\pi}{2}$, but there is a discontinuity at $x = 0$. He found that this discontinuity from a limit of continuous functions could only arise when the convergence was 'indefinitely slow'. Cauchy himself examined the same issue in 1853, articulated a 'Cauchy criterion' for uniform convergence, and completed the theorem (qn 20) in its modern form. It is to Weierstrass in his lectures in Berlin in the 1860s that we owe the theorems of this chapter. He had read the papers of Abel many years before and written an unpublished article in 1841 about uniform convergence. Weierstrass was able to prove that every continuous function was the uniform limit of a sequence of polynomials (1885) and believed that this theorem legitimated the study of continuous but non-differentiable functions stimulated by a function proposed by Riemann in 1861, namely $f(x) = \sum_{n=1}^\infty \frac{\sin n^2 x}{n^2}$, which is continuous everywhere and differentiable only at some isolated points.

Answers

1 $\lim_{n\to\infty} 1/(n^2 + x^2) = 0$ for all x.

3 $f(0) = 1,\ f(x) = 0$ when $x \neq 0$.

4 $f(x) = 0$ when $0 \le x < 1,\ f(1) = 1$.

5 $f(x) = -1$ when $0 \le x < 1,\ f(1) = 0,\ f(x) = 1$ when $1 < x$.

6 Not possible with qns 1 or 2.
Question 3 (i) $x^2 < 1/10^2$, (ii) $x^2 < 1/100^2$.
Question 4 (i) $1/\sqrt[10]{2} < x < 1$, (ii) $1/\sqrt[100]{2} < x < 1$.
Question 5 (i) $1/\sqrt[10]{3} < x < 1$ or $1 < x < \sqrt[10]{3}$,
(ii) $1\sqrt[100]{3} < x < 1$ or $1 < x < \sqrt[100]{3}$.

7 $1/(n^2 + x^2) \le 1/n^2$.

8 $|(\sin x)/n| \le 1/n$.

9 $0 < |x| < 1/n$.

10 $\sqrt[n]{(0.99)} < x < 1$.

11 $1/\sqrt[n]{3} < x < 1$ or $1 < x < \sqrt[n]{3}$.

12 $f_n(x) \to 1$ as $x \to 0$, so
$|f_n(x) - f(x)| \to 1 = \sup\{|f_n(x) - f(x)|: x \in \mathbb{R}\}$.

13 $f_n(x) \to 1$ as $x \to 1^-$, so
$|f_n(x) - f(x)| \to 1 = \sup\{|f_n(x) - f(x)|: 0 \le x \le 1\}$.

14 $f_n(x) \to 0$ as $x \to 1^-$, so
$|f_n(x) - f(x)| \to 1 = \sup\{|f_n(x) - f(x)|: 0 \le x \le 1\}$.
$f_n(x) \to 0$ as $x \to 1^+$, so
$|f_n(x) - f(x)| \to 1 = \sup\{|f_n(x) - f(x)|: 1 \le x\}$.

17 The pointwise limit function is $f(x) = 0$ whatever the domain.
(i) On $[-a, a]$, $|f_n(x) - f(x)| \le |a|/n \to 0$ as $n \to \infty$.
(ii) For given n, $|f_n(x) - f(x)|$ is unbounded for $x \in \mathbb{R}$.

The critical question is whether you can use n to control the greatest values of $|f_n(x) - f(x)|$.

18 (i) On $[0, a]$, $|f_n(x) - f(x)| \le a^n \to 0$ as $n \to \infty$. Convergence uniform.
(ii) On $[0, 1)$, $f_n(x) \to 1$ as $x \to 1^-$, so

$|f_n(x) - f(x)| \to 1 = \sup\{|f_n(x) - f(x)| : 0 \le x < 1\}$. Convergence not uniform.

This is a tricky and uncomfortable example, but it shows what a subtle business uniform convergence may be.

19 The pointwise limit function is given by $f(x) = 0$.
$\sup\{|f_n(x) - f(x)| : x \in \mathbb{R}\} = \frac{1}{2}$ irrespective of n. Convergence not uniform.

20 By the uniform convergence of $(f_n) \to f$,
$(\sup\{|f_n(x) - f(x)| : x \in A\}) \to 0$ as $n \to \infty$.
So for some N, $|f(x) - f_n(x)|$ and $|f_n(a) - f(a)| \le \frac{1}{3}\varepsilon$ for $n > N$.
Since each f_n is continuous a suitable δ may be found.
Now, if $|x - a| < \delta$, $|f(x) - f(a)| < \varepsilon$ since we know the three inequalities hold for some value of n. Thus $\lim_{x\to a} f(x) = f(a)$, and f is continuous at a.

22 First equality by continuity of f_n at a.
Second equality by definition of pointwise limit function.
Third equality by uniform convergence and qn 20 which establishes that f is continuous.
Fourth equality by definition of pointwise limit function.

23 In qn 3,

$$\lim_{n\to\infty} \lim_{x\to 0} f_n(x) = \lim_{n\to\infty} 1 = 1.$$

$$\lim_{x\to 0} \lim_{n\to\infty} f_n(x) = \lim_{x\to 0} 0 = 0.$$

24 For sufficiently large n, $|f_n - f| < \varepsilon$. So $\int |f_n - f| < \varepsilon(b - a)$. Use qn 10.35.

25 $\lim_{n\to\infty} \int_0^1 f_n = \frac{1}{2}$. $\int_0^1 \lim_{n\to\infty} f_n = 0$. $f_n'(1/\sqrt{(2n+1)}) = 0$.
$$\sup\{|f_n(x) - f(x)| : 0 \le x \le 1\} = \frac{n}{\sqrt{(1+2n)} \cdot (1 + 1/(2n))^n} \to \infty$$
as $n \to \infty$.

26 The pointwise limit function is given by $f(x) = 0$. $\int f_n = 1$. $\int f = 0$.

27 The function f_n is integrable as in qn 10.14. The pointwise limit function is Dirichlet's function, qn 6.20.

28 $|f_n(x_{n+1}) - f(x_{n+1})| = 1$. Convergence not uniform.

29 Let S_n be an upper sum for f_n. Then $S_n + \varepsilon_1(b - a)$ is an upper sum for f. Let s_n be a lower sum for f_n. Then $s_n - \varepsilon_1(b - a)$ is a lower sum for f. It is possible to take n sufficiently large so that $\varepsilon_1 < \varepsilon/4(b - a)$. Since f_n is integrable there are upper and lower sums such that $S_n - s_n < \frac{1}{2}\varepsilon$ from qn 10.23. Then $(S_n + \varepsilon_1(b - a)) - (s_n - \varepsilon_1(b - a)) < \varepsilon$.
So f is integrable by qn 6.24.

30 The argument in qn 24 does not appeal to continuity.

31 Let $(f_n) \to f$, then by qn 30,

$$\lim_{n\to\infty} F_n(x) = \int_a^x \lim_{n\to\infty} f_n = \int_a^x f = F(x) \text{ (say)}.$$

$$F_n(x) - F(x) = \int_a^x (f_n - f)$$

and, by qn 10.35,

$$|F_n(x) - F(x)| \le \int_a^x |f_n - f|.$$

If $n > N \Rightarrow \sup\{|f_n(x) - f(x)| : a \le x \le b\} < \varepsilon/(b-a)$, then $n > N \Rightarrow |F_n(x) - F(x)| < \varepsilon$.

32 The pointwise limit function is given by $f(x) = 0$.
$f_n'(0) = 1,\ f'(0) = 0$. Yes.

33 $f(x) = 0$. $|f_n(x) - f(x)| \le 1/n$, so convergence is uniform.
$f_n'(x) = -\sin nx$ which is not pointwise convergent for any $x \ne k\pi$.
$\int_0^x f_n = (\sin nx)/n^2$. Limit of integral = integral of limit by qn 24.

34 (i) Question 20, (ii) qn 10.39, (iii) qn 24, (iv) qn 10.51, (v) qn 3.54(v), the difference rule, (vi) Fundamental Theorem of Calculus, qn 10.54, (vii) qns 24 and 31.

35 (i) Question 5.86.
(ii)
$$\begin{aligned} |e_{n+m}(x) - e_n(x)| &= |x^{n+1}/(n+1)! + \ldots + x^{n+m}/(n+m)!| \\ &\le |x^{n+1}/(n+1)!| + \ldots + |x^{n+m}/(n+m)!| \\ &\le a^{n+1}/(n+1)! + \ldots + a^{n+m}/(n+m)! \\ &\le e_{n+m}(a) - e_n(a). \end{aligned}$$
(iii) Question 3.76, the inequality rule.
(iv) $(e(a) - e_n(a)) \to 0$ as $n \to \infty$.
(v) Let $f_n = e_n$, then $f_n' = e_{n-1}$. $(f_n') \to e$ uniformly and $(f_n) \to e$ uniformly. From qn 34(vi), $e = e'$.

36 $\Sigma\, u_n$ is absolutely convergent by (i), (ii) and the first comparison test, qn 5.26. So by qn 5.67 (absolute convergence) the series is convergent for all $x \in A$ and f is well defined. The inequality with m terms is obtained by repeated application of the triangle inequality (qn 2.61), the second inequality is only a rewrite of the first, and the third inequality comes from qn 3.76, the inequality rule. As $N \to \infty$, the right-hand side of the last inequality tends to 0; see qn 5.17.

37 (i) Take $M_n = 1/n^2$, (ii) $M_n = (\frac{1}{2})^n$, (iii) $M_n = 1/n^2$, (iv) $M_n = a/n^2$, (v) $M_n = 1/n^2 + a^2/n^3$, (vi) $M_n = (n+1)a^n$ using Cauchy's nth root test, qn 5.35, or d'Alembert, qn 5.43, (vii) $M_n = 1/2n^2$ using differentiation and qn 2.45.

38 $f(x) = 1/(1-x)$. As $x \to 1^-$, $|f_n(x) - f(x)| \to \infty$.
If $-a \le x \le a < 1$, $|x^n| \le a^n = M_n$.

39 (i) When $x = n^2$, there is a term in the series, after the nth, which is greater than $\frac{1}{2}$. So although the series is convergent for each value of x by comparison with $\Sigma\ 1/n^2$, the greatest difference between the nth partial sum (i.e. f_n) and the limit function is not null.
(ii) Likewise when $x = 1/\sqrt[2n]{(2n)}$, there is a term in the series, after the nth, which is greater than 1. So although the series is convergent for each value of x in $(-1, 1)$, the greatest difference between the nth partial sum and the limit function is greater than 1.

40 The limit function is continuous by qn 20.

41 A partial sum of a power series is a polynomial which is continuous. The uniform convergence of the series on $[0, x]$, from qn 40, gives the limit of the integral equal to the integral of the limit from qn 24.
The radius of convergence of the two series is the same by qn 5.107.

42 Partial sums are polynomials, so derivatives of partial sums are also polynomials. Both series have the same circle of convergence by qn 5.107 so, within that circle, both converge uniformly. This is the context in which qn 34(vi) may be applied.

43
$$(1-x)^{-1} = 1 + x + x^2 + \ldots + x^n + \ldots$$
$$= 1 + \binom{-1}{1}(-x) + \binom{-1}{2}(-x)^2 + \ldots + \binom{-1}{n}(-x)^n + \ldots$$
$$= \sum_{n=0}^{\infty} \binom{-1}{n}(-x)^n.$$
The radius of convergence $= 1$.

Suppose $(1-x)^{-m} = \sum_{n=0}^{\infty} \binom{-m}{n}(-x)^n$.

Then $m(1-x)^{-m-1} = \sum_{n=0}^{\infty} -(n+1)\binom{-m}{n+1}(-x)^n$

and hence $(1-x)^{-m-1} = \sum_{n=0}^{\infty} -\frac{n+1}{m}\binom{-m}{n+1}(-x)^n$

thus $(1-x)^{-m-1} = \sum_{n=0}^{\infty} \binom{-m-1}{n}(-x)^n$.

This establishes the result by induction. Compare with qn 5.113.

44 Radius of convergence for both series $|x| = 1$ by qn 5.107.

45 By qn 5.98, the series is convergent for $|x| < 1$, so f is a pointwise limit function. Within the circle of convergence the convergence is uniform (qn 40) so, using qn 42,

$$f'(x) = \sum_{n=0}^{\infty}(n+1)\binom{a}{n+1}x^n.$$

Now

$$n\frac{a(a-1)\ldots(a-n+1)}{n!} + (n+1)\frac{a(a-1)\ldots(a-n)}{(n+1)!} = \frac{a(a-1)\ldots(a-n+1)}{(n-1)!}\left(1+\frac{a-n}{n}\right).$$

So $(x+1)f'(x) = af(x)$.

$$\begin{aligned} g'(x) &= -a(1+x)^{-a-1}f(x) + (1+x)^{-a}f'(x), \text{ using qn 11.34,} \\ &= (1+x)^{-a-1}(-af(x) + (1+x)f'(x)) \\ &= 0. \end{aligned}$$

By the Mean Value Theorem (qn 9.17) g is a constant function. Since $f(0) = 1,\ g(0) = 1$, so $f(x) = (1+x)^a$.

46 (i) $\lfloor x+1 \rfloor = \lfloor x \rfloor + 1$, so $f(x) = f(x+1)$. $\lfloor m \rfloor = m$. $0 \le f(x) < 1$.

(ii) s_2 and s_3 are half- and quarter-sized versions of s.

(iii) When $\frac{1}{2} < f(x) < 1,\ 0 < 1 - f(x) < \frac{1}{2}$.
$s_1(x) + s_2(x) \le 3/4$, $s_1(x) + s_2(x) + s_3(x) \le 7/8$; though these are not least upper bounds.

(iv) s is linear on segments and contiguous, so s is continuous for all x. Now the composite of two continuous functions and a scalar multiple gives the continuity of s_n.

(v) $s_n(x) = (1/2^{n-1})s(2^{n-1}x) \le 1/2^n$. b is uniformly convergent by the Weierstrass M-test, and continuous since the components are continuous.

(vi)
$$\begin{aligned} b(2^N x) &= \sum_{n=1}^{\infty} s_n(2^N x) = \sum_{n=1}^{\infty} \frac{s(2^{n-1}\cdot 2^N x)}{2^{n-1}} \\ &= 2^N \sum_{n=1}^{\infty} \frac{s(2^{N+n-1}x)}{2^{N+n-1}} = 2^N\left(\sum_{n=1}^{\infty} s_n(x) - \sum_{n=1}^{n=N} s_n(x)\right). \end{aligned}$$

(vii) s_1 is linear on intervals $[m/2, (m+1)/2]$ for any integer m.
s_2 is linear on intervals $[m/4, (m+1)/4]$ for any integer m.
s_n is linear on intervals $[m/2^n, (m+1)/2^n]$ for any integer m.

(viii) $s(1/3) = 1/3$. $s_n(1/3) = (1/2^{n-1})s(2^{n-1}/3) = (1/2^{n-1})(1/3)$.
$b(1/3) = 1/3 + (1/3)(1/2) + (1/3)(1/4) + \ldots$.

(ix) $b(c) = L_N(c)$ and $b(d) = L_N(d)$ from (vii).
$b(k) = L_N(k) + (1/2^N)b(2^N(m+1/3)/2^N) = L_N(k) + (1/2^N)b(1/3)$.
Now use (viii).

(x) Consider

$$\left(\frac{b(k)-b(a)}{k-a} - \frac{b(c)-b(a)}{c-a}\right) - \left(\frac{L(k)-L(a)}{k-a} - \frac{L(c)-L(a)}{c-a}\right)$$
$$= \left(\frac{b(k)-L(k)}{k-a}\right) + (b(a)-L(a))\left(\frac{1}{c-a} - \frac{1}{k-a}\right).$$

Also, $0 < k - a \leq (1 + 1/3)/2^N$.

(xi) Like (x).

(xii) Every neighbourhood of a contains an interval of the form $(a, a + 1/2^{N-1}]$ and therefore points corresponding to c, k and d as in (ix), (x) and (xi). So there can be no limit g such that
$\left|\frac{b(x)-b(a)}{x-a} - g\right| < \frac{1}{5}$,
for all x inside *any* neighbourhood of a.

Appendices

Appendix 1

Properties of the real numbers

Algebraic properties of a field of numbers – chapter 1 onwards

If a, b, and c are numbers, then

$a + b$ is a number;
$a + b = b + a$;
$a + (b + c) = (a + b) + c$;
there is a number 0 such that
$a + 0 = a$, for all a;
for each a there is a number $-a$
such that $a + (-a) = 0$;
$a + (-b)$ is usually written $a - b$;

$a \cdot b$ is a number;
$a \cdot b = b \cdot a$;
$a \cdot (b \cdot c) = (a \cdot b) \cdot c$;
there is a number $1 \neq 0$, such that
$a \cdot 1 = a$, for all $a \neq 0$;
for each $a \neq 0$, there is a number $1/a$
such that $a \cdot (1/a) = 1$;
$a \cdot (1/b)$ is usually written a/b;

$$a \cdot (b + c) = a \cdot b + a \cdot c.$$

From these algebraic properties it follows that $-(-a) = a$, $1/(1/a) = a$, $a \cdot 0 = 0$, $(-a) \cdot (-b) = a \cdot b$ and also $a \cdot b = 0$ only when $a = 0$ or $b = 0$.

While these algebraic properties hold for the rational numbers, $\mathbb{Q}$, the real numbers, $\mathbb{R}$, the complex numbers, $\mathbb{C}$, and some finite systems such as arithmetic modulo 2, $\mathbb{Z}_2$, or modulo 3, $\mathbb{Z}_3$, they do not hold universally for the number system on a pocket calculator. Let $a = -b$ where a is as large a number as your calculator will show, then if c is as small a number as you can key in, the machine will calculate $a + (b + c)$ as 0, and $(a + b) + c$ as c.

Order properties of a field of numbers – chapter 2 onwards

The trichotomy law

If a is a number, then *either* $a = 0$, *or* a is positive *or* $-a$ is positive, and only one of these is true. When $-a$ is positive, a is said to be negative.

Positive closure under addition

The sum of two positive numbers is positive.

Positive closure under multiplication
The product of two positive numbers is positive.

Definition of 'less than'
We say $a < b$ if and only if $b - a$ is positive.

A field of numbers with order properties has to contain the positive numbers $1, 1+1, 1+1+1, 1+1+1+1$, etc. as an increasing sequence and so it must be infinite and contain versions of $\mathbb{N}$, $\mathbb{Z}$ and $\mathbb{Q}$. A field of numbers with order properties can be modelled by an indefinitely long line. The numbers are then matched with a dense set of points on the line. Because squares cannot be negative (qn 2.15) the complex numbers $\mathbb{C}$ do not have the properties of order.

Property of Archimedean order – chapter 3 onwards

Every number is exceeded by an integer.

If all the numbers are marked on a line, and the integers are marked as a special row of pegs, then the property of Archimedean order says every number lies between two pegs. This rules out both infinite numbers and infinitesimal numbers, both of which can be tolerated in a 'non-standard' number system. An equivalent property, which is given in many texts, is that if two positive numbers a and b are given, then some multiple of the first will exceed the second, $b < na$. Yet another equivalent property, due to Euclid, is that if from a quantity, a half or more is removed, and then from what is left, half or more is removed, and so on, then at some stage, what is left will be less than any given quantity, however small.

Principle of completeness – chapter 4 onwards

Every infinite decimal sequence is convergent.

The property of Archimedean order implies that every number is the limit of an infinite decimal sequence. The rational numbers are the limits of terminating or recurring decimals, and the rational numbers form an Archimedean ordered field. But there are points on a line not corresponding to rational numbers. The completeness principle is adopted so that the real number system matches precisely the points on a line.

Propositions equivalent to completeness in the context of Archimedean order are

I. Every bounded monotonic sequence of real numbers is convergent.
II. The intersection of a set of nested closed intervals is not empty.
III. Every bounded sequence of real numbers has a convergent subsequence.

IV. Every infinite bounded set of real numbers has a cluster point.
V. Every Cauchy sequence of real numbers is convergent.
VI. Every non-empty set of real numbers which is bounded above has a least upper bound.

Of these propositions, I, III, IV and VI are sufficient to imply Archimedean order in an ordered field. In Appendix 3, qn 10 sets up the non-Archimedean ordered field of Laurent series, in which Cauchy sequences converge.

It is with completeness that we can prove the Intermediate Value Theorem, and can guarantee the existence of nth roots, logarithms, exponentials and trigonometric functions.

Any one of the properties of the following theorems implies completeness in an Archimedean ordered field.

(i) D'Alembert's ratio test of series
(ii) The Intermediate Value Theorem
(iii) $f' = 0$ implies f is constant
(iv) The Mean Value Theorem
(v) Continuous functions on a closed interval are bounded.

See Teismann (2013) and Propp (2013).

Appendix 2

Geometry and intuition

Sometimes authors and lecturers on analysis insist that students must not use geometrical intuition in developing the fundamental concepts of analysis or in constructing proofs in analysis. Such a prohibition appears to be consonant with Felix Klein's description in 1895 of the developments due to Weierstrass, Cantor and Dedekind, namely the *Arithmetisation of Analysis*.

However, such advice is impossible to implement and is in any case untrue to the origins of the subject. We can hardly conceive of a Dedekind cut, for example, without imagining a 'real line', and such imagining was certainly part of Dedekind's own thought. We have eyes, and we have imaginations with which to visualise, and such visualisation is central to much of the development of analysis. Every development of the real number system is a way of formalising our intuitions of the points on an endless straight line. We cannot conceive how the theory of real functions could have developed had there been no graphs drawn.

However, geometric intuition is not always reliable, and knowing when it should be trusted and when it should not is part of the mathematical maturity which should develop during an analysis course.

There are contexts in which geometrical intuition is misleading.

1 When comparing infinities: because there is a one-to-one correspondence between the points of the segment $[0, 1]$ and the points on the segment $[0, 2]$, there appear to be the 'same' number of points on both segments. The conflict with intuition here is simply to do with infinity, not to do with rationals and irrationals, because the same paradox arises if we restrict our attention to rational points.

2 When comparing denseness with completeness: because there is an infinity of rationals between any two points on the line there are rationals as close as we like to any point. That most of the cluster points of $\mathbb{Q}$ are not in $\mathbb{Q}$ again seems paradoxical. Even the terminating decimals are dense on the line and will give us measurements as accurate as we may wish, yet they do not even include all the rational numbers.

3 Continuity: when using a pencil to make a 'continuous' line the Intermediate Value Theorem appears to be unfalsifiable. This is closely related to the denseness–completeness issue in 2.

4 What a continuous function may be like is not obvious. A function may be constructed, somewhat unexpectedly, from the segment [0, 1] onto the unit square $\{(x, y) | 0 \leqslant x \leqslant 1, 0 \leqslant y \leqslant 1\}$ by mapping the number $0.a_1a_2a_3a_4a_5a_6\ldots$ to the point $(0.a_1a_3a_5\ldots,\ 0.a_2a_4a_6\ldots)$.
This function is well defined provided terminating decimals in [0, 1] are always represented by recurring 9s. The function is continuous and one-to-one except where one of the image coordinates terminates.

5 The connection between continuity and differentiability seems straightforward enough (with non-differentiability at the occasional sharp point on the curve) until one considers a function which is everywhere continuous and nowhere differentiable, such as David Tall's *Blancmange Function* in qn 12.46.

6 In drawing graphs, unless both domain and range are bounded, only a part of the graph of a function can be exhibited; under changes of scale, gradients may appear to change; an open interval and a closed interval do not look different.

The examples 2–5 above all show the inadequacy of considering a mark made by a pencil, without lifting it from the paper, as the illustrative model of the graph of a continuous function. It is tantalising to contrast these illustrations of suspect visualisation with an example where the intuitive, pencil and paper, point of view is sound enough, as at the beginning of chapter 7. Try finding the kind of set A which can be the range of a continuous function $\mathbb{R} \to A \subseteq \mathbb{R}$. Here even quite rough work with pencil and paper leads to a precise and accurate formulation.

I would certainly concur with the judgement of J. E. Littlewood who wrote in 1953 (*Bollobás*, 1986, p. 54):

> My pupils *will* not use pictures, even unofficially and when there is no question of expense. This practice is increasing. I have lately discovered that it has existed for 30 years or more, and also why. A heavy warning used to be given that pictures are not rigorous; this has never had its bluff called and has permanently frightened its victims into playing for safety. Some pictures, of course, are not rigorous, but I should say most are (and I use them wherever possible myself). An obviously legitimate case is to use a graph to define an awkward function (e.g. behaving differently in successive stretches).

Appendix 3

Questions for student investigation and discussion

Multiple-choice questions for student discussion

1 $a^2 + b^2 + c^2 \geq bc + ca + ab$ is true:

A. always;
B. sometimes;
C. never.

2 If $(a_n) \to a$, must $\left(\left(\sum a_n\right)/n\right) \to a$?

A. Yes
B. No

If $\left(\left(\sum a_n\right)/n\right) \to a$, must $(a_n) \to a$?

A. Yes
B. No

3 The sequence (a_n) is bounded. Which of the following propositions are sufficient to guarantee its convergence?

(i) $(a_n) \to 0$;
(ii) (a_n) is positive and bounded below;
(iii) (a_n) is bounded above and increasing;
(iv) for all n, $a_n^2 < a_n$;
(v) none of the above.

4 Between two rationals there is an irrational. Between two irrationals there is a rational. So the rationals and the irrationals alternate on the line.

A. This is a valid argument.
B. The result is true but the argument is invalid.
C. The result is false.

5 True or false?
Whatever the values of a or b, the set $[a, b] \cap \mathbb{Q}$ contains its own supremum and infimum.
Whatever the values of a or b, the set $(a, b) \cap \mathbb{Q}$ never contains its own supremum or infimum.

6 A real function f is continuous on $(0, 1)$ and takes positive values. Which **one** of the following statements must be true?

(i) f is bounded on $(0, 1)$;
(ii) $f(x)$ tends to a non-negative limit as $x \to 0^+$;
(iii) f attains a minimum, though not necessarily a maximum, value;
(iv) the Riemann integral $\int_0^1 f$ exists;
(v) none of the four above.

7 A real function f is continuous on the open interval $(0, 1)$.

(i) If f is uniformly continuous, must f be bounded?
(ii) If f is bounded, must f be uniformly continuous?
(iii) May f be neither bounded, nor uniformly continuous?

8 A differentiable function $f : \mathbb{R} \to \mathbb{R}$ is strictly monotonic increasing. Must its derivative be positive?

Problems for corporate or individual investigation

9 (i) Find the $(2n - 1)$th partial sum and the $2n$th partial sum of the series

$$1 - \frac{2}{3} + \frac{2}{3} - \frac{3}{5} + \frac{3}{5} - \frac{4}{7} + \frac{4}{7} - \dots$$

and their limits. Explain the difference.
(ii) What is wrong with this argument?

$$\frac{1}{\sqrt{8}} + \frac{1}{\sqrt{27}} + \frac{1}{\sqrt{64}} + \frac{1}{\sqrt{125}} + \dots > \frac{1}{\sqrt{9}} + \frac{1}{\sqrt{36}} + \frac{1}{\sqrt{81}} + \frac{1}{\sqrt{144}} + \dots$$
$$= \frac{1}{3} + \frac{1}{6} + \frac{1}{9} + \frac{1}{12} + \dots$$
$$= \frac{1}{3}\left(1 + \frac{1}{2} + \frac{1}{3} + \frac{1}{4} + \dots\right)$$

which is divergent.

10 On the set of formal power series $\sum_{n=-m}^{\infty} a_n x^n$ for some integer m, and $a_n \in \mathbb{R}$, can you construct $+$, $\times$ and $<$ so that the set is an ordered field? You will need to take $x^2 < x < 1 < 1/x < 1/x^2$. Is the field then Archimedean ordered?
Must Cauchy sequences converge?

11 Find the cluster points of the bounded sets

(i) $\{\sin n | n \in \mathbb{N}\}$,
(ii) $\{n\sqrt{2} - \lfloor n\sqrt{2} \rfloor | n \in \mathbb{N}\}$.
Is there a subsequence of $(\sin n)$ which tends to 0? If so, can you find one?

12 Does $(\sqrt[n]{|\sin n|})$ have a limit?

13 Can you give meaning to the following?

(i) $1 + \cfrac{1}{2 + \cfrac{1}{3 + \cfrac{1}{4 + \dots}}}$,

(ii) $\sqrt{(1 + \sqrt{(2 + \sqrt{(3 + \dots)})})}$,

(iii) $2\sqrt{2}, 4\sqrt{(2 - \sqrt{2})}, 8\sqrt{(2 - \sqrt{(2 + \sqrt{2})})}$,
$16\sqrt{(2 - \sqrt{(2 + \sqrt{(2 + \sqrt{2})})})}, \dots$

14 (*Cauchy*, 1821) If for a sequence of positive terms $(a_{n+1}/a_n) \to k$, prove that $(\sqrt[n]{a_n}) \to k$. What about the converse?

15 (*Cantor*, 1874) A real number x is said to be *algebraic* if it is the solution of a polynomial equation with integer coefficients:

$$a_0 + a_1 x + a_2 x^2 + \dots + a_n x^n = 0.$$

Calling $|a_0| + |a_1| + |a_2| + \dots + |a_n| + n$ the 'weight' of the polynomial, show that the set of algebraic numbers is countable. A real number which is not algebraic is said to be *transcendental.*

16 Does a real sequence (a_n) necessarily converge if, given $\varepsilon > 0$, there exists an integer N (depending on p and ε), such that $|a_{n+p} - a_n| < \varepsilon$ when $n > N$, for each integer p?

17 (a) Is there a non-empty set of real numbers which is both closed (contains all its cluster points) and open (contains a neighbourhood of each of its points)?
(b) What subsets of the real numbers can be the set of cluster points of some subset of the real numbers?

18 Under what circumstances does

$$\lim_{x\to\infty}(\sqrt{(ax^2+bx+c)}-\sqrt{(Ax^2+Bx+C)})$$

exist?

19 (*Pringsheim*, 1899) Find the values of the function f defined by $f(x)=\lim_{n\to\infty}\lim_{m\to\infty}(\cos n!\,\pi x)^{2m}$.

20 If a, b, c and d are irrational numbers with $a<b$ and $c<d$, can you construct a bijection with domain $[a,b]\cap\mathbb{Q}$ and range $[c,d]\cap\mathbb{Q}$? A continuous bijection? A differentiable bijection?

21 Let $a_1<a_2<a_3<\ldots<a_n$.
A function $f:[a_1,a_n]\to\mathbb{R}$ is defined by

$$f(x)=|x-a_1|+|x-a_2|+\ldots+|x-a_n|.$$

Is f continuous? Is f bounded? Where does it attain its bounds?
A function $g:[a_1,a_n]\to\mathbb{R}$ is defined by

$$g(x)=(x-a_1)^2+(x-a_2)^2+\ldots+(x-a_n)^2.$$

Is g continuous? Is g bounded? Where does it attain its bounds?

22 *The waterfall function*
If the function $f:[a,b]\to\mathbb{R}$ is continuous at every point of its domain, must the function $g:[a,b]\to\mathbb{R}$ defined by

$$g(t)=\sup\{f(x)|\,a\le x\le t\}$$

be continuous at every point?

23 A function f is defined on $[a,b]$. $x_i=a+i(b-a)/n$. Under what conditions will the sequence with nth term $(\sum_{i=1}^{i=n}f(x_i))/n$ be convergent as $n\to\infty$?

24 (*Newton–Raphson*) Under what circumstances will a sequence (a_n) defined by the relation $a_{n+1}=a_n-f(a_n)/f'(a_n)$ converge? Assume $f(a)=0$, $f'(a)\neq 0$ and f'' is continuous. Define $g(x)=x-f(x)/f'(x)$. Show that in some neighbourhood of a, $|g'(x)|<L<1$, for some L. Deduce that if a_1 is in this neighbourhood $(a_n)\to a$. The calculus format for this iteration is due to Thomas Simpson (1740). Newton (1669) and Raphson (1702) worked algebraically with polynomials. The application of this iteration to the quadratic function $f(x)=x^2-a$ coincides with Heron's method for finding the square root of a. See qn 2.37.

25 (*Cauchy*, 1821) What properties can be established for a function $f:\mathbb{R}\to\mathbb{R}$ such that $f(x+y)=f(x)\cdot f(y)$? What further properties can you claim if f is continuous with $f(1)=2$?

26 For each of the following functions $f : (0, 1] \to \mathbb{R}$, determine whether

(i) $\lim_{x \to 0+} f(x)$ exists,
(ii) whether f can be extended to a continuous function at 0,
(iii) whether f can be extended to a differentiable function at 0.

$f(x) = x\lfloor 1/x \rfloor,\ x^2\lfloor 1/x \rfloor,\ x\lfloor 1/x^2 \rfloor$ and $x^2\lfloor 1/x^2 \rfloor$.

27 A monk starts at daybreak at the bottom of a mountain and makes a pilgrimage to the shrine at the top. He arrives just as the sun is setting. The following day he starts at daybreak and makes his way down, again arriving at sunset. Show that at some time he was at the same place at the same time on both days. His rational colleague, not yet a monk, made the same pilgrimage, but only taking rational steps and noting rational times. Must there be a corresponding matching point for him?

Bibliography

Armitage, J. V. and Griffiths, H. B., 1969, *Companion to Advanced Mathematics*, vol. 1, Cambridge University Press.
Part II of this book contains a concise but illuminating overview of first-year undergraduate analysis from a second-year/metric space point of view.

Artmann, B., 1988, *The Concept of Number*, Ellis Horwood.
An advanced and thorough treatment.

Ausubel, D. P., 1968, *Educational Psychology: A Cognitive View*, Holt, Rinehart and Winston.

Baylis, J. and Haggarty, R., 1988, *Alice in Numberland*, Macmillan Education.
A prosy, jokey, but none the less serious introduction to the real numbers.

Beckenbach, E. F. and Bellman, R., 1961, *An Introduction to Inequalities*, Mathematical Association of America.
Inequalities for the sixth-former.

Boas, R. P., 1960, *A Primer of Real Functions*, Mathematical Association of America.
Excellent further reading.

Bollobás, B. (ed.), 1986, *Littlewood's Miscellany*, Cambridge University Press.

Bolzano, B., 1950, *Paradoxes of the Infinite*, Routledge and Kegan Paul.
First published posthumously in 1851; a remarkable testament to one of the finest analytical thinkers of the period 1800–1850.

Bonar, D. D. and Khoury, M., 2006, *Real Infinite Series*, Mathematical Association of America.
Illuminating and enriching our chapter 5. Clear arguments and excellent visuals.

Boyer, C. B., 1959, *The History of the Calculus*, Dover reprint of *The Concepts of the Calculus*, 1939.

Bressoud, D., 1994, *A Radical Approach to Real Analysis*, Mathematical Association of America.
An historically inspired (1807–1829) and trail-blazing course, for users of the software *Mathematica*.

Bryant, V., 1990, *Yet Another Introduction to Analysis*, Cambridge University Press.
A good collection of expository ideas.

Burkill, J. C., 1960, *An Introduction to Mathematical Analysis*, Cambridge University Press.
A concise accessible treatment which establishes the irrationality of π in the last exercise of chapter 7.

Burkill, J. C. and Burkill, H., 1970, *A Second Course in Mathematical Analysis*, Cambridge University Press.
Contains a very good introduction to uniform convergence.

Burn, R. P., 1990, Filling holes in the real line, *Math. Gaz.*, **74**, 228–32.

Burn, R. P., 1997, *A Pathway to Number Theory*, Cambridge University Press.
Gives the Fundamental Theory of Arithmetic in chapter 1.

Cauchy, A. L., 1821, *Cours d'Analyse de l'École Royale Polytechnique: Analyse Algébrique*, Debure frères. Reprinted 1989, Jacques Gabay. English translation 2009, Springer.
In later publications Cauchy referred to this as *Analyse Algébrique*. It was the first book to use modern methods of proof for convergence and continuity.

Cohen, D., 1988, *Calculus By and For Young People*. Published by the author at 809 Stratford Drive, Champaign, Ill. 61821, USA.
A numerical approach to sequences, series and graphs, said to be for those aged 7 and upwards. Just right for the sixth form!

Cohen, L. W. and Ehrlich, G., 1963, *The Structure of the Real Number System*, van Nostrand Reinhold.
A detailed, thorough and modern account of the number system from the Peano postulates to the various forms of completeness.

Courant, R. and John, F., 1965, *Introduction to Calculus and Analysis*, vol. 1, New York: Wiley. Reprinted 1989, Berlin: Springer.
Excellent text and diagrams. Style of discussion consonant with a good sixth-form text. Gives integration before differentiation. Strong on applications.

Dedekind, R., 1963, *Essays on the Theory of Numbers*, Dover.
First published 1872 and 1887, these are two of the historically seminal documents in the arithmetisation of analysis.

Dieudonné, J., 1960, *Foundations of Modern Analysis*, Academic Press.
For 'further reading' only.

du Bois-Reymond, P., 1882, *Die Allgemeine Funktionentheorie*, Laupp, Tübingen. French translation 1887, *Théorie Générale des Fonctions*, Niçoise, Nice. Reprint 1995, Jacques Gabay, Paris.
This is not a textbook but a discussion of numbers and functions favouring infinite decimals. It has a fine and full discussion of completeness.

Dugac, P., 1978, Sur les fondements de l'Analyse de Cauchy à Baire, doctoral thesis, Université Pierre et Marie Curie, Paris.
A masterly survey of the foundation of analysis during the nineteenth century, particularly informative on the contribution of Weierstrass.

Dugac, P., 2003, *Histoire de l'Analyse*, Vuibert.
A comprehensive study of the notion of limit.

Edwards, C. H., 1979, *The Historical Development of the Calculus*, Springer.
Strong on the seventeenth and eighteenth centuries. Written to generate interest in mathematics.

Ferrar, W. L., 1938, *Convergence*, Oxford University Press.

Gardiner, A., 1982, *Infinite Processes*, Springer.
An attempt to bridge the gap between sixth-form and undergraduate thinking on three fronts: numbers, geometry and functions.

Gelbaum, B. R. and Olmsted, J. M. H., 1964, *Counterexamples in Analysis*, Holden-Day.
A rich collection of correctives to the untutored intuition.

Grabiner, J. V., 1981, *The Origins of Cauchy's Rigorous Calculus*, MIT.
The definitive introduction to Cauchy's analysis and calculus. Strong on the late eighteenth and early nineteenth century.

Grattan-Guinness, I. (ed.), 1980, *From the Calculus to Set Theory, 1630–1910*, Duckworth.
Strong on the nineteenth century, but you must know the mathematics before you start reading this history book.

Hairer, E. and Wanner, G., 1995, *Analysis by its History*, Springer.
The historical remarks in this book are useful, and the diagrams quite outstanding. The references to primary texts are invaluable.

Hardy, G. H., 1908, *Pure Mathematics* (1st edition), Cambridge University Press.
Until about 1960, this (2nd edition, 1914, onwards) was the leading English language text on analysis. The text is hard going but the exercises offer a rich and broad diet.

Hardy, G. H., Littlewood, J. E., and Pólya, G., 1952, *Inequalities*, Cambridge University Press.
A graduate text.

Hardy, G. H. and Wright, E. M., 1960, *An Introduction to the Theory of Numbers*, Oxford University Press.
The standard reference on number theory. Contains a full discussion of decimals in chapter 9.

Hart, F. M., 1988, *Guide to Analysis*, Macmillan.
A first course with many answers worked out in full.

Hauchart, C. and Rouche, N., 1987, *Apprivoiser l'Infini*, Ciaco.
A thesis on the beginnings of analysis in the secondary school.

Heath, T. L. (trans.), 1956, *The Thirteen Books of Euclid's Elements*, Dover.

Heijenoort, J. van, 1967, *From Frege to Gödel*, Harvard University Press.
Historical source material.

Hemmings, R. and Tahta, D., 1984, *Images of Infinity*, Leapfrogs – Tarquin.
A rich stimulus to the intuition. Ideal for a sixth-former.

Hight, D. W., 1977, *A Concept of Limits*, Dover.
Unique for its geometrical illustrations of limits.

Ivanov, O. A., 1998, *Easy as π?*, Springer.

Kazarinoff, N. D., 1961, *Analytic Inequalities*, Holt, Rinehart and Winston.
Suitable for a sixth-former.

Klambauer, G., 1979, *Problems and Propositions in Analysis*, Marcel Dekker.
For the lecturer's bookshelf: substantial problems with solutions.

Klein, F., 1924, *Elementary Mathematics from an Advanced Standpoint*. Part I. *Arithmetic, Algebra and Analysis*, 3rd edition trans. E. R. Hedrick and C. A. Noble, Dover reprint.
Contains an interesting discussion of the elementary functions and proofs of the transcendence of e and π.

Knopp, K., 1928, *Theory and Application of Infinite Series*, Blackie. Dover reprint.
Thorough, humane and with a wealth of historical reference.

Kopp, P. E., 1996, *Analysis*, Arnold.

Körner, T. W., 1991, Differentiable functions on the rationals. *Bull. Lond. Math. Soc.*, **23**, 557–62.

Korovkin, P. P., 1961, *Inequalities*, Pergamon Press (contained in *Popular Lectures in Mathematics*, vols 1–6, trans. H. Moss).
Very good for sixth-formers.

Landau, E., 1960, *Foundations of Analysis*, Chelsea, New York.
First published in German 1930; this was the first published account of the development of the number system from the Peano postulates to Dedekind sections for undergraduates.

Leavitt, T. C. J., 1967, *Limits and Continuity*, McGraw-Hill.
Useful diagrams for beginners, a partly programmed text.

Ledermann, W. and Weir, A. J., 1996, *Introduction to Group Theory* (2nd edition), Addison-Wesley.
The appendix contains a proof of the equivalence of induction and well-ordering.

Levi, H., 1961, *Elements of Algebra*, Chelsea, New York.
The number system developed from notions of cardinality.
Lieber, L. P., 1953, *Infinity*, Holt, Rinehart and Winston.
An artistic and poetic education for the mathematical intuition.
Maor, E., 1994, *e: the story of a number*, Princeton University Press.
An excellent historical introduction to analysis at the advanced high school level.
Moise, E. E., 1982, *Introductory Problem Courses in Analysis and Topology*, Springer.
Mason, J. and Klymchuk, S., 2009, *Using Counter-examples in Calculus*, Imperial College, London.
A remarkable collection of student conjectures, each of which demands an imaginative response to generate a counter-example.
Nelsen, R. B., 1993, *Proofs Without Words*, Mathematical Association of America.
Particularly interesting on series.
Niven, I., 1961, *Numbers: Rational and Irrational*, Mathematical Association of America.
A superb introduction for the sixth-former.
Northrop, E. P., 1960, *Riddles in Mathematics: A Book of Paradoxes*, Pelican. (1st UK edition, 1945, English Universities Press).
Excellent high-school material. Fascinating on series.
O'Brien, K. E., 1966, *Sequences*, Houghton-Mifflin.
For beginners.
Osgood, W.F., 1907, *Lehrbuch der Funktionentheorie*, Teubner.
Propp, J., 2013, Real analysis in reverse. *Amer. Math. Monthly*, **120**, 5, 392–408.
A careful analysis of properties equivalent to completeness.
Pólya, G., 1954, *Induction and Analogy in Mathematics* (*Mathematics and Plausible Reasoning*, vol. 1), Princeton University Press.
Rich with insight into how mathematics works. Gives historical illumination.
Pólya, G. and Szegö, G., 1976, *Problems and Theorems in Analysis*, vol. 1, Springer.
Hard problems with solutions: a very rich diet.
Quadling, D. A., 1955, *Mathematical Analysis*, Oxford University Press.
Reade, J. B., 1986, *An Introduction to Mathematical Analysis*, Oxford University Press.
As readable as a conventional text can be.
Rosenbaum, L. J., 1966, *Induction in Mathematics*, Houghton-Mifflin.
Good sixth-form material.
Rudin, W., 1953, *Principles of Mathematical Analysis*, McGraw-Hill.
Concise and purposeful. A good second course.
Scott, D. B. and Tims, S. R., 1966, *Mathematical Analysis*, Cambridge University Press.
Smith, W. K., 1964, *Limits and Continuity*, Macmillan.
A concrete development of the neighbourhood definition of limit.
Sominskii, I. S., 1961, *The Method of Mathematical Induction*, Pergamon Press (contained in *Popular Lectures in Mathematics*, vols 1–6 trans. H. Moss).
Excellent sixth-form material for the student willing to do some work!
Spivak, M., 2006, *Calculus*, Cambridge.
The modern lecturer's favourite text: beautifully written, copiously illustrated, and with an extensive collection of exercises.
Stewart, I. and Tall, D., 1977, *The Foundations of Mathematics*, Oxford University Press.
A serious introduction to university mathematics.
Swann, H. and Johnson, J., 1977, *Prof. E. McSquared's Original, Fantastic and Highly Edifying Calculus Primer*, William Kaufmann.
A comic-strip approach to functions and limits (by neighbourhoods).

Tall, D., 1982, The blancmange function: continuous everywhere but differentiable nowhere. *Math. Gaz.*, **66**, 11–22.

Tall, D., 1992, The transition to advanced mathematical thinking: functions, limits, infinity and proof. Article in Grouws, D. A. (ed.), *Handbook of Research on Mathematics Teaching and Learning*, Macmillan.
A useful survey of recent research.

Teismann, H., 2013, Toward a more complete list of completeness axioms. *Amer. Math. Monthly*, **120**, 2, 99–114.

Thurston, H. A., 1967, *The Number-system*, Dover.
A useful combination of logic and rationale in the development of the number system.

Toeplitz, O., 1963, *The Calculus: A Genetic Approach*, University of Chicago Press.
History: the key to a humane approach to analysis.

Wheeler, D., 1974, ***R** is for Real*, Open University Press.
A problem-motivated introduction to the real numbers.

Yarnelle, J. E., 1964, *An Introduction to Transfinite Mathematics*, D. C. Heath.
The most readable account of countability that I know.

Zippin, L., 1962, *Uses of Infinity*, Mathematical Association of America.

Index

2.53 means the reference is to be found in question 53 of chapter 2.
2.53^- means that the reference is to be found before that question.
2.53^+ means that the reference is to be found after that question.
2.53–58 refers to questions 53 to 58 of chapter 2.
2.53,64 refers to questions 53 and 64 of chapter 2.
2H means that the reference is to be found in the historical note to chapter 2.

Printed in the United States
By Bookmasters